SOIL PROPERTIES

Testing, Measurement, and Evaluation

Fifth Edition

CHENG LIU

JACK B. EVETT

The University of North Carolina at Charlotte

Upper Saddle River, New Jersey
Columbus, Ohio

To
Kimmie, Jonathon, and Michele Liu
and
Linda, Susan, Scott, Sarah, and Sallie Evett

Library of Congress Cataloging-in-Publication Data
Liu, Cheng
  Soil properties: testing, measurement, and evaluation / Cheng
Liu, Jack B. Evett. — 5th ed.
    p. cm.
  ISBN 0-13-093005-9
  1. Soil mechanics. 2. Soils—Testing. I. Evett, Jack B.
    II. Title.
TA710.L547   2003
624.1'5136'0287—dc21   2002017088

Editor in Chief: Stephen Helba
Editor: Ed Francis
Production Editor: Holly Shufeldt
Production Coordination: Carlisle Publishers Services
Design Coordinator: Diane Ernsberger
Cover Designer: Heather Miller
Cover Photo: Corbis Stock Market
Production Manager: Matthew Ottenweller
Marketing Manager: Mark Marsden

This book was set in New Century Schoolbook by Carlisle Communications, Ltd. It was printed and bound by
Banta Book Company. The cover was printed by Phoenix Color Corp.

Photo Credits: page 42 by Cheng Liu; pages 91, 104, 106, 177, 185, 204, 234, 278, 312, 328, 330, 362, 387 by
Soiltest, Inc.; pages 330 and 363 by Wykeham Farrance, Inc.; page 218 by Troxler Electronic Laboratories, Inc.

Pearson Education Ltd.
Pearson Education Australia Pty. Limited
Pearson Education Singapore, Pte. Ltd.
Pearson Education North Asia Ltd.
Pearson Education Canada, Ltd.
Pearson Educación de Mexico S.A. de C.V.
Pearson Education—Japan
Pearson Education Malaysia Pte. Ltd.
Pearson Education, *Upper Saddle River, New Jersey*

10 9 8 7 6 5 4 3 2 1
ISBN 0-13-093005-9

Contents

Preface

The preface to the first edition of this book, expresses our purpose in writing the book and describes 11 specific features of it. We believe these features are still valid for this edition. We have updated the book to conform with the very latest information from the American Society for Testing and Materials (ASTM).

We wish to express our sincere appreciation to Carlos G. Bell of The University of North Carolina at Charlotte and to W. Kenneth Humphries of the University of South Carolina, who read our original manuscript and offered many helpful suggestions. Also, we thank Renda Gwaltney, who typed the entire original manuscript.

We thank our colleague, Professor Alan Stadler of The University of North Carolina at Charlotte, for reviewing Chapter 15 for us. We also express our appreciation to our colleague, Professor Ambrose "Bo" Barry of The University of North Carolina at Charlotte, for his assistance in preparing the CD and the CD Installation Tips at the back of the book. These allow the user to more easily analyze data collected in a laboratory experiment.

As mentioned in the preface to the first edition, we believe the features cited for that edition, as well as the expansion and improvements provided in succeeding editions, distinguish our book from other soils laboratory manuals and make it more helpful and more useful. We hope you will enjoy using it, and we would be pleased to receive your comments, suggestions, and/or criticisms.

Cheng Liu
Jack B. Evett
Charlotte, North Carolina

Acknowledgments

The authors wish to thank the following for their review of the fifth edition.

- Charles A. Matrosic, P. E., Ferris State University
- Timothy W. Zeigler, Southern Polytechnic State University
- Mohammad Najafi, Ph. D., P. E., Director, Trenchless Research and Development Center, Visiting Associate Professor, University of Missouri-Columbia

Sources

- Material used in this book from 1998 *Annual Book of ASTM Standards*. Copyright © American Society for Testing and Materials, 100 Barr Harbor Drive, West Conshohocken, PA 19428-2959. Reprinted with permission.
- Material used in this book from *Highway Materials* by Robert D. Krebs and Richard D. Walker. Copyright © 1971, McGraw-Hill, Inc. Used with the permission of McGraw-Hill Book Company. (Material used on pages 168 and 169)
- Material used in this book from *Engineering Properties of Soils and Their Measurements*, by Joseph E. Bowles. Copyright © 1970, McGraw-Hill, Inc. Used with the permission of McGraw-Hill Book Company. (Material used on pages 262 and 264)
- Material used in this book from *Engineering Properties of Soils and Their Measurements*, by Joseph E. Bowles. Copyright © 1978, McGraw-Hill, Inc. Used with the permission of McGraw-Hill Book Company. (Material used on pages 52, 250, 277, 302, 322, 348, 362, 363, 375, 376, and 401)
- Material used in this book from *Soils and Foundations*, 5th edition, by Cheng Liu and Jack B. Evett. Copyright © 2001 by Pearson Education, Inc. Used with the permission of Pearson Education, Inc. (Figures 19–4, 19–5, 19–6, 21–5, 21–8, 22–5 and 22–7)
- Material used in this book from Troxler Electronic Laboratories, Inc. (Figures 15–1 and 15–2).

Preface to the
First Edition

We have attempted to prepare a fundamental as well as a practical soils laboratory manual to complement our textbook Soils and Foundations, also published by Prentice Hall [1998]. We truly believe that this manual will prove to be extremely useful for the beginning student in soil engineering. To back up this claim, we offer the following helpful features of our book.

1. One of the major features is the simple and direct style of writing, which will, we believe, make it easy for the user to understand.

2. We have included for each chapter an introduction that includes a "definition," "scope," and "objective" for each experiment. This should give the reader an initial understanding of what he or she is attempting to do and for what purpose.

3. We have included one (or more) numerical examples together with each test in every chapter. These are completely worked examples, showing step by step the computations required for the analysis and evaluation of the test data collected. Charts and graphs are also included, if needed. Thus the reader has access to a completely worked example to study prior to and after performing the test and thus will know better what is to be done and how.

4. In addition to presenting step-by-step computations, we have provided data reporting forms and necessary graph papers for most tests. These forms and graph papers provide a convenient means of recording test data, carrying out required computations, and plotting required curves as well as displaying the test results. At the end of each chapter blank copies of all such forms are included for the user. Blank copies of all needed graph papers are included at the end of the text. The appropriate graph paper for each experiment can be photocopied as needed.

5. We have used a convention, which we have not seen before, of presenting all data collected during a test in boldface type. All other values (primarily computed values) are presented in regular type. This convention is used throughout the numerical examples. We believe this feature will be extremely helpful to the reader, as it will always be obvious which data were "measured" and which were "computed."

6. We have given, where appropriate, "typical values" for various tests. Inclusion of typical values should help the user determine if his or her test results are reasonable.

7. For most tests, we have included in the conclusion section information on "method of presentation" and "engineering uses of the test results." We believe that this will help the beginning student understand better what data, results, and other information should be presented in the test report and what the test results will be used for in practical engineering problems.

8. Most of our test procedures follow closely those of the American Society for Testing and Materials (ASTM) and the American Association of State Highway and Transportation Officials (AASHTO). As practicing engineers and architects almost always cite ASTM and/or AASHTO in contracts and specifications, these standards should be followed.

9. In addition to giving the exact step-by-step test procedure for each experiment, we have included first a general overview of the entire procedure. This will give the user an overall preview of the entire process of testing prior to tackling the sometimes laborious step-by-step procedure.

10. The presentation of the three consecutive sections "Procedure," "Data," and "Calculations" should be very useful. Immediately following "Procedure" is the section "Data," which lists explicitly the data that must be collected during the performance of the test. Immediately following "Data" is the section "Calculations," which shows precisely how the collected data are evaluated to obtain the desired test results.

11. The inclusion of "Determining the Moisture Content of a Soil (Calcium Carbide Gas Moisture Tester)" (Chapter 12) is not only significant but essential. This field procedure for determining the moisture content of a soil is quite practical and is well accepted in conjunction with the in-place density test. (There are other methods for determining the moisture content of soils, but they generally require drying overnight. When checking soil compaction in the field, results are needed almost immediately, and the calcium carbide gas moisture tester gives the required results very quickly.)

We believe the features cited above distinguish our book from other soils laboratory manuals and make it more helpful and more useful. We hope you will enjoy using it, and we would be pleased to receive your comments, suggestions, and/or criticisms.

We wish to express our sincere appreciation to Carlos G. Bell of the University of North Carolina at Charlotte and to W. Kenneth Humphries of the University of South Carolina, who read our manuscript and offered many helpful suggestions. Also, we thank Renda Gwaltney, who typed the entire manuscript.

Cheng Liu
Jack B. Evett
Charlotte, North Carolina

CHAPTER ONE

Introduction

Structures of all types (buildings, bridges, highways, etc.) rest directly on, in, or against soil; hence, proper analysis of soil and design of foundations are necessary to ensure that these structures remain safe and free of undue settling and collapse. A comprehensive knowledge of the soil in a specific location may also be important in other contexts, including the use of soil as a source of construction material. In order to obtain such knowledge, soil samples must be collected from a job site and tested in a soils laboratory to evaluate the soil's engineering properties quantitatively. This laboratory textbook deals specifically with collecting, and especially testing, soil.

Soil testing is an extremely important step in an overall construction design project. Soil conditions vary from one location to another; hence, virtually no construction site presents soil conditions exactly like any other. It can be extremely important that properties may vary, even profoundly, within one site. As a result, soil conditions at every site must be thoroughly investigated prior to preparing detailed designs.

Experienced soils engineers can obtain a fairly good idea of the soil conditions at a given location by examining soil samples obtained from exploratory borings. However, quantitative results of laboratory tests on the samples are necessary to analyze the soil conditions and effect an appropriate design that is based on actual data. The importance of securing sufficient and accurate soil property data can hardly be overemphasized.

1

This book deals with soil laboratory testing procedures, as well as the collection and evaluation of test data. However, to provide a more complete picture concerning individual experiments, each experiment is introduced together with its definition, scope, and objective. A conclusion is given at the end of each chapter explaining the method of presentation, typical values, and engineering uses of test results. In addition, a complete numerical example, including necessary graphs, is presented in each chapter so that students can grasp not only the procedures and principles of each test, but also the entire process of evaluating applicable laboratory data.

Twenty-two different soil tests are included in this book, with a single chapter devoted to each. These include virtually all soil tests that are required and done on a routine basis. All except a very few procedures follow those of the American Society for Testing and Materials (ASTM) and/or the American Association of State Highway and Transportation Officials (AASHTO).

COMMON FORMAT

In order to standardize the presentation of each laboratory experiment in the remainder of the book, the following format is used for each chapter:

Introduction
> Definitions
> Scope of Test
> Object of Test

Apparatus and Supplies

Description of Soil Sample

Preparation of Samples and Test Specimens

Adjustment and Calibration of Mechanical Device

Procedure

Data

Calculations

Numerical Example

Charts and Graphs

Results

Conclusions
> Method of Presentation
> Typical Values
> Engineering Uses of Test Results

References

Blank Forms

This format is intended to be all-inclusive; hence, some of the foregoing headings will not apply in some cases and are therefore omitted from such cases. In other instances, additional headings not included here may be used.

LABORATORY REPORTS

Most, if not all, soil testing of any value culminates in a written report. The reason for conducting tests is to evaluate certain soil properties quantitatively. To be useful, results must generally be made a matter of record and also communicated to whoever is to use them. This invariably calls for a written report. For college laboratory experiments, written reports are required to communicate results to the laboratory instructor. With commercial laboratory tests, written reports are needed to communicate results to clients, project engineers within the company, and the like.

It should not be assumed that a single format exists for all written laboratory reports. The purpose of a report, company policy, and individual style, among other things, are factors that may affect report format. Reports from commercial laboratories to clients may consist simply of a letter transmitting a single laboratory-determined parameter. More often, however, reports constitute extensive documents that relate in considerable detail all factors bearing on a test.

The authors suggest that the format presented in the preceding section be adopted as a guide for students to follow in preparing laboratory reports to be submitted to their instructors. Reports should be typed on 8½- by 11-inch plain white paper. They should be submitted in a folder, with the following information appearing on the front of the folder:

1. Title of experiment

2. Author of report

3. Course number and section

4. Names of laboratory partners

5. Date of experiment and date of report

Much has been written on how to write technical reports. In reality, some individuals write very well, some write very poorly, and many fall somewhere between these extremes. Readers who need assistance in writing are referred to one of many books available on technical writing. Suffice it to say here that reports should be written with correct grammar, punctuation, and spelling. Use of personal pronouns should be avoided. (Instead of "I tested the sample," use "The sample was tested.") Finally, the report should be coherent. It should be readable, easily and logically, from the first page to the last.

LABORATORY GRAPHS

Often it is helpful to use graphical displays to present experimental data. In some cases, it is necessary in soil testing to plot certain experimentally determined data on graph paper in order to evaluate test results correctly. Good graphing techniques include the use of appropriate scales, proper labeling of axes, accurate plotting of points, and so on. All graphs should include the following:

1. Title

2. Date of plotting

3. Person preparing the graph

4. Type of soil tested

5. Project and/or laboratory number, if appropriate

As a final comment concerning graph preparation, when experimentally determined points have been plotted and it is necessary to draw a line through the points, in most cases a smooth curve should be drawn (using a French curve), rather than connecting adjacent data points by straight-line segments.

Nowadays more and more graphs are plotted by computer. When computers are used, it is still the responsibility of the person presenting a plot to verify its authenticity. Just because something comes from a computer does not mean it is automatically correct!

SAFETY   Whenever one enters a laboratory or performs field testing, the potential for an accident exists, even if it is no worse than being cut by a dropped glass container. It could, of course, be much worse.

In performing soil testing, hazardous materials, operations, and equipment may be encountered. This book does not purport to cover all safety considerations applying to the procedures presented. It is the responsibility of the reader and user of these procedures to consult and establish necessary safety and health practices and determine the applicability of any regulatory limitations.

Extreme care and caution must always be exercised in performing soil laboratory and field tests. Certainly, the laboratory is no place for horseplay.

CHAPTER TWO

Description and Identification of Soils (Visual-Manual Procedure)

(Referenced Document: ASTM D 2488)

INTRODUCTION

Soil exists throughout the world in a wide variety of types. Different types of soil exhibit diverse behavior and physical properties. Inasmuch as the engineering properties and behavior of soils are governed by their physical properties, it is important to describe and identify soils in terms that will convey their characteristics clearly and accurately to soils engineers.

The remainder of this book covers a number of tests that are performed to provide explicit evaluations of soil properties. However, before tests on soil samples from a given area are conducted, a prudent visual, tactile, and perhaps olfactory inspection along with a few simple tests can be performed to provide an initial appraisal of the soil in the area. Such information can be helpful in preliminary planning and in relating field observations to subsequent test results. Of course, any initial appraisal should be described clearly using appropriate and recognizable terminology.

This chapter covers procedures for describing soils for engineering purposes and gives a procedure for identifying soils based on visual, tactile, and olfactory examinations and manual tests. Of course, the results obtained through these procedures are merely a rough appraisal of a soil. When precise classification is needed for engineering purposes, the procedures of Chapter 10 should be used.

DEFINITIONS   The American Society for Testing and Materials (ASTM) gives the following definitions for various types of soil [1]:*

(1) For particles retained on a 3-in. (75-mm) US standard sieve, the following definitions are suggested:

Cobbles—particles of rock that will pass a 12-in. (300-mm) square opening and be retained on a 3-in. (75-mm) sieve, and

Boulders—particles of rock that will not pass a 12-in. (300-mm) square opening.

(2) *Clay*—soil passing a No. 200 (75-μm) sieve that can be made to exhibit plasticity (putty-like properties) within a range of water contents and that exhibits considerable strength when air dry. For classification, a clay is a fine-grained soil, or the fine-grained portion of a soil, with a plasticity index equal to or greater than 4, and the plot of plasticity index versus liquid limit falls on or above the "A" line (see Figure 10–3).

(3) *Gravel*—particles of rock that will pass a 3-in. (75-mm) sieve and be retained on a No. 4 (4.75-mm) sieve with the following subdivisions:

coarse—passes a 3-in. (75-mm) sieve and is retained on a ¾-in. (19-mm) sieve.

fine—passes a ¾-in. (19-mm) sieve and is retained on a No. 4 (4.75-mm) sieve.

(4) *Organic clay*—a clay with sufficient organic content to influence the soil properties. For classification, an organic clay is a soil that would be classified as a clay, except that its liquid limit value after oven drying is less than 75% of its liquid limit value before oven drying.

(5) *Organic silt*—a silt with sufficient organic content to influence the soil properties. For classification, an organic silt is a soil that would be classified as a silt except that its liquid limit value after oven drying is less than 75% of its liquid limit value before oven drying.

(6) *Peat*—a soil comprised primarily of vegetable tissue in various stages of decomposition, usually with an organic odor, a dark brown to black color, a spongy consistency, and a texture ranging from fibrous to amorphous.

(7) *Sand*—particles of rock that will pass a No. 4 (4.75-mm) sieve and be retained on a No. 200 (75-μm) sieve with the following subdivisions:

coarse—passes a No. 4 (4.75-mm) sieve and is retained on a No. 10 (2.00-mm) sieve.

---

*Numbers in brackets refer to references listed at the end of each chapter.

medium—passes a No. 10 (2.00-mm) sieve and is retained on a No. 40 (425-μm) sieve.

fine—passes a No. 40 (425-μm) sieve and is retained on a No. 200 (75-μm) sieve.

(8) *Silt*—soil passing a No. 200 (75-μm) sieve that is nonplastic or very slightly plastic and that exhibits little or no strength when air dry. For classification, a silt is a fine-grained soil, or the fine-grained portion of a soil, with a plasticity index less than 4, and the plot of plasticity index versus liquid limit falls below the "A" line (see Figure 10–3).

The visual-manual procedure covered in this chapter for describing and identifying soils utilizes the following group symbols (see Chapter 10 for more details):

G	gravel
S	sand
M	silt
C	clay
O	organic
PT	peat
W	well graded
P	poorly graded

Normally, two group symbols are used to classify a soil; for example, SW indicates well-graded sand. ASTM D 2488-00 provides "Standard Practice for Description and Identification of Soils (Visual-Manual Procedure)."

APPARATUS AND SUPPLIES [1]

Pocketknife or small spatula

Small test tube and stopper (or jar with lid)

Small hand lens

Water

Hydrochloric acid (HCl), small bottle, dilute, one part HCl (10 N) to three parts distilled water

Note—Unless otherwise indicated, references to water mean water from a city water supply or natural source, including nonpotable water. When preparing the dilute HCl solution, slowly add acid to the water, following necessary safety precautions. Handle with caution and store safely. If the solution comes in contact with the skin, rinse thoroughly with water. Do not add water to acid.

PREPARATION OF SAMPLES AND TEST SPECIMENS [1]

(1) The sample shall be considered to be representative of the stratum from which it was obtained by an appropriate, accepted, or standard procedure.

Note 1—Preferably, the sample procedure should be identified as having been conducted in accordance with ASTM Practices D 1452, D 1587, or D 2113, or Method D 1586.

(2) The sample shall be carefully identified as to origin.

Note 2—Remarks as to the origin may take the form of a boring number and sample number in conjunction with a job number, a geologic stratum, a pedologic horizon or a location description with respect to a permanent monument, a grid system, or a station number and offset with respect to a stated centerline and a depth or elevation.

(3) For accurate description and identification, the minimum amount of the specimen to be examined shall be in accordance with the following schedule:

Maximum Particle Size, Sieve Opening	Minimum Specimen Size, Dry Weight
4.75 mm (No. 4)	100 g (0.25 lb)
9.5 mm (⅜ in.)	200 g (0.5 lb)
19.0 mm (¾ in.)	1.0 kg (2.2 lb)
38.1 mm (1½ in.)	8.0 kg (18 lb)
75.0 mm (3 in.)	60.0 kg (132 lb)

Note 3—If random isolated particles are encountered that are significantly larger than the particles in the soil matrix, the soil matrix can be accurately described and identified in accordance with the preceding schedule.

(4) If the field sample or specimen being examined is smaller than the minimum recommended amount, the report shall include an appropriate remark.

PROCEDURE [1]

Description Information for Soils

(1) *Angularity*—Describe the angularity of the sand (coarse sizes only), gravel, cobbles, and boulders as angular, subangular, subrounded, or rounded in accordance with the criteria in Table 2–1. A range of angularity may be stated, such as subrounded to rounded.

TABLE 2–1 Criteria for Describing Angularity of Coarse-Grained Particles [1]

Description	Criteria
Angular	Particles have sharp edges and relatively plane sides with unpolished surfaces
Subangular	Particles are similar to angular description but have rounded edges
Subrounded	Particles have nearly plane sides but have well-rounded corners and edges
Rounded	Particles have smoothly curved sides and no edges

(2) *Shape*—Describe the shape of the gravel, cobbles, and boulders as flat, elongated, or flat and elongated if they meet the criteria in Table 2–2 and Figure 2–1. Otherwise, do not mention the shape. Indicate the fraction of the particles that have the shape, such as: one-third of the gravel particles are flat.

(3) *Color*—Describe the color. Color is an important property in identifying organic soils, and within a given locality it may also be useful in identifying materials of similar geologic origin. If the sample contains layers or patches of varying color, this shall be noted and all representative colors shall be described. The color shall be described for moist samples. If the color represents a dry condition, this shall be stated in the report.

TABLE 2–2 Criteria for Describing Particle Shape (see Figure 2–1) [1]

The particle shape shall be described as follows where length, width, and thickness refer to the greatest, intermediate, and least dimensions of a particle, respectively.

Flat	Particles with width/thickness > 3
Elongated	Particles with length/width > 3
Flat and elongated	Particles meet criteria for both flat and elongated

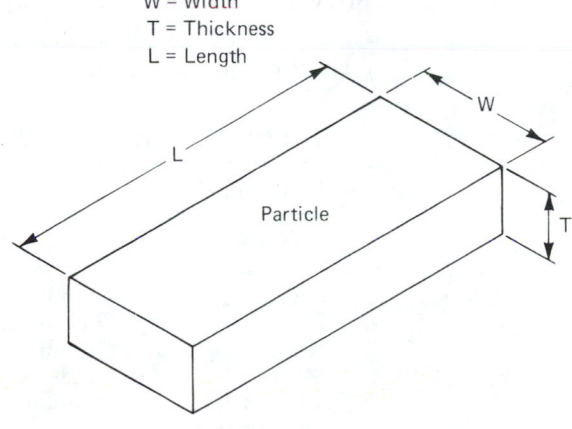

PARTICLE SHAPE

W = Width
T = Thickness
L = Length

Flat: W/T > 3
Elongated: L/W > 3
Flat and elongated: meets both criteria

FIGURE 2–1 Criteria for Particle Shape [1]

(4) *Odor*—Describe the odor if organic or unusual. Soils containing a significant amount of organic material usually have a distinctive odor of decaying vegetation. This is especially apparent in fresh samples, but if the samples are dried, the odor may often be revived by heating a moistened sample. If the odor is unusual (petroleum product, chemical, and the like), it shall be described.

(5) *Moisture Condition*—Describe the moisture condition as dry, moist, or wet, in accordance with the criteria in Table 2–3.

TABLE 2–3   Criteria for Describing Moisture Condition [1]

Description	Criteria
Dry	Absence of moisture, dusty, dry to the touch
Moist	Damp but no visible water
Wet	Visible free water, usually soil is below water table

(6) *HCl Reaction*—Describe the reaction with HCl as none, weak, or strong, in accordance with the criteria in Table 2–4. Since calcium carbonate is a common cementing agent, a report of its presence on the basis of the reaction with dilute hydrochloric acid is important.

TABLE 2–4   Criteria for Describing the Reaction with HCl [1]

Description	Criteria
None	No visible reaction
Weak	Some reaction with bubbles forming slowly
Strong	Violent reaction with bubbles forming immediately

(7) *Consistency*—For intact fine-grained soil, describe the consistency as very soft, soft, firm, hard, or very hard, in accordance with the criteria in Table 2–5. This observation is inappropriate for soils with significant amounts of gravel.

TABLE 2–5   Criteria for Describing Consistency [1]

Description	Criteria
Very soft	Thumb will penetrate soil more than 1 in. (25 mm)
Soft	Thumb will penetrate soil about 1 in. (25 mm)
Firm	Thumb will indent soil about ¼ in. (6 mm)
Hard	Thumb will not indent soil but readily indented with thumbnail
Very hard	Thumbnail will not indent soil

(8) *Cementation*—Describe the cementation of intact coarse-grained soils as weak, moderate, or strong, in accordance with the criteria in Table 2–6.

TABLE 2-6   Criteria for Describing Cementation [1]

Description	Criteria
Weak	Crumbles or breaks with handling or little finger pressure
Moderate	Crumbles or breaks with considerable finger pressure
Strong	Will not crumble or break with finger pressure

(9) *Structure*—Describe the structure of intact soils in accordance with the criteria in Table 2–7.

TABLE 2-7   Criteria for Describing Structure [1]

Description	Criteria
Stratified	Alternating layers of varying material or color with layers at least 6 mm thick; note thickness
Laminated	Alternating layers of varying material or color with the layers less than 6 mm thick; note thickness
Fissured	Breaks along definite planes of fracture with little resistance to fracturing
Slickensided	Fracture planes appear polished or glossy, sometimes striated
Blocky	Cohesive soil that can be broken down into small angular lumps which resist further breakdown
Lensed	Inclusion of small pockets of different soils, such as small lenses of sand scattered through a mass of clay; note thickness
Homogeneous	Same color and appearance throughout

(10) *Range of Particle Sizes*—For gravel and sand components, describe the range of particle sizes within each component as defined in (3) and (7) in the "Definitions" section. For example, about 20% fine to coarse gravel, about 40% fine to coarse sand.

(11) *Maximum Particle Size*—Describe the maximum particle size found in the sample in accordance with the following information:

(11.1) *Sand Size*—If the maximum particle size is a sand size, describe as fine, medium, or coarse as defined in (7) in the "Definitions" section. For example, maximum particle size, medium sand.

(11.2) *Gravel Size*—If the maximum particle size is a gravel size, describe the maximum particle size as the smallest sieve opening that the particle will pass. For example, maximum particle size, 1½ in. (will pass a 1½-in. square opening but not a ¾-in. square opening).

(11.3) *Cobble or Boulder Size*—If the maximum particle size is a cobble or boulder size, describe the maximum dimension of the largest particle. For example, maximum dimension, 18 in. (450 mm).

(12) *Hardness*—Describe the hardness of coarse sand and larger particles as hard, or state what happens when the particles are hit

by a hammer; for example, gravel-size particles fracture with considerable hammer blow, some gravel-size particles crumble with hammer blow. "Hard" means particles do not crack, fracture, or crumble under a hammer blow.

(13) Additional comments shall be noted, such as the presence of roots or root holes, difficulty in drilling or augering hole, caving of trench or hole, or the presence of mica.

(14) A local or commercial name or a geologic interpretation of the soil, or both, may be added if identified as such.

(15) A classification or identification of the soil in accordance with other classification systems may be added if identified as such.

Identification of Peat

(1) A sample composed primarily of vegetable tissue in various stages of decomposition that has a fibrous to amorphous texture, usually a dark brown to black color, and an organic odor shall be designated as a highly organic soil and shall be identified as peat, PT, and not subjected to the identification procedures described hereafter.

Preparation for Identification

(1) The soil identification portion of this practice is based on the portion of the soil sample that will pass a 3-in. (75-mm) sieve. The larger than 3-in. (75-mm) particles must be removed manually, for a loose sample, or mentally, for an intact sample before classifying the soil.

(2) Estimate and note the percentage of cobbles and the percentage of boulders. Performed visually, these estimates will be on the basis of volume percentage.

> *Note 4*—Since the percentages of the particle-size distribution in ASTM Test Method D 2487 are by dry weight, and the estimates of percentages of gravel, sand, and fines in this practice are by dry weight, it is recommended that the report state that the percentages of cobbles and boulders are by volume.

(3) Of the fraction of the soil smaller than 3 in. (75 mm), estimate and note the percentage, by dry weight, of the gravel, sand, and fines (see suggested procedures at the end of the chapter).

> *Note 5*—Since the particle-size components appear visually on the basis of volume, considerable experience is required to estimate the percentages on the basis of dry weight. Frequent comparison with laboratory particle-size analyses should be made.

(3.1) The percentages shall be estimated to the closest 5%. The percentages of gravel, sand, and fines must add up to 100%.

(3.2) If one of the components is present but not in sufficient quantity to be considered 5% of the smaller than 3-in. (75-mm) portion, indicate its presence by the term *trace,* for example, trace of fines. A trace is not to be considered in the total of 100% for the components.

Preliminary Identification

(1) The soil is *fine grained* if it contains 50% or more fines. Follow the procedures for identifying fine-grained soils.

(2) The soil is *coarse grained* if it contains less than 50% fines. Follow the procedures for identifying coarse-grained soils.

Procedure for Identifying Fine-Grained Soils

(1) Select a representative sample of the material for examination. Remove particles larger than the No. 40 sieve (medium sand and larger) until a specimen equivalent to about a handful of material is available. Use this specimen for performing the dry strength, dilatancy, and toughness tests.

(2) *Dry Strength:*

(2.1) From the specimen, select enough material to mold into a ball about 1 in. (25 mm) in diameter. Mold the material until it has the consistency of putty, adding water if necessary.

(2.2) From the molded material, make at least three test specimens. A test specimen shall be a ball of material about ½ in. (12 mm) in diameter. Allow the test specimens to dry in air or sun or by artificial means, as long as the temperature does not exceed 60°C.

(2.3) If the test specimen contains natural dry lumps, those that are about ½ in. (12 mm) in diameter may be used in place of the molded balls.

> Note 6—The process of molding and drying usually produces higher strengths than are found in natural dry lumps of soil.

(2.4) Test the strength of the dry balls or lumps by crushing between the fingers. Note the strength as none, low, medium, high, or very high in accordance with the criteria in Table 2–8. If natural dry lumps are used, do not use the results of any of the lumps that are found to contain particles of coarse sand.

TABLE 2–8   Criteria for Describing Dry Strength [1]

Description	Criteria
None	The dry specimen crumbles into powder with mere pressure of handling
Low	The dry specimen crumbles into powder with some finger pressure
Medium	The dry specimen breaks into pieces or crumbles with considerable finger pressure
High	The dry specimen cannot be broken with finger pressure Specimen will break into pieces between thumb and a hard surface
Very high	The dry specimen cannot be broken between the thumb and a hard surface

(2.5) The presence of high-strength, water-soluble cementing materials, such as calcium carbonate, may cause exceptionally high dry strengths. The presence of calcium carbonate can usually be detected from the intensity of the reaction with dilute hydrochloric acid.

(3) *Dilatancy:*

(3.1) From the specimen, select enough material to mold into a ball about ½ in. (12 mm) in diameter. Mold the material, adding water if necessary, until it has a soft, but not sticky, consistency.

(3.2) Smooth the soil ball in the palm of one hand with the blade of a knife or small spatula. Shake horizontally, striking the side of the hand vigorously against the other hand several times. Note the reaction of water appearing on the surface of the soil. Squeeze the sample by closing the hand or pinching the soil between the fingers, and note the reaction as none, slow, or rapid, in accordance with the criteria in Table 2–9. The reaction is the speed with which water appears while shaking and disappears while squeezing.

TABLE 2–9   Criteria for Describing Dilatancy [1]

Description	Criteria
None	No visible change in the specimen
Slow	Water appears slowly on the surface of the specimen during shaking and does not disappear or disappears slowly upon squeezing
Rapid	Water appears quickly on the surface of the specimen during shaking and disappears quickly upon squeezing

(4) *Toughness:*

(4.1) Following the completion of the dilatancy test, the test specimen is shaped into an elongated pat and rolled by hand on a smooth surface or between the palms into a thread about ⅛ in. (3 mm) in diameter. (If the sample is too wet to roll easily, it should be spread into a thin layer and allowed to lose some water by evaporation.) Fold the sample threads and reroll repeatedly until the thread crumbles at a diameter of about ⅛ in. The thread will crumble at a diameter of ⅛ in. when the soil is near the plastic limit. Note the pressure required to roll the thread near the plastic limit. Also, note the strength of the thread. After the thread crumbles, the pieces should be lumped together and kneaded until the lump crumbles. Note the toughness of the material during kneading.

(4.2) Describe the toughness of the thread and lump as low, medium, or high, in accordance with the criteria in Table 2–10.

TABLE 2–10   Criteria for Describing Toughness [1]

Description	Criteria
Low	Only slight pressure is required to roll the thread near the plastic limit. The thread and the lump are weak and soft
Medium	Medium pressure is required to roll the thread to near the plastic limit. The thread and the lump have medium stiffness
High	Considerable pressure is required to roll the thread to near the plastic limit. The thread and the lump have very high stiffness

(5) *Plasticity*—On the basis of observations made during the toughness test, describe the plasticity of the material in accordance with the criteria given in Table 2–11.

TABLE 2–11   Criteria for Describing Plasticity [1]

Description	Criteria
Nonplastic	An ⅛-in. (3-mm) thread cannot be rolled at any water content
Low	The thread can barely be rolled and the lump can not be formed when drier than the plastic limit
Medium	The thread is easy to roll and not much time is required to reach the plastic limit. The thread cannot be rerolled after reaching the plastic limit. The lump crumbles when drier than the plastic limit
High	It takes considerable time rolling and kneading to reach the plastic limit. The thread can be rerolled several times after reaching the plastic limit. The lump can be formed without crumbling when drier than the plastic limit

(6) Decide whether the soil is an *inorganic* or an *organic* fine-grained soil. If inorganic, follow the steps given next.

(7) *Identification of Inorganic Fine-Grained Soils:*

(7.1) Identify the soil as a *lean clay,* CL, if the soil has medium to high dry strength, none or slow dilatancy, and medium toughness and plasticity (see Table 2–12).

TABLE 2–12   Identification of Inorganic Fine-Grained Soils from Manual Tests [1]

Soil Symbol	Dry Strength	Dilatancy	Toughness
ML	None to low	Slow to rapid	Low or thread cannot be formed
CL	Medium to high	None to slow	Medium
MH	Low to medium	None to slow	Low to medium
CH	High to very high	None	High

(7.2) Identify the soil as a *fat clay,* CH, if the soil has high to very high dry strength, no dilatancy, and high toughness and plasticity (see Table 2–12).

(7.3) Identify the soil as a *silt,* ML, if the soil has no to low dry strength, slow to rapid dilatancy, and low toughness and plasticity, or is nonplastic (see Table 2–12).

(7.4) Identify the soil as an *elastic silt,* MH, if the soil has low to medium dry strength, no to slow dilatancy, and low to medium toughness and plasticity (see Table 2–12).

> *Note 7*—These properties are similar to those for a lean clay. However, the silt will dry quickly on the hand and have a smooth, silky feel when dry. Some soils that would classify as MH in accordance with the criteria in Test Method D 2487 are visually difficult to distinguish from lean clays, CL. It may be necessary to perform laboratory testing for proper identification.

(8) *Identification of Organic Fine-Grained Soils:*

(8.1) Identify the soil as an *organic soil,* OL/OH, if the soil contains enough organic particles to influence the soil properties. Organic soils usually have a dark brown to black color and may have an organic odor. Often, organic soils will change color, for example, black to brown, when exposed to the air. Some organic soils will lighten in color significantly when air dried. Organic soils normally will not have a high toughness or plasticity. The thread for the toughness test will be spongy.

> *Note 8*—In some cases, through practice and experience, it may be possible to further identify the organic soils as organic silts or organic clays, OL or OH. Correlations between the dilatancy, dry strength, and toughness tests and laboratory tests can be made to identify organic soils in certain deposits of similar materials of known geologic origin.

(9) If the soil is estimated to have 15% to 25% sand or gravel, or both, the words "with sand" or "with gravel" (whichever is more predominant) shall be added to the group name. For example: "lean clay with sand, CL" or "silt with gravel, ML" (see Figures 2–2 and 2–3). If the percentage of sand is equal to the percentage of gravel, use "with sand."

(10) If the soil is estimated to have 30% or more sand or gravel, or both, the words "sandy" or "gravelly" shall be added to the group name. Add the word "sandy" if there appears to be more sand than gravel. Add the word "gravelly" if there appears to be more gravel than sand. For example: "sandy lean clay, CL," "gravelly fat clay, CH," or "sandy silt, ML" (see Figures 2–2 and 2–3). If the percentage of sand is equal to the percentage of gravel, use "sandy."

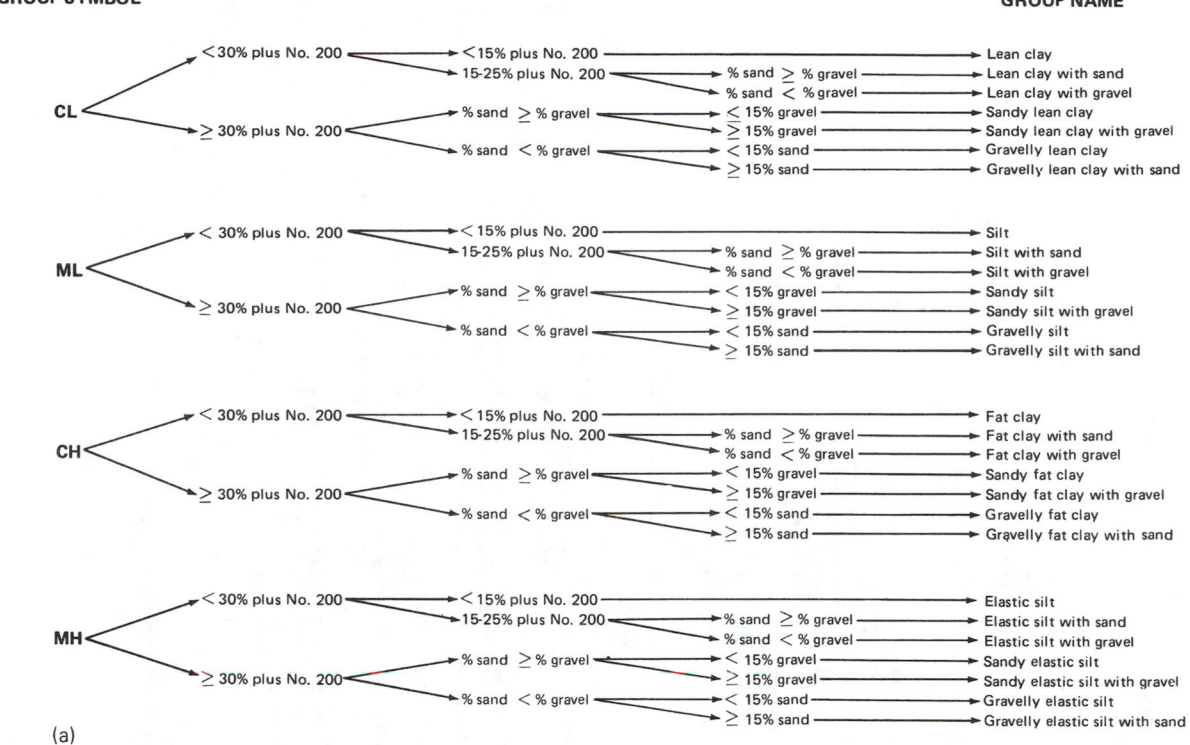

(a)

NOTE: Percentages are based on estimating amounts of fines, sand, and gravel to the nearest 5%.

GROUP SYMBOL   **GROUP NAME**

(b)

NOTE: Percentages are based on estimating amounts of fines, sand, and gravel to the nearest 5%.

FIGURE 2–2   Flowchart for Identifying Inorganic Fine-Grained Soil (50% or more fines) [1]

Procedures for Identifying Coarse-Grained Soils (Contains less than 50% fines)

(1) The soil is a *gravel* if the percentage of gravel is estimated to be more than the percentage of sand.

(2) The soil is a *sand* if the percentage of gravel is estimated to be equal to or less than the percentage of sand.

(3) The soil is a *clean gravel* or *clean sand* if the percentage of fines is estimated to be 5% or less.

(3.1) Identify the soil as a *well-graded gravel,* GW, or as a *well-graded sand,* SW, if it has a wide range of particle sizes and substantial amounts of the intermediate particle sizes.

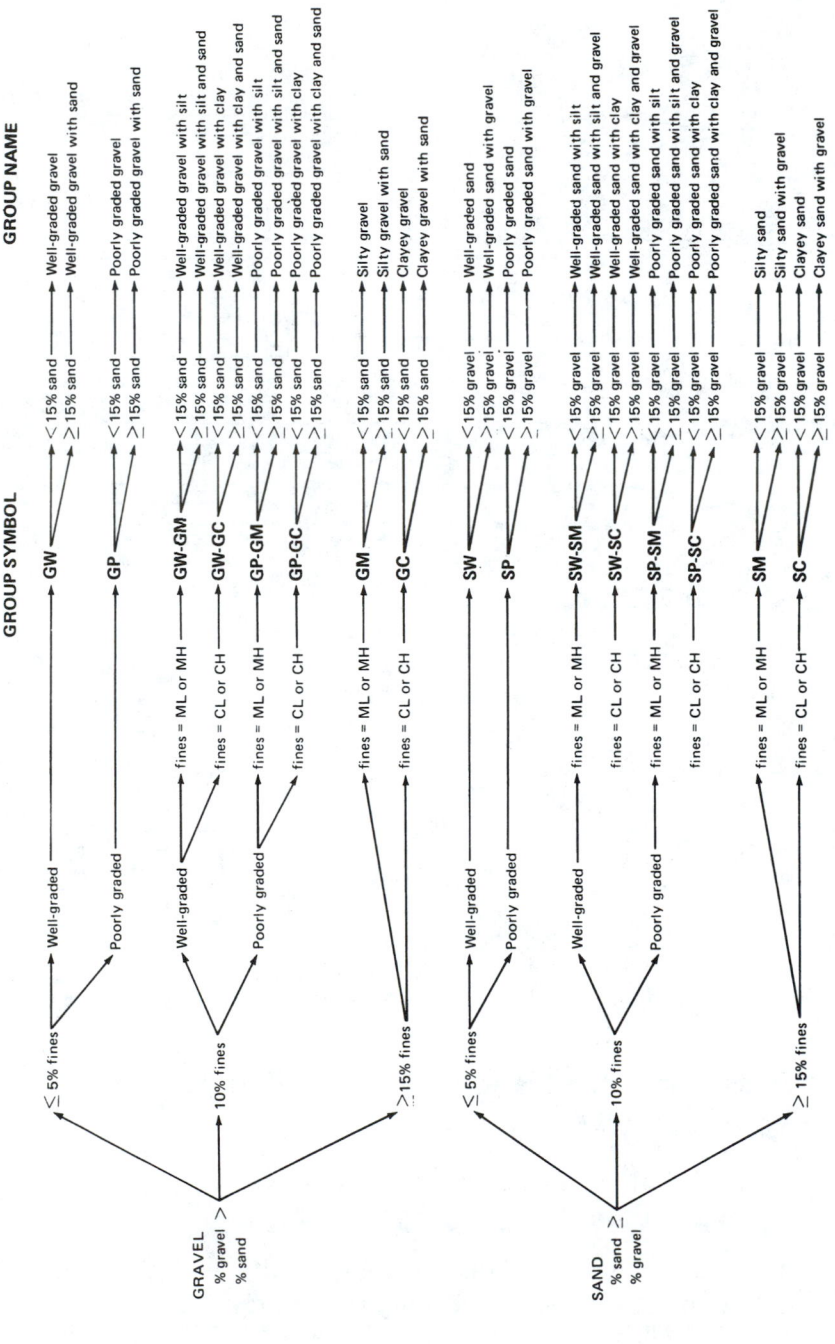

FIGURE 2–3 Flowchart for Identifying Coarse-Grained Soil (less than 50% fines) [1]

NOTE: Percentages are based on estimating amounts of fines, sand, and gravel to the nearest 5%.

(3.2) Identify the soil as a *poorly graded gravel,* GP, or as a *poorly graded sand,* SP, if it consists predominantly of one size (uniformly graded), or it has a wide range of sizes with some intermediate sizes obviously missing (gap or skip graded).

(4) The soil is either a *gravel with fines* or a *sand with fines* if the percentage of fines is estimated to be 15% or more.

(4.1) Identify the soil as a *clayey gravel,* GC, or a *clayey sand,* SC, if the fines are clayey as determined by the procedures in the previous section.

(4.2) Identify the soil as a *silty gravel,* GM, or a *silty sand,* SM, if the fines are silty as determined by the procedures in the previous section.

(5) If the soil is estimated to contain 10% fines, give the soil a dual identification using two group symbols.

(5.1) The first group symbol shall correspond to a clean gravel or sand (GW, GP, SW, SP), and the second symbol shall correspond to a gravel or sand with fines (GC, GM, SC, SM).

(5.2) The group name shall correspond to the first group symbol plus the words "with clay" or "with silt" to indicate the plasticity characteristics of the fines. For example: "well-graded gravel with clay, GW-GC" or "poorly graded sand with silt, SP-SM" (see Figure 2–3).

(6) If the specimen is predominantly sand or gravel but contains an estimated 15% or more of the other coarse-grained constituent, the words "with gravel" or "with sand" shall be added to the group name. For example: "poorly graded gravel with sand, GP" or "clayey sand with gravel, SC" (see Figure 2–3).

(7) If the field sample contains any cobbles or boulders, or both, the words "with cobbles" or "with cobbles and boulders" shall be added to the group name. For example: "silty gravel with cobbles, GM."

REPORT [1]

(1) The report shall include the information as to origin and the items indicated in Table 2–13.

> *Note 9—Example: Clayey Gravel with Sand and Cobbles, GC—* About 50% fine to coarse, subrounded to subangular gravel; about 30% fine to coarse, subrounded sand; about 20% fines with medium plasticity, high dry strength, no dilatancy, medium toughness; weak reaction with HCl; original field sample had about 5% (by volume) subrounded cobbles, maximum dimension, 150 mm.

> *In-Place Conditions—*Firm, homogeneous, dry, brown
> *Geologic Interpretation—*Alluvial fan

TABLE 2-13   Checklist for Description of Soils [1]

1. Group name
2. Group symbol
3. Percent of cobbles or boulders, or both (by volume)
4. Percent of gravel, sand, or fines, or all three (by dry weight)
5. Particle-size range:
    Gravel—fine, coarse
    Sand—fine, medium, coarse
6. Particle angularity: angular, subangular, subrounded, rounded
7. Particle shape (if appropriate): flat, elongated, flat and elongated
8. Maximum particle size or dimension
9. Hardness of coarse sand and larger particles
10. Plasticity of fines: nonplastic, low, medium, high
11. Dry strength: none, low, medium, high, very high
12. Dilatancy: none, slow, rapid
13. Toughness: low, medium, high
14. Color (in moist condition)
15. Odor (mention only if organic or unusual)
16. Moisture: dry, moist, wet
17. Reaction with HCl: none, weak, strong
For intact samples:
18. Consistency (fine-grained soils only): very soft, soft, firm, hard, very hard
19. Structure: stratified, laminated, fissured, slickensided, lensed, homogeneous
20. Cementation: weak, moderate, strong
21. Local name
22. Geologic interpretation
23. Additional comments: presence of roots or root holes, presence of mica, gypsum, etc., surface coatings on coarse-grained particles, caving or sloughing of auger hole or trench sides, difficulty in augering or excavating, etc.

Note 10—If desired, the percentages of gravel, sand, and fines may be stated in terms indicating a range of percentages, as follows:

Trace—Particles are present, but estimated to be less than 5%
Few—5% to 10%
Little—15% to 25%
Some—30% to 45%
Mostly—50% to 100%

(2) If, in the soil description, the soil is identified using a classification group symbol and name as described in Test Method D 2487, it must be distinctly and clearly stated in log forms, summary tables, reports, and the like that the symbol and name are based on visual-manual procedures.

EXAMPLES OF VISUAL SOIL DESCRIPTIONS [1]

(1) The following examples show how the information required can be reported. The information that is included in descriptions should be based on individual circumstances and need.

(1.1) *Well-Graded Gravel with Sand (GW)*—About 75% fine to coarse, hard, subangular gravel; about 25% fine to coarse, hard, subangular sand; trace of fines; maximum size, 75 mm, brown, dry; no reaction with HCl.

(1.2) *Silty Sand with Gravel (SM)*—About 60% predominantly fine sand; about 25% silty fines with low plasticity, low dry strength, rapid dilatancy, and low toughness; about 15% fine, hard, sub-rounded gravel, a few gravel-size particles fractured with hammer blow; maximum size, 25 mm; no reaction with HCl (Note—Field sample size smaller than recommended).

In-Place Conditions—Firm, stratified, and contains lenses of silt 1 to 2 in. (25 to 50 mm) thick, moist, brown to gray; in-place density 106 lb/ft^3; in-place moisture 9%.

(1.3) *Organic Soil (OL/OH)*—About 100% fines with low plasticity, slow dilatancy, low dry strength, and low toughness; wet, dark brown, organic odor; weak reaction with HCl.

(1.4) *Silty Sand with Organic Fines (SM)*—About 75% fine to coarse, hard, subangular reddish sand; about 25% organic and silty dark brown nonplastic fines with no dry strength and slow dilatancy; wet; maximum size, coarse sand; weak reaction with HCl.

(1.5) *Poorly Graded Gravel with Silt, Sand, Cobbles and Boulders (GP-GM)*—About 75% fine to coarse, hard, subrounded to subangular gravel; about 15% fine, hard, subrounded to subangular sand; about 10% silty nonplastic fines; moist, brown; no reaction with HCl; original field sample had about 5% (by volume) hard, subrounded cobbles and a trace of hard, subrounded boulders, with a maximum dimension of 18 in. (450 mm).

SUGGESTED PROCEDURES FOR ESTIMATING THE PERCENTAGES OF GRAVEL, SAND, AND FINES IN A SOIL SAMPLE [1]

(1) *Jar Method*—The relative percentage of coarse- and fine-grained material may be estimated by thoroughly shaking a mixture of soil and water in a test tube or jar and then allowing the mixture to settle. The coarse particles will fall to the bottom and successively finer particles will be deposited with increasing time; the sand sizes will fall out of suspension in 20 to 30 s. The relative proportions can be estimated from the relative volume of each separate size. This method should be correlated to particle-size laboratory determinations.

(2) *Visual Method*—Mentally visualize the gravel size particles placed in a sack (or other container) or sacks. Then do the same with the sand-size particles and the fines. Mentally compare the number of sacks to estimate the percentage of plus No. 4 sieve size and minus No. 4 sieve size present. The percentages of sand and fines in the minus sieve size No. 4 material can then be estimated from the wash test.

(3) *Wash Test (for relative percentages of sand and fines)*—Select and moisten enough minus No. 4 sieve size material to form a 1-in. (25-mm) cube of soil. Cut the cube in half, set one-half to the side, and place the other half in a small dish. Wash and decant the fines out of the material in the dish until the wash water is clear and then compare the two samples and estimate the percentage of sand

and fines. Remember that the percentage is based on weight, not volume. However, the volume comparison will provide a reasonable indication of grain size percentages.

(3.1) While washing, it may be necessary to break down lumps of fines with the finger to get the correct percentages.

REFERENCE   [1]   ASTM, *2001 Annual Book of ASTM Standards,* West Conshohocken, PA, 2001. Copyright, American Society for Testing and Materials, 100 Barr Harbor Drive, West Conshohocken, PA 19428-2959. Reprinted with permission.

CHAPTER THREE

Determining the Moisture Content of Soil (Conventional Oven Method)

(Referenced Document: ASTM 2216)

INTRODUCTION The *moisture content of soil* (also referred to as *water content*) is an indicator of the amount of water present in soil. By definition, moisture content is the ratio of the mass of water in a sample to the mass of solids in the sample, expressed as a percentage. In equation form,

$$w = \frac{M_w}{M_s} \times 100 \qquad (3\text{--}1)$$

where:

w = moisture content of soil (expressed as a percentage)

M_w = mass of water in soil sample (i.e., initial mass of moist soil minus mass of oven-dried soil)

M_s = mass of soil solids in sample (i.e., the soil's "oven-dried mass")

M_w and M_s may be expressed in any units of mass, but both should be expressed in the same unit.

It might be noted that the moisture content could be *mistakenly* defined as the ratio of mass of water to total mass of moist soil (rather than to the mass of oven-dried soil). Because the total mass of moist soil is the sum of the mass of water and oven-dried soil, this incorrect definition would give a fraction in which both numerator and denominator vary (but not in the same proportion) according to the amount of moisture present. Such a definition would be undesirable, because moisture content would then be based on a varying quantity of moist mass of soil rather than a constant quantity of oven-dried soil. Stated another way, with the incorrect definition, the moisture content would not be directly proportional to the mass of water present. With the correct definition given by Eq. (3–1), moisture content is directly proportional to the mass of water present. This characteristic makes moisture content, as defined by Eq. (3–1), one of the most useful and important soil parameters.

APPARATUS AND SUPPLIES

Drying oven (with accurate temperature control and temperature gage)

Balance (with accuracy to 0.01 g)

Containers (e.g., tin or aluminum moisture cans with lids)

Desiccator

Container-handling apparatus: gloves, tongs, or suitable holder for moving and handling hot containers after drying

Miscellaneous: knives, spatulas, scoops, quartering cloth, sample splitters, etc.

SAMPLES [1]

(1) Samples shall be preserved and transported in accordance with ASTM Test Method D 4220 Groups B, C, or D soils. Keep the samples that are stored prior to testing in noncorrodible airtight containers at a temperature between approximately 3° and 30°C and in an area that prevents direct contact with sunlight. Disturbed samples in jars or other containers shall be stored in such a way as to prevent or minimize moisture condensation on the insides of the containers.

(2) The water content determination should be done as soon as practicable after sampling, especially if potentially corrodible containers (such as thin-walled steel tubes, paint cans, etc.) or plastic sample bags are used.

TEST SPECIMEN [1]

(1) For water contents being determined in conjunction with another ASTM method, the specimen mass requirement stated in that method shall be used if one is provided. If no minimum specimen mass is provided in that method, then the values given below shall apply.

(2) The minimum mass of moist material selected to be representative of the total sample shall be in accordance with the following:

Maximum Particle Size (100% passing)	Standard Sieve Size	Recommended Minimum Mass of Moist Test Specimen for Water Content Reported to ±0.1%	Recommended Minimum Mass of Moist Test Specimen for Water Content Reported to ±1%
2 mm or less	No. 10	20 g	20 g*
4.75 mm	No. 4	100 g	20 g*
9.5 mm	⅜-in.	500 g	50 g
19.0 mm	¾-in.	2.5 kg	250 g
37.5 mm	1½-in.	10 kg	1 kg
75.0 mm	3-in.	50 kg	5 kg

NOTE—*To be representative not less than 20 g shall be used.

(2.1) The minimum mass used may have to be increased to obtain the needed significant digits for the mass of water when reporting water contents to the nearest 0.1%.

(3) Using a test specimen smaller than the minimum indicated in (2) requires discretion, though it may be adequate for the purposes of the test. Any specimen used not meeting these requirements shall be noted on the test data forms or test data sheets.

(4) When working with a small (less than 200 g) specimen containing a relatively large gravel particle, it is appropriate not to include this particle in the test specimen. However, any discarded material shall be described and noted on the test data forms or test data sheets.

(5) For those samples consisting entirely of intact rock, the minimum specimen mass shall be 500 g. Representative portions of the sample may be broken into smaller particles, depending on the sample's size, the container, and balance being used, and to facilitate drying to constant mass, see Section (4) under "Procedure."

TEST SPECIMEN SELECTION [1]

(1) When the test specimen is a portion of a larger amount of material, the specimen must be selected to be representative of the water condition of the entire amount of material. The manner in which the test specimen is selected depends on the purpose and application of the test, type of material being tested, the water condition, and the type of sample (from another test, bag, block, and the likes).

(2) For disturbed samples, such as trimmings, bag samples, and the like, obtain the test specimen by one of the following methods (listed in order of preference):

(2.1) If the material is such that it can be manipulated and handled without significant moisture loss and segregation, the material should be mixed thoroughly and then select a representative portion using a scoop of a size that no more than a few scoopfuls are required to obtain the proper size of specimen defined by No. (2) under the Section "Test Specimen."

(2.2) If the material is such that it cannot be thoroughly mixed and/or split, form a stockpile of the material, mixing as much as possible. Take at least five portions of material at random locations using a sampling tube, shovel, scoop, trowel, or similar device appropriate to the maximum particle size present in the material. Combine all the portions for the test specimen.

(2.3) If the material or conditions are such that a stockpile cannot be formed, take as many portions of the material as practical using random locations that will best represent the moisture condition. Combine all the portions for the test specimen.

(3) Intact samples, such as block, tube, split barrel, and the like, obtain the test specimen by one of the following methods depending on the purpose and potential use of the sample.

(3.1) Using a knife, wire saw, or other sharp cutting device, trim the outside portion of the sample a sufficient distance to see if the material is layered and to remove material that appears more dry or more wet than the main portion of the sample. If the existence of layering is questionable, slice the sample in half. If the material is layered, see No. (3.3).

(3.2) If the material is not layered, obtain the specimen meeting the mass requirements in No. (2) under the Section "Test Specimen" by: (1) taking all or one-half of the interval being tested; (2) trimming a representative slice from the interval being tested; or (3) trimming the exposed surface of one-half or from the interval being tested.

> *Note 1*—Migration of moisture in some cohesionless soils may require that the full section be sampled.

(3.3) If a layered material (or more than one material type) is encountered, select an average specimen, or individual specimens, or both. Specimens must be properly identified as to location or what they represent and appropriate remarks entered on the test data forms or test data sheets.

PROCEDURE   Determination of moisture content of soil is actually quite simple. As indicated by Eq. (3–1), it is necessary only to determine the (1) mass of water in the soil sample and (2) mass of soil solids in the same sample. This is easily done by determining the mass of the moist soil sample, drying the sample to remove moisture, and then measuring the mass of the remaining oven-dried sample. The mass of the remaining oven-dried sample is, of course, the mass of soil solids in the sample. The difference between that mass and the mass of the original moist sample is the mass of water in the original sample. Substituting these values into Eq. (3–1) will give the desired moisture content of the soil.

The actual step-by-step procedure is as follows (ASTM D 2216-98 [1]):

(1) Determine and record the mass of the clean and dry specimen container (and its lid, if used).

(2) Select representative test specimens in accordance with "Test Specimen Selection."

(3) Place the moist test specimen in the container and, if used, set the lid securely in position. Determine the mass of the container and moist material using a balance selected on the basis of the specimen mass. Record this value.

> *Note 2*—To prevent mixing of specimens and yielding of incorrect results, all containers and lids, if used, should be numbered and the container numbers shall be recorded on the laboratory data sheets. The lid numbers should match the container numbers to eliminate confusion.

> *Note 3*—To assist in the oven-drying of large test specimens, they should be placed in containers having a large surface area (such as pans) and the material broken up into smaller aggregations.

(4) Remove the lid (if used) and place the container with moist material in the drying oven. Dry the material to a constant mass. Maintain the drying oven at 110 ± 5°C unless otherwise specified. The time required to obtain constant mass will vary depending on the type of material, size of specimen, oven type and capacity, and other factors. The influence of these factors generally can be established by good judgment and experience with the materials being tested and the apparatus being used.

> *Note 4*—In most cases, drying a test specimen overnight (about 12 to 16 h) is sufficient. In cases where there is doubt concerning the adequacy of drying, drying should be continued until the change in mass after two successive periods (greater than 1 h) of drying is an insignificant amount (less than about 0.1%). Specimens of sand may often be dried to constant mass in a period of about 4 h when a forced-draft oven is used.

> *Note 5*—Since some dry materials may absorb moisture from moist specimens, dried specimens should be removed before placing moist specimens in the same oven. However, this would not be applicable if the previously dried specimens will remain in the drying oven for an additional time period of about 16 h.

(5) After the material has dried to constant mass, remove the container from the oven (and replace the lid if used). Allow the material and container to cool to room temperature or until the container can be handled comfortably with bare hands and the operation of the balance will not be affected by convection currents and/or its being heated. Determine the mass of the container and oven-dried material using the same balance as used in (3). Record this value. Tight-fitting lids shall be used if it appears that the specimen is absorbing moisture from the air prior to determination of its dry mass.

> *Note 6*—Cooling in a desiccator is acceptable in place of tight-fitting lids since it greatly reduces absorption of moisture from the atmosphere during cooling, especially for containers without tight-fitting lids.

DATA Data collected in this test should include the following:

Mass of container, M_c

Mass of container and wet specimen, M_{cws}

Mass of container and oven-dried specimen, M_{cs}

Calculate the water content of the material as follows:

$$w = [(M_{cws} - M_{cs})(M_{cs} - M_c)] \times 100 = \frac{M_w}{M_s} \times 100 \qquad \textbf{(3–2)}$$

where:
w = water content, %
M_{cws} = mass of container and wet specimen, g
M_{cs} = mass of container and oven-dried specimen, g
M_c = mass of container, g
M_w = mass of water ($M_w = M_{cws} - M_{cs}$), g
M_s = mass of solid particles ($M_s = M_{cs} - M_c$), g

NUMERICAL EXAMPLE A laboratory test was conducted according to the procedure described previously. The following data were obtained:

Mass of container, M_c = **59.85 g**

Mass of container and wet specimen, M_{cws} = **241.25 g**

Mass of container and oven-dried soil, M_{cs} = **215.43 g**

Note: The data above are shown in boldface type to differentiate them from other values that are not collected during the test. In other words, boldface numbers indicate data collected during the test; all other numbers appear in regular type. This distinction is observed in the numerical examples throughout this book.

$$M_w = M_{cws} - M_{cs}$$

$$M_w = \mathbf{241.25} - \mathbf{215.43} = 25.82 \text{ g}$$

$$M_s = M_{cs} - M_c$$

$$M_s = \mathbf{215.43} - \mathbf{59.85} = 155.58 \text{ g}$$

Equation (3–1) can now be used to determine the desired moisture content.

$$w = \frac{M_w}{M_s} \times 100 \tag{3–1}$$

$$w = \frac{25.82}{155.58} \times 100 = 16.6\%$$

These results, together with the initial data, are summarized in the form on the following page. At the end of the chapter, two blank copies of this form are included for the reader's use.

REFERENCE   [1] ASTM, *2001 Annual Book of ASTM Standards,* West Conshohocken, PA, 2001. Copyright, American Society for Testing and Materials, 100 Barr Harbor Drive, West Conshohocken, PA 19428-2959. Reprinted with permission.

Soils Testing Laboratory
Moisture Content Determination

Sample No. _____15_____   Project No. _____SR2828_____

Boring No. _____B-7_____   Location _____Newell, N.C._____

Depth _____4 ft_____

Description of Sample _____Brown silty clay_____

Tested by _____John Doe_____   Date _____1/15/02_____

Determination No.:	1	2	3
Container (can) no.	A-1		
Mass of container + wet specimen, M_{cws} (g)	241.25		
Mass of container + oven-dried specimen, M_{cs} (g)	215.43		
Mass of container, M_c (g)	59.85		
Mass of water, M_w (g)	25.82		
Mass of solid particles, M_s (g)	155.58		
Moisture content, w (%)	16.6		

Soils Testing Laboratory
Moisture Content Determination

Sample No. _____   Project No. _____

Boring No. _____   Location _____

Depth _____

Description of Sample _____

Tested by _____   Date _____

Determination No.:	1	2	3
Container (can) no.			
Mass of container + wet specimen, M_{cws} (g)			
Mass of container + oven-dried specimen, M_{cs} (g)			
Mass of container, M_c (g)			
Mass of water, M_w (g)			
Mass of solid particles, M_s (g)			
Moisture content, w (%)			

Soils Testing Laboratory
Moisture Content Determination

Sample No. _____   Project No. _____

Boring No. _____   Location _____

Depth _____

Description of Sample _____

Tested by _____   Date _____

Determination No.:	1	2	3
Container (can) no.			
Mass of container + wet specimen, M_{cws} (g)			
Mass of container + oven-dried specimen, M_{cs} (g)			
Mass of container, M_c (g)			
Mass of water, M_w (g)			
Mass of solid particles, M_s (g)			
Moisture content, w (%)			

CHAPTER FOUR

Determining the Moisture Content of Soil (Microwave Oven Method)

(Referenced Document: ASTM D 4643)

INTRODUCTION

Chapter 3 presented the conventional oven method for determining moisture content of soil. That method has been used for many years in soils laboratories everywhere. It has one disadvantage, however; the time required to completely dry a soil sample in a conventional oven can be rather lengthy, while an assessment of moisture content may be needed quickly.

Soil can be dried faster in a modern microwave oven, and in 1987 ASTM published for the first time a standard test method (D 4643-87) for determining moisture content of soil using a microwave oven. This method is much quicker than the conventional oven method and has been found to give reliable results for most soil types. (The method may not give reliable results for (1) soils containing significant amounts of mica, gypsum, halloysite, montmorillonite, or other hydrated materials; (2) highly organic soils; or (3) soils in which the pore water contains dissolved solids.)

The microwave oven method generally gives results comparable to those obtained using conventional ovens, but ASTM states that if there are questions of accuracy between the two methods, the conventional oven method shall be the referee method. Furthermore, the microwave oven method is intended not as a replacement for the conventional oven method, but rather as a supplement when rapid results are needed to expedite other phases of testing.

In using microwave ovens for drying, care must be exercised to ensure that soil is not overheated, thereby causing a spuriously high reading of moisture content. To avoid this happening, the method presented uses an incremental drying procedure. Using microwave ovens having settings at less than full power can also be helpful in reducing overheating.

It should also be noted that when soil is subjected to microwave energy, its behavior depends on its mineralogical composition. Therefore, no one procedure is applicable to all types of soil. The procedure discussed in this chapter serves only as a guide when using microwave ovens to determine moisture content of soils.

APPARATUS AND SUPPLIES [1]

Microwave oven (preferably with a vented chamber and variable power controls; input power ratings of 700 W are adequate)

Balance (with accuracy of 0.01 g)

Containers (must be suitable for microwave ovens—i.e., nonmetallic and resistant to sudden and extreme temperature change; porcelain, glass, and even paper containers are generally satisfactory)

Glove or potholder

Desiccator (a cabinet or jar of suitable size containing silica gel, anhydrous calcium phosphate, or equivalent)

Heat sink (a material or liquid placed in the microwave to absorb energy after the moisture has been driven from the test specimen; the heat sink reduces the possibility of overheating the specimen and damage to the oven; glass beakers filled with water and materials that have a boiling point above water, such as nonflammable oils, have been used successfully; moistened bricks have also been used)

Stirring tools (spatulas, putty knives, and glass rods for cutting and stirring the test specimen before and during the test; short lengths of glass rods have been found useful for stirring and may be left in the specimen container during testing, reducing the possibility of specimen loss due to adhesion to the stirring tool)

HAZARDS [1]

(1) Handle hot containers with a container holder. Some soil types can retain considerable heat, and serious burns could result from improper handling.

(2) Suitable eye protection is recommended due to the possibility of particle shattering during the heating, mixing, or mass determinations.

(3) Safety precautions supplied by the manufacturer of the microwave should be observed. Particular attention should be paid to keeping the door sealing gasket clean and in good working condition.

Note 1—The use of a microwave oven for the drying of soils may be considered abusive by the manufacturers and constitute voiding of warranties. Microwave drying of soils containing metallic materials may cause arcing in the oven. Highly organic soils and soils containing oils and coal may ignite and burn during microwave drying. Continued operation of the oven after the soil has reached constant weight may also cause damage or premature failure of the microwave oven.

Note 2—When first introduced, microwave ovens were reported to affect heart pacemakers, primarily because of the operating frequencies of the two devices. Since that time, pacemakers have been redesigned, and the microwave oven is not regarded as the health hazard it once was. However, it is advisable to post warnings that a microwave is in use.

(4) Highly organic soils and soils containing oil or other contaminants may ignite into flames during microwave drying. Means for smothering flames to prevent operator injury or oven damage should be available during testing. Fumes given off from contaminated soils or wastes may be toxic, and the oven should be vented accordingly.

(5) Due to the possibility of steam explosions, or thermal stress shattering porous or brittle aggregates, a covering over the sample container may be appropriate to prevent operator injury or oven damage. A cover of heavy paper toweling has been found satisfactory for this purpose. This also prevents scattering of the test sample in the oven during the drying cycle.

(6) Do not use metallic containers in a microwave oven because arcing and oven damage may result.

(7) Observe manufacturer's operating instructions when installing and using the oven.

(8) The placement of the test specimen directly on the glass liner tray provided with some ovens is strongly discouraged. The concentrated heating of the specimen may result in the glass tray shattering, possibly causing injury to the operator.

TEST SAMPLE

A representative sample of the moist soil to be tested should be taken in an amount as indicated in Table 4–1. If cohesive soil samples are tested, break them to approximately ¼-in. (6-mm) particles in order to speed drying and prevent overheating of the surface while drying the interior.

TABLE 4–1 Test Specimen Masses [1]

Sieve Retaining Not More Than about 10% of Sample	*Recommended Mass of Moist Specimen, g*
2.0 mm (No. 10)	100 to 200
4.75 mm (No. 4)	300 to 500
19 mm (¾ in.)	500 to 1,000

PROCEDURE   The procedure for finding moisture content using microwave ovens is virtually the same as that given in Chapter 3 for conventional ovens, except for the manner in which soil specimens are dried.

The actual step-by-step procedure is as follows (ASTM D 4643-00[1]):

(1) Determine the mass of a clean, dry container or dish, and record.

(2) Place the soil specimen in the container, and immediately determine and record the mass.

(3) Place the soil and container in a microwave oven with the heat sink and turn the oven on for 3 min. If experience with a particular soil type and specimen size indicates shorter or longer initial drying times can be used without overheating, the initial and subsequent drying times may be adjusted.

> *Note 3*—The 3-min initial setting is for a minimum sample mass of 100 g, as indicated in Table 4–1. Smaller samples are not recommended when using the microwave oven because drying may be too rapid for proper control. When very large samples are needed to represent soil containing large gravel particles, the sample may need to be split into segments and dried separately to obtain the dry mass of the total sample.

> *Note 4*—Most ovens have a variable power setting. For the majority of soils tested, a setting of "high" should be satisfactory; however, for some soils such a setting may be too severe. The proper setting can be determined only through the use of and experience with a particular oven for various soil types and sample sizes. The energy output of microwave ovens may decrease with age and usage; therefore, power settings and drying times should be established for each oven.

(4) After the set time has elapsed, remove the container and soil from the oven, either weigh the specimen immediately, or place in desiccator to cool to allow handling and to prevent damage to the balance. Determine and record the mass.

(5) With a small spatula, knife, or short length of glass rod, carefully mix the soil, taking special precaution not to lose any soil.

(6) Return the container and soil to the oven and reheat in the oven for 1 min.

(7) Repeat (4) through (6), until the change between two consecutive mass determinations would have an insignificant effect on the calculated moisture content. A change of 0.1% or less of the initial wet mass of the soil should be acceptable for most specimens.

(8) Use the final mass determination in calculating the water content. Obtain this value immediately after the heating cycle, or, if the mass determination is to be delayed, after cooling in desiccator.

(9) When routine testing of similar soils is contemplated, the drying times and number of cycles may be standardized for each oven. When standardized drying times and cycles are utilized, periodic verification to assure that the results of the final dry mass determination are equivalent to the procedure in (7) should be performed.

> *Note 5*—Incremental heating, together with stirring, will minimize overheating and localized drying of the soil, thereby yielding results more consistent with results obtained by Method D 2216. The recommended time increments have been suitable for most specimens having particles smaller than a No. 4 sieve and with a mass of approximately 200 g; however, they may not be appropriate for all soils and ovens, and adjustment may be necessary.

> *Note 6*—Water content specimens should be discarded after testing and not used in any other tests due to particle breakdown, chemical changes or losses, melting, or losses of organic constituencies.

DATA Inasmuch as the microwave oven method for determining moisture content differs from the conventional oven method only in the manner in which soil specimens are dried, data collected are the same as indicated in Chapter 3.

CALCULATIONS See Chapter 3.

NUMERICAL EXAMPLE See Chapter 3.

REFERENCE [1] ASTM, *2001 Annual Book of ASTM Standards,* West Conshohocken, PA, 2001. Copyright, American Society for Testing and Materials, 100 Barr Harbor Drive, West Conshohocken, PA 19428-2959. Reprinted with permission.

CHAPTER FIVE

Determining the Specific Gravity of Soil

(Referenced Document: ASTM D 854)

INTRODUCTION   In general, the term *specific gravity* is defined as the ratio of the mass of a given volume of material to the mass of an equal volume of water. In effect, it tells how much the material is heavier (or lighter) than water. The particular geotechnical term *specific gravity of soil* actually denotes the specific gravity of the solid matter of the soil and refers, therefore, to the ratio of the mass of solid matter of a given soil sample to the mass of an equal volume (i.e., equal to the volume of the solid matter) of water. Alternatively, specific gravity of soil may be defined as the ratio of the unit mass of solids (mass of solids divided by volume of solids) in the soil to the unit mass of water. In equation form,

$$G_s = \frac{M_s/V_s}{\rho_w} = \frac{M_s}{V_s\,\rho_w} \tag{5-1}$$

where:

G_s = specific gravity of soil (dimensionless)
M_s = mass of solids, g
V_s = volume of solids, cm^3
ρ_w = unit mass (mass density) of water (1 g/cm^3)

The method given in this chapter covers determination of specific gravity of soils that pass the 4.75-mm (No. 4) sieve, using a pycnometer. When the soil contains particles larger than the 4.75-mm sieve, ASTM Test Method C 127 shall be used for the material retained on the 4.75-mm sieve and this test method shall be used for the material passing the 4.75-mm sieve. When the specific gravity value is to be used in calculations in connection with the hydrometer portion of ASTM Test Method D 422 (Chapter 9), it is intended that the specific gravity test be made on the portion of the sample that passes the 2.00-mm (No. 10) sieve. [1]

APPARATUS AND SUPPLIES

Pycnometer (volumetric flask), with a minimum capacity of 250 mL, preferably with a volume of 500 mL; the volume of the pycnometer must be two to three times greater than the volume of the soil-water mixture used during the deairing portion of the test (see Figure 5–1)

Balance (with accuracy to 0.01 g)

Thermometer

Desiccator

Drying oven

Vacuum pump

Evaporating dish

FIGURE 5–1   Apparatus for Determining Specific Gravity of Soil
(Courtesy of Cheng Liu.)

Spatula

Mechanically operated stirring device (malt mixer) (see Figure 5–1)

Large beaker

Distilled water is used in this test method. This water may be purchased and is readily available at most grocery stores; hereafter, distilled water will be referred to as water.

CALIBRATION OF PYCNOMETER

One parameter that is required in order to compute the specific gravity of soil is the mass of the pycnometer when filled with water, M_{pw}. The value of M_{pw} is not constant; it varies slightly as a function of water temperature (for the same volume of water). Furthermore, M_{pw} must be known at a temperature that equals the temperature of the same pycnometer when filled later with a mixture of water and soil sample. The value of M_{pw} at any desired temperature can be obtained by proper calibration of the pycnometer. The following procedure may be used to calibrate the device:

1. The pycnometer must be cleaned and dried, and its mass (M_p) determined and recorded.

2. The pycnometer is then filled with water that is approximately at room temperature, and the mass of the pycnometer plus water, M_{pw}, is accurately determined and recorded.

3. The water temperature T_i must be determined (and recorded) to the nearest 0.1°C (0.2°F) by inserting a thermometer in the water.

4. The value of M_{pw} can then be computed for any other water temperature, T_x, from the equation [1]

$$M_{pw} \text{ (at } T_x) = \frac{\text{density of water at } T_x}{\text{density of water at } T_i} [M_{pw} \text{ (at } T_i) - M_p] + M_p \qquad \textbf{(5–2)}$$

where:

M_{pw} = mass of pycnometer and water, g
M_p = mass of pycnometer, g
T_i = observed temperature of water, °C
T_x = any other desired temperature, °C

The required densities of water may be obtained from Table 5–1.

If many determinations of specific gravity are to be made using a particular pycnometer, it would be helpful to prepare a table (or graph) from which values of M_{pw} could be read for any desired water temperatures. Such tables (or graphs) can be developed by using Eq. (5–2) to determine values of M_{pw} corresponding to a sufficient number of different temperatures T_x.

Table 5–1   Density of Water and Temperature Coefficient (K) for Various Temperatures[A][1]

Temperature (°C)	Density (g/mL)[B]	Temperature Coefficient (K)	Temperature (°C)	Density (g/mL)[B]	Temperature Coefficient (K)	Temperature (°C)	Density (g/mL)[B]	Temperature Coefficient (K)	Temperature (°C)	Density (g/mL)[B]	Temperature Coefficient (K)
15.0	0.99910	1.00090	16.0	0.99895	1.00074	17.0	0.99878	1.00057	18.0	0.99860	1.00039
.1	0.99909	1.00088	.1	0.99893	1.00072	.1	0.99876	1.00055	.1	0.99858	1.00037
.2	0.99907	1.00087	.2	0.99891	1.00071	.2	0.99874	1.00054	.2	0.99856	1.00035
.3	0.99906	1.00085	.3	0.99890	1.00069	.3	0.99872	1.00052	.3	0.99854	1.00034
.4	0.99904	1.00084	.4	0.99888	1.00067	.4	0.99871	1.00050	.4	0.99852	1.00032
.5	0.99902	1.00082	.5	0.99886	1.00066	.5	0.99869	1.00048	.5	0.99850	1.00030
.6	0.99901	1.00080	.6	0.99885	1.00064	.6	0.99867	1.00047	.6	0.99848	1.00028
.7	0.99899	1.00079	.7	0.99883	1.00062	.7	0.99865	1.00045	.7	0.99847	1.00026
.8	0.99898	1.00077	.8	0.99881	1.00061	.8	0.99863	1.00043	.8	0.99845	1.00024
.9	0.99896	1.00076	.9	0.99879	1.00059	.9	0.99862	1.00041	.9	0.99843	1.00022
19.0	0.99841	1.00020	20.0	0.99821	1.00000	21.0	0.99799	0.99979	22.0	0.99777	0.99957
.1	0.99839	1.00018	.1	0.99819	0.99998	.1	0.99797	0.99977	.1	0.99775	0.99954
.2	0.99837	1.00016	.2	0.99816	0.99996	.2	0.99795	0.99974	.2	0.99773	0.99952
.3	0.99835	1.00014	.3	0.99814	0.99994	.3	0.99793	0.99972	.3	0.99770	0.99950
.4	0.99833	1.00012	.4	0.99812	0.99992	.4	0.99791	0.99970	.4	0.99768	0.99947
.5	0.99831	1.00010	.5	0.99810	0.99990	.5	0.99789	0.99968	.5	0.99766	0.99945
.6	0.99829	1.00008	.6	0.99808	0.99987	.6	0.99786	0.99966	.6	0.99764	0.99943
.7	0.99827	1.00006	.7	0.99806	0.99985	.7	0.99784	0.99963	.7	0.99761	0.99940
.8	0.99825	1.00004	.8	0.99804	0.99983	.8	0.99782	0.99961	.8	0.99759	0.99938
.9	0.99823	1.00002	.9	0.99802	0.99981	.9	0.99780	0.99959	.9	0.99756	0.99936
23.0	0.99754	0.99933	24.0	0.99730	0.99909	25.0	0.99705	0.99884	26.0	0.99679	0.99858
.1	0.99752	0.99931	.1	0.99727	0.99907	.1	0.99702	0.99881	.1	0.99676	0.99855
.2	0.99749	0.99929	.2	0.99725	0.99904	.2	0.99700	0.99879	.2	0.99673	0.99852
.3	0.99747	0.99926	.3	0.99723	0.99902	.3	0.99697	0.99876	.3	0.99671	0.99850
.4	0.99745	0.99924	.4	0.99720	0.99899	.4	0.99694	0.99874	.4	0.99666	0.99847
.5	0.99742	0.99921	.5	0.99717	0.99897	.5	0.99692	0.99871	.5	0.99665	0.99844
.6	0.99740	0.99919	.6	0.99715	0.99894	.6	0.99689	0.99868	.6	0.99663	0.99842
.7	0.99737	0.99917	.7	0.99712	0.99892	.7	0.99687	0.99866	.7	0.99660	0.99839
.8	0.99735	0.99914	.8	0.99710	0.99889	.8	0.99684	0.99863	.8	0.99657	0.99836
.9	0.99732	0.99912	.9	0.98707	0.99887	.9	0.99681	0.99860	.9	0.99654	0.99833
27.0	0.99652	0.99831	28.0	0.99624	0.99803	29.0	0.99595	0.99774	30.0	0.99585	0.99744
.1	0.99649	0.99828	.1	0.99621	0.99800	.1	0.99592	0.99771	.1	0.99562	0.99741
.2	0.99646	0.99825	.2	0.99618	0.99797	.2	0.99589	0.99768	.2	0.99559	0.99738
.3	0.99643	0.99822	.3	0.99615	0.99794	.3	0.99586	0.99765	.3	0.99556	0.99735
.4	0.99641	0.99820	.4	0.99612	0.99791	.4	0.99583	0.99762	.4	0.99553	0.99732
.5	0.99638	0.99817	.5	0.99609	0.99788	.5	0.99580	0.99759	.5	0.99550	0.99729
.6	0.99635	0.99814	.6	0.99607	0.99785	.6	0.99577	0.99756	.6	0.99547	0.99726
.7	0.99632	0.99811	.7	0.99604	0.99783	.7	0.99574	0.99753	.7	0.99544	0.99723
.8	0.99629	0.99808	.8	0.99601	0.99780	.8	0.99571	0.99750	.8	0.99541	0.99720
.9	0.99627	0.99806	.9	0.99598	0.99777	.9	0.99568	0.99747	.9	0.99538	0.99716

[A]Reference: *CRC Handbook of Chemistry and Physics*, David R. Lide, Editor-in-Chief, 74th Edition, 1993–1994.
[B]mL = cm^3.

TEST SPECIMEN [1]

(1) The test specimen may be moist or oven-dry soil and shall be representative of the soil solids that passes the U.S. Standard No. 4 sieve in the total sample. Table 5–2 gives guidelines on recommended dry soil mass versus soil type and pycnometer size.

(1.1) Two important factors concerning the amount of soil solids being tested are as follows. First, the mass of the soil solids divided by its specific gravity will yield four-significant digits. Secondly, the mixture of soil solids and water is a slurry not a highly viscous fluid (thick paint) during the deairing process.

Table 5–2 Recommended Mass for Test Specimen

Soil Type	Specimen Dry Mass (g) When Using 250 mL Pycnometer	Specimen Dry Mass (g) When Using 500 mL Pycnometer
SP, SP-SM	60 ± 10	100 ± 10
SP-SC, SM, SC	45 ± 10	75 ± 10
Silt or Clay	35 ± 5	50 ± 10

PROCEDURE The initial step of the procedure is to put the soil sample in a pycnometer, which is then filled with water, taking care to eliminate air bubbles (by methods described in detail shortly). The next step is to determine the mass of the pycnometer when filled with water and soil, M_{pws}, and then measure the temperature of the soil-and-water mixture. If samples containing natural moisture content are used, the soil-and-water mixture must be poured onto an evaporating dish and then dried in an oven to determine the mass of solids, M_s. With the temperature of the soil-and-water mixture known, the mass of the pycnometer when filled with water, M_{pw}, can be found from the calibration of the pycnometer. With these data now known, the specific gravity of soil can be computed by dividing the mass of solids, M_s, by the mass of an equal volume of water, $M_s + M_{pw} - M_{pws}$.

The actual step-by-step procedure is as follows (ASTM D 854-00 [1]):

(1) *Pycnometer Mass*—Using the same balance used to calibrate the pycnometer, verify that the mass of the pycnometer is within 0.06 g of the average calibrated mass. If it is not, re-calibrate the dry mass of the pycnometer.

(2) *Method A—Procedure for Moist Specimens:*

(2.1) Determine the water content of a portion of the sample in accordance with Test Method D 2216. Using this water content, calculate the range of wet masses for the specific gravity specimen in accordance with No. (1) under the Section "Test Specimen." From the sample, obtain a specimen within this range. Do not sample to obtain an exact predetermined mass.

(2.2) Disperse the soil using a blender or equivalent device to disperse the soil. Add the soil to about 100 mL of water. The minimum volume of slurry that can be prepared by this equipment will typically require using a 500-mL pycnometer.

(2.3) Using the funnel, pour the slurry into the pycnometer. Rinse any soil particles remaining on the funnel into the pycnometer using a wash/spray squirt bottle.

(2.4) Proceed as described in (4).

(3) *Method B—Procedure for Oven-Dried Specimens:*

(3.1) Dry the specimen to a constant mass in an oven maintained at $110 \pm 5°C$. Break up any clods of soil using a mortar and pestle. If the soil will not easily disperse after drying or has changed composition, use Test Method A. [Method A shall be used for organic soils; highly plastic soils; fine grained soils; tropical soils; and soils containing halloysite.]

(3.2) Place the funnel into the pycnometer. The stem of the funnel must extend past the calibration mark or stopper seal. Spoon the soil solids directly into the funnel. Rinse any soil particles remaining on the funnel into the pycnometer using a wash/spray squirt bottle.

(4) *Preparing the Soil Slurry*—Add water until the water level is between ⅓ and ½ of the depth of the main body of the pycnometer. Agitate the water until slurry is formed. Rinse any soil adhering to the pycnometer into the slurry.

(4.1) If slurry is not formed, but a viscous paste, use a pycnometer having a larger volume. See No. (1.1) under the Section "Test Specimen."

Note 1—For some soils containing a significant fraction of organic matter, kerosene is a better wetting agent than water and may be used in place of water for oven-dried specimens. If kerosene is used, the entrapped air should only be removed by use of an aspirator. Kerosene is a flammable liquid that must be used with extreme caution.

(5) *Deairing the Soil Slurry*—Entrapped air in the soil slurry can be removed using either heat (boiling), vacuum or combining heat and vacuum.

(5.1) When using the heat-only method (boiling), use a duration of at least 2 h after the soil-water mixture comes to a full boil. Use only enough heat to keep the slurry boiling. Agitate the slurry as necessary to prevent any soil from sticking to or drying onto the glass above the slurry surface.

(5.2) If only a vacuum is used, the pycnometer must be continually agitated under vacuum for at least 2 h. Continually agitated means the silt/clay soil solids will remain in suspension, and the slurry is in constant motion. The vacuum must remain relatively constant and be sufficient to cause bubbling at the beginning of the deairing process.

(5.3) If a combination of heat and vacuum is used, the pycnometers can be placed in a warm water bath (not more than 40°C) while applying the vacuum. The water level in the bath should be slightly below the water level in the pycnometer. If the pycnometer glass becomes hot, the soil will typically stick to or dry onto the glass. The duration of vacuum and heat must be at least 1 h after the initiation of boiling. During the process, the slurry should be agitated as necessary to maintain boiling and prevent soil from drying onto the pycnometer.

(6) *Filling the Pycnometer with Water*—Fill the pycnometer with deaired water by introducing the water through a piece of small-diameter flexible tubing with its outlet end kept just below the surface of the slurry in the pycnometer or by using the pycnometer filling tube. If the pycnometer filling tube is used, fill the tube with water, and close the valve. Place the tube such that the drainage holes are just at the surface of the slurry. Open the valve slightly to allow the water to flow over the top of the slurry. As the clear water layer develops, raise the tube and increase the flow rate. If the added water becomes cloudy, do not add water above the calibration mark or into the stopper seal area. Add the remaining water the next day.

(6.1) If using the stoppered iodine flask, fill the flask, such that the base of the stopper will be submerged in water. Then rest the stopper at an angle on the flared neck to prevent air entrapment under the stopper. If using a volumetric or stoppered flask, fill the flask to above or below the calibration mark depending on preference.

(7) If heat has been used, allow the specimen to cool to approximately room temperature.

(8) *Thermal Equilibrium*—Put the pycnometer(s) into the insulated container. The thermometer (in a beaker of water), and some deaired water in a bottle along with either an eyedropper or pipette should also be placed in the insulated container. Keep these items in the closed container overnight to achieve thermal equilibrium.

(9) *Pycnometer Mass Determination*—If the insulated container is not positioned near a balance, move the insulated container near the balance or vice versa. Open the container and remove the pycnometer. Only touch the rim of the pycnometer because the heat from hands can change the thermal equilibrium. Place the pycnometer on an insulated block (Styrofoam or equivalent).

(9.1) If using a volumetric flask, adjust the water to the calibration mark as follows: If using a volumetric flask as a pycnometer, adjust the water to the calibration mark, with the bottom of the meniscus level with the mark. If water has to be added, use the thermally equilibrated water from the insulated container. If water has to be removed, use a small suction tube or paper towel. Check for and remove any water beads on the pycnometer stem or on the exterior of the flask. Measure and record the mass of pycnometer and water to the nearest 0.01 g.

(9.2) If a stoppered flask is used, place the stopper in the bottle while removing the excess water using an eyedropper. Dry the rim using a paper towel. Be sure the entire exterior of the flask is dry.

(10) Measure and record the mass of pycnometer, soil, and water to the nearest 0.01 g using the same balance used for pycnometer calibration.

(11) *Pycnometer Temperature Determination*—Measure and record the temperature of the slurry/soil-water mixture to the nearest

0.1°C using the thermometer and method used during calibration. This is the test temperature, T_x.

(12) *Mass of Dry Soil*—Determine the mass of a tare or pan to the nearest 0.01 g. Transfer the soil slurry to the tare or pan. It is imperative that all of the soil be transferred. Water can be added. Dry the specimen to a constant mass in an oven maintained at 110 ± 5°C and cool it in a desiccator. If the tare can be sealed so that the soil cannot absorb moisture during cooling, a desiccator is not required. Measure the dry mass of soil solids plus tare to the nearest 0.01 g using the designated balance. Calculate and record the mass of dry soil solids to the nearest 0.01 g.

Note 2—This method has been proven to provide more consistent, repeatable results than determining the dry mass prior to testing. This is most probably due to the loss of soil solids during the deairing phase of testing.

DATA   Data collected in this test should include the following:

[A] Calibration of Pycnometer

Mass of pycnometer, M_p (g)

Mass of pycnometer plus water, M_{pw} (g)

Observed temperature of water, T_i (°C)

[B] Specific Gravity Determination

Mass of pycnometer plus water and soil, M_{pws} (g)

Temperature of contents of pycnometer when M_{pws} was determined, T_x (°C)

Mass of large evaporating dish or beaker, M_d (g)

Mass of large evaporating dish or beaker plus oven-dried soil, M_{ds} (g)

CALCULATIONS   Calculate the specific gravity of soil to the nearest 0.01 using the equation [1]

$$G_{s\,@\,20°C} = \frac{KM_s}{M_s + M_{pw}\,(\text{at } T_x) - M_{pws}} \qquad (5\text{–}3)$$

where:

$G_{s\,@\,20°C}$ = specific gravity of soil based on water at 20°C

K = conversion factor used to report specific gravity based on water at 20°C, which is the usual practice; the value of K can be determined from Table 5–1

$$
\begin{aligned}
M_s =&\ \text{mass of sample of oven-dried soil (i.e., } M_{ds} - M_d\text{), g} \\
M_{pw}(\text{at } T_x) =&\ \text{mass of pycnometer filled with water at temperature} \\
&\ T_x \text{ [see Eq. (5--2)], g} \\
T_x =&\ \text{temperature of contents of pycnometer when } M_{pws} \\
&\ \text{was determined, °C} \\
M_{pws} =&\ \text{mass of pycnometer plus water and soil (at } T_x\text{), g}
\end{aligned}
$$

The weighted average specific gravity for soils containing particles both larger and smaller than the 4.75-mm sieve can be calculated using the equation [1]

$$
G_{\text{avg @ 20°C}} = \frac{1}{\dfrac{R_1}{100 G_{1\,\text{@ 20°C}}} + \dfrac{P_1}{100 G_{2\,\text{@ 20°C}}}} \tag{5--4}
$$

where:

$$
\begin{aligned}
G_{\text{avg @ 20°C}} =&\ \text{weighted average specific gravity of soils composed of} \\
&\ \text{particles larger and smaller than the 4.75-mm sieve} \\
R_1 =&\ \text{percent of soil particles retained on 4.75-mm sieve} \\
P_1 =&\ \text{percent of soil particles passing the 4.75-mm sieve} \\
G_{1\,\text{@ 20°C}} =&\ \text{apparent specific gravity of soil particles retained on} \\
&\ \text{the 4.75-mm sieve as determined by Test Method C 127} \\
G_{2\,\text{@ 20°C}} =&\ \text{specific gravity of soil particles passing the 4.75-mm} \\
&\ \text{sieve as determined by this test method}
\end{aligned}
$$

NUMERICAL EXAMPLE

A laboratory test was conducted according to the procedure described previously. The following data were obtained:

[A] Calibration of Pycnometer

Mass of pycnometer, $M_p = $ **158.68 g**

Mass of pycnometer plus water, $M_{pw} = $ **656.43 g**

Observed temperature of water, $T_i = $ **24°C**

[B] Specific Gravity Determination

Mass of pycnometer plus water and soil, $M_{pws} = $ **718.52 g**

Temperature of contents of pycnometer when M_{pws} was determined, $T_x = $ **22°C**

Mass of evaporating dish, $M_d = $ **289.14 g**

Mass of evaporating dish plus oven-dried soil, $M_{ds} = $ **387.15 g**

Equation (5--2) can be used to evaluate M_{pw} (at T_x):

$$
M_{pw}\,(\text{at } T_x) = \frac{\text{density of water at } T_x}{\text{density of water at } T_i}\,[M_{pw}\,(\text{at } T_i) - M_p] + M_p \tag{5--2}
$$

$$M_{pw} \text{ (at } T_x) = \frac{0.99777}{0.99730} [656.43 - 158.68] + 158.68 = 656.66 \text{ g}$$

(Note that the density values were determined from Table 5–1 for a value of T_x equal to **22°C** and a value of T_i equal to **24°C**.)

The mass of the solids, M_s, can be computed by subtracting the mass of the evaporating dish, M_d, from the mass of the dish plus oven-dried soil, M_{ds}:

$$M_s = 387.15 - 289.14 = 98.01 \text{ g}$$

Equation (5–3) can now be used to compute the specific gravity of the soil, $G_{s @ 20°C}$:

$$G_{s @ 20°C} = \frac{KM_s}{M_s + M_{pw} \text{ (at } T_x) - M_{pws}} \tag{5–3}$$

$$G_{s @ 20°C} = \frac{(0.99957)(98.01)}{98.01 + 656.66 - 718.52} = 2.71$$

(Note that the value of K was determined from Table 5–1 for a value of T_x equal to **22°C**.)

These results, together with the initial data, are summarized in the form on the following page. At the end of the chapter, two blank copies of this form are included for the reader's use.

Soils Testing Laboratory
Specific Gravity Determination

Sample No. _____16_____   Project No. _____SR2828_____

Boring No. _____B-7_____   Location _____Newell, N.C._____

Depth _____4 ft_____

Description of Sample _____Brown silty clay_____

Tested by _____John Doe_____   Date _____1/23/02_____

[A] Calibration of Pycnometer

(1) Mass of dry, clean pycnometer, M_p __158.68__ g
(2) Mass of pycnometer + water, M_{pw} __656.43__ g
(3) Observed temperature of water, T_i __24__ °C

[B] Specific Gravity Determination

Test Method used (check one)   ☐ Method A
                                ☒ Method B

Maximum particle size of test specimen (check one)
                                ☒ No. 4 (4.75 mm)
                                ☐ No. 10 (2 mm)

Determination No.:	1	2	3
Mass of pycnometer + soil + water, M_{pws} (g)	718.52		
Temperature, T_x (°C)	22		
Mass of pycnometer + water at T_x, M_{pw} (at T_x) (g)	656.66		
Evaporating dish no.	1A		
Mass of evaporating dish, M_d (g)	289.14		
Mass of evaporating dish + oven-dried soil, M_{ds} (g)	387.15		
Mass of solids, M_s (g)	98.01		
Conversion factor, K	0.99957		
Specific gravity of soil, $G_{s@20°C} = \dfrac{KM_s}{M_s + M_{pw} \text{ (at } T_x) - M_{pws}}$	2.71		

51

CONCLUSIONS   The single value determined (and therefore reported) in this test procedure is the specific gravity of soil, G_s. For most soils, the value of G_s falls within the range 2.65 to 2.80. Table 5–3 gives some typical values of G_s for different types of soil.

Table 5–3   Expected Value for G_s [2]

Type of Soil	G_s
Sand	2.65–2.67
Silty sand	2.67–2.70
Inorganic clay	2.70–2.80
Soils with mica or iron	2.75–3.00
Organic soils	Variable, but may be under 2.00

Although the procedure for determining specific gravity of soils is relatively simple, several warnings are in order. To begin with, all data measurements—particularly the mass measurements—must be done extremely accurately in order to get an accurate value of G_s. It is good practice to use the same instrument for all mass determinations in a specific gravity test in order to eliminate any variations among instruments. The accuracy of the test is also highly dependent on how well air has been removed from the soil-and-water mixture in the pycnometer. Both soil and water can contain air, and if not completely removed from the soil-and-water mixture, any air entrapped inside the pycnometer will decrease the value of M_{pws}. This will, in turn, decrease the computed value of the specific gravity [see Eq. (5–3)]. Finally, care must be exercised to prevent (or at least minimize) the effects of nonuniform temperature of the soil-and-water mixture in the pycnometer.

The specific gravity of soil is an important parameter that is used together with other soil parameters (such as void ratio and degree of saturation) to compute additional useful information. Of particular note among the latter are the density and unit weight of soil, which are used in many soil engineering problems.

REFERENCES   [1]   ASTM, *2001 Annual Book of ASTM Standards*, West Conshohocken, PA, 2001. Copyright, American Society for Testing and Materials, 100 Barr Harbor Drive, West Conshohocken, PA 19428-2959. Reprinted with permission.

[2]   Joseph E. Bowles, *Engineering Properties of Soils and Their Measurement*, 2d ed., McGraw-Hill Book Company, New York, 1978.

Soils Testing Laboratory
Specific Gravity Determination

Sample No. _____ Project No. _____

Boring No. _____ Location _____

Depth _____

Description of Sample _____

Tested by _____ Date _____

[A] Calibration of Pycnometer

    (1) Mass of dry, clean pycnometer, M_p _____ g

    (2) Mass of pycnometer + water, M_{pw} _____ g

    (3) Observed temperature of water, T_i _____ °C

[B] Specific Gravity Determination

    Test Method used (check one)   ☐ Method A

                                                     ☐ Method B

    Maximum particle size of test specimen (check one)

                                                         ☐ No. 4 (4.75 mm)

                                                         ☐ No. 10 (2 mm)

Determination No.:	1	2	3
Mass of pycnometer + soil + water, M_{pws} (g)			
Temperature, T_x (°C)			
Mass of pycnometer + water at T_x, M_{pw} (at T_x) (g)			
Evaporating dish no.			
Mass of evaporating dish, M_d (g)			
Mass of evaporating dish + oven-dried soil, M_{ds} (g)			
Mass of solids, M_s (g)			
Conversion factor, K			
Specific gravity of soil, $G_{s @ 20°C} = \dfrac{KM_s}{M_s + M_{pw} \text{ (at } T_x) - M_{pws}}$			

Soils Testing Laboratory
Specific Gravity Determination

Sample No. _____   Project No. _____

Boring No. _____   Location _____

Depth _____

Description of Sample _____

Tested by_____   Date _____

[A] Calibration of Pycnometer

(1) Mass of dry, clean pycnometer, M_p _____ g
(2) Mass of pycnometer + water, M_{pw} _____ g
(3) Observed temperature of water, T_i _____ °C

[B] Specific Gravity Determination

Test Method used (check one)   ☐ Method A
                                ☐ Method B

Maximum particle size of test specimen (check one)
                                ☐ No. 4 (4.75 mm)
                                ☐ No. 10 (2 mm)

Determination No.:	1	2	3
Mass of pycnometer + soil + water, M_{pws} (g)			
Temperature, T_x (°C)			
Mass of pycnometer + water at T_x, M_{pw} (at T_x) (g)			
Evaporating dish no.			
Mass of evaporating dish, M_d (g)			
Mass of evaporating dish + oven-dried soil, M_{ds} (g)			
Mass of solids, M_s (g)			
Conversion factor, K			
Specific gravity of soil, $G_{s@20°C} = \dfrac{KM_s}{M_s + M_{pw} \text{ (at } T_x) - M_{pws}}$			

CHAPTER SIX

Determining the Liquid Limit of Soil

(Referenced Document: ASTM D 4318)

INTRODUCTION Atterberg [1, 2] established four states of consistency—i.e., degree of firmness—for fine-grained soils: *liquid, plastic, semisolid,* and *solid* (Figure 6–1). The dividing line between liquid and plastic states is the *liquid limit,* the dividing line between plastic and semisolid states is the *plastic limit,* and the dividing line between semisolid and solid states is the *shrinkage limit.* If a soil in the liquid state is gradually dried out, it will pass through the liquid limit, plastic state, plastic limit, semisolid state, and shrinkage limit and reach the solid state. The liquid, plastic, and shrinkage limits are therefore quantified in terms of water content. For example, the liquid limit is reported in terms of the water content at which a soil changes from the liquid to the plastic state. The difference between the liquid limit and plastic limit is the *plasticity index.*

The three limits and the index just defined are useful numbers in classifying soils and making judgments in regard to their applications. This chapter gives the laboratory procedure for determining liquid limit, whereas Chapters 7 and 8 give the procedures for determining (1) plastic limit and plasticity index and (2) shrinkage limit, respectively.

As explained, the liquid limit is the dividing line between the liquid and plastic states. It is quantified for a given soil as a specific water content; from a physical standpoint, it is the water content at which the shear strength of the soil becomes so small that the soil "flows" to close a standard groove cut in a sample of soil when it is jarred in a standard

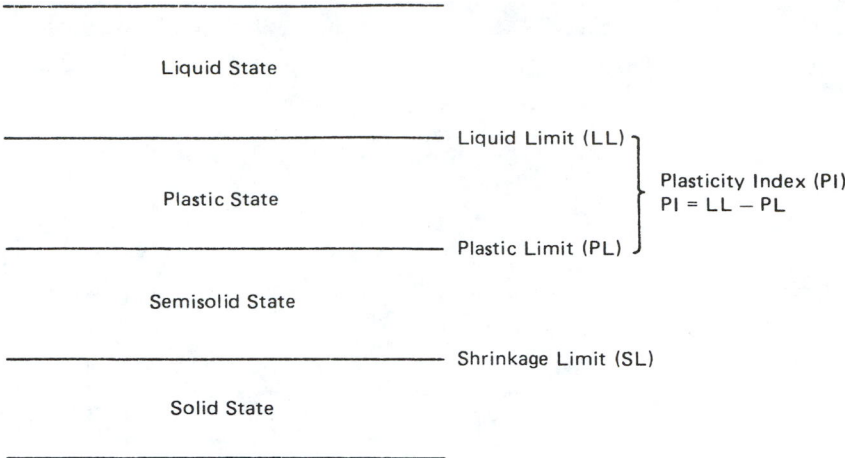

FIGURE 6–1   Atterberg Limits [2]

manner. The liquid limit is identified in the laboratory as that water content at which the groove is closed a distance of ½ in. when the soil sample is jarred in the standard manner by exactly 25 drops (or blows) from a height of 1 cm in a standardized liquid limit device.

In addition to being useful in identifying and classifying soils, the liquid limit can also be used to compute an approximate value of the compression index, C_c, for normally consolidated clays by equation [3, 4]

$$C_c = 0.009\,(LL - 10) \tag{6-1}$$

where LL, the liquid limit, is expressed as a percentage. The compression index is used in determining expected consolidation settlement of load on clay.

APPARATUS AND SUPPLIES

Liquid limit device (see Figure 6–2)

Flat grooving tool (see Figure 6–3)

Drop gage—a metal block for adjusting the height of drop of the cup (see Figure 6–4)

Balance (with accuracy to 0.01 g)

Evaporating dish

Spatula

Containers

Oven

No. 40 sieve

The liquid limit device and flat grooving tool are special equipment designed and built solely for determining liquid limits of soils. In simple terms, the liquid limit device consists of a brass cup that is alternately raised and dropped as a handle is turned. Figures 6–2 and 6–3 illustrate

Dimensions

Letter	A△	B△	C△	E△	F	G	H	J△	K△	L△	M△
MM	54 ± 0.5	2 ± 0.1	27 ± 0.5	56 ± 2.0	32	10	16	60 ± 1.0	50 ± 2.0	150 ± 2.0	125 ± 2.0

Letter	N	P	R	T	U△	V	W	Z
MM	24	28	24	45	47 ± 1.0	3.8	13	6.5

△ Essential dimensions

Cam Angle Degrees	Cam Radius
0	0.742 R
30	0.753 R
60	0.764 R
90	0.773 R
120	0.784 R
150	0.796 R
180	0.818 R
210	0.854 R
240	0.901 R
270	0.945 R
300	0.974 R
330	0.995 R
360	1.000 R

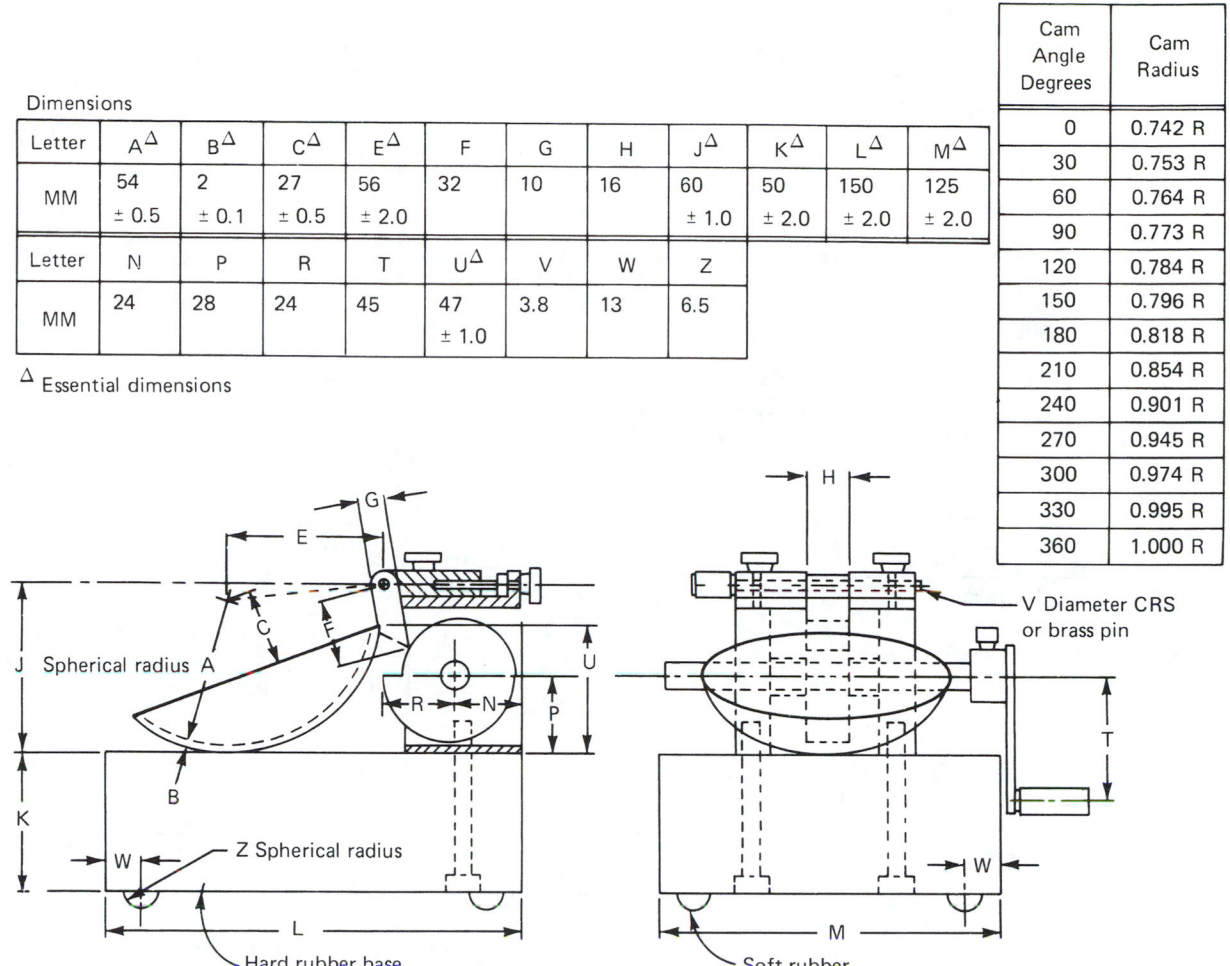

FIGURE 6–2 Hand-Operated Liquid Limit Device [5]

schematically the liquid limit device and flat grooving tool and also give their specified dimensions. Figure 6–4 shows the dimensions of a drop gage.

PREPARATION OF TEST SPECIMENS [5]

Obtain a representative portion from the total sample sufficient to provide 150 to 200 g of material passing the 425-μm (No. 40) sieve. Free flowing samples (materials) may be reduced by the methods of quartering or splitting. Non-free flowing or cohesive materials shall be mixed thoroughly in a pan with a spatula or scoop and a representative portion scooped from the total mass by making one or more sweeps with a scoop through the mixed mass.

Two procedures for preparing test specimens are provided as follows: *Wet preparation procedure,* as described in (1), and *dry preparation procedure,* as described in (2). The procedure to be used shall be specified by the requesting authority. If no procedure is specified, use the wet preparation procedure.

Dimensions

Letter	A$^\Delta$	B$^\Delta$	C$^\Delta$	D$^\Delta$	E$^\Delta$	F$^\Delta$
MM	2 ± 0.1	11 ± 0.2	40 ± 0.5	8 ± 0.1	50 ± 0.5	2 ± 0.1
Letter	G	H	J	K$^\Delta$	L$^\Delta$	N
MM	10 minimum	13	60	10 ± 0.05	60 deg ± 1 deg	20

$^\Delta$ Essential dimensions

$^\square$ Back at least 15 mm from tip

Note: Dimension A should be 1.9-2.0 and dimension D
should be 8.0-8.1 when new to allow for adequate service life

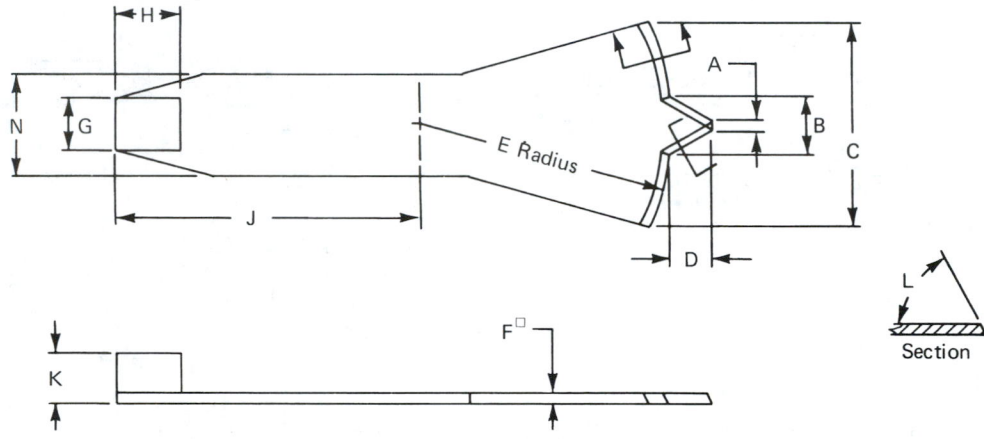

FIGURE 6–3   Grooving Tool (Optional Height-of-Drop Gage Attached) [5]

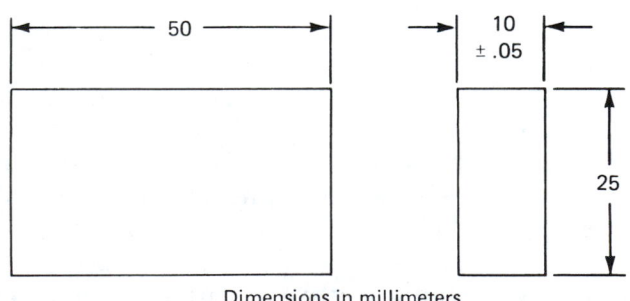

Dimensions in millimeters

FIGURE 6–4   Height of Drop Gage [5]

(1) *Wet Preparation*—Except where the dry method of specimen preparation is specified, prepare specimens for test as described in the following sections.

(1.1) *Material Passes the 425-μm (No. 40) Sieve:*

(1.1.1) Determine by visual and manual methods that the specimen obtained according to the first paragraph in this section has little or no material retained on a 425-μm (No. 40) sieve. If this is the case, prepare

150 to 200 g of material by mixing thoroughly with distilled or demineralized water on the glass plate or mixing dish using the spatula. If desired, soak the material in a mixing/storage dish with a small amount of water to soften the material before the start of mixing. If using Method A (see page 65), adjust the water content of the material to bring it to a consistency that would require about 25 to 35 blows of the liquid limit device to close the groove (Note 1). For Method B (see page 66), the number of blows should be between about 20 and 30 blows.

(1.1.2) If, during mixing, a small percentage of material is encountered that would be retained on a 425-μm (No. 40) sieve, remove these particles by hand (if possible). If it is impractical to remove the coarser material by hand, remove small percentages (less than about 15%) of coarser material by working the material (having the above consistency) through a 425-μm sieve. During this procedure, use a piece of rubber sheeting, rubber stopper, or other convenient device provided the procedure does not distort the sieve or degrade material that would be retained if the washing method described in (1.2) were used. If larger percentages of coarse material are encountered during mixing, or it is considered impractical to remove the coarser material by the procedures just described, wash the sample as described in (1.2). When the coarse particles found during mixing are concretions, shells or other fragile particles, do not crush these particles to make them pass a 425-μm sieve, but remove by hand or by washing.

(1.1.3) Place the prepared material in the mixing/storage dish, check its consistency (adjust if required), cover to prevent loss of moisture, and allow to stand (cure) for at least 16 h (overnight). After the standing period and immediately before starting the test, thoroughly remix the soil.

> *Note 1*—The time taken to adequately mix a soil will vary greatly, depending on the plasticity and initial water content. Initial mixing times of more than 30 min may be needed for stiff, fat clays.

(1.2) *Material Containing Particles Retained on a 425-μm (No. 40) Sieve:*

(1.2.1) Place the specimen (see first paragraph in this section) in a pan or dish and add sufficient water to cover the material. Allow the material to soak until all lumps have softened and the fines no longer adhere to the surfaces of the coarse particles (Note 2).

> *Note 2*—In some cases, the cations of salts present in tap water will exchange with the natural cations in the soil and significantly alter the test results if tap water is used in the soaking and washing operations. Unless it is known that such cations are not present in the tap water, distilled or demineralized water should be used. As a general rule, water containing more than 100 mg/L of dissolved solids should not be used for washing operations.

(1.2.2) When the material contains a large percentage of material retained on the 425-μm (No. 40) sieve, perform the following washing operation in increments, washing no more than 0.5 kg (1 lb) of material at one time. Place the 425-μm (No. 40) sieve in the bottom of the clean pan. Transfer without any loss of material the soil-water mixture onto the sieve. If gravel or coarse sand particles are present, rinse as many of these as possible with small quantities of water from a wash bottle, and discard. Alternatively, transfer the soil-water mixture over a 2.00-mm (No. 10) sieve nested atop the 425-μm (No. 40) sieve, rinse the fine material through, and remove the 2.00-mm (No. 10) sieve. After washing and removing as much of the coarser material as possible, add sufficient water to the pan to bring the level to about 13 mm (½ in.) above the surface of the 425-μm (No. 40) sieve. Agitate the slurry by stirring with the fingers while raising and lowering the sieve in the pan and swirling the suspension so that fine material is washed from the coarser particles. Disaggregate fine soil lumps that have not slaked by gently rubbing them over the sieve with the fingertips. Complete the washing operation by raising the sieve above the water surface and rinsing the material retained with a small amount of clean water. Discard material retained on the 425-μm (No. 40) sieve.

(1.2.3) Reduce the water content of the material passing the 425-μm (No. 40) sieve until it approaches the liquid limit. Reduction of water content may be accomplished by one or a combination of the following methods: (a) exposing to air currents at ordinary room temperature, (b) exposing to warm air currents from a source such as an electric hair dryer, (c) decanting clear water from the surface of suspension, (d) filtering in a buchner funnel or using filter candles, or (e) draining in a colander or plaster of paris dish lined with high retentivity, high wet-strength filter paper. If a plaster of paris dish is used, take care that the dish never becomes sufficiently saturated that it fails to absorb water into its surface. Thoroughly dry dishes between uses. During evaporation and cooling, stir the sample often enough to prevent overdrying of the fringes and soil pinnacles on the surface of the mixture. For soil samples containing soluble salts, use a method of water reduction such as (a) or (b) that will not eliminate the soluble salts from the test specimen.

(1.2.4) If applicable, remove the material retained on the filter paper. Thoroughly mix this material or the above material on the glass plate or in the mixing dish using the spatula. Adjust the water content of the mixture, if necessary, by adding small increments of distilled or demineralized water or by allowing the mixture to dry at room temperature while mixing on the glass plate. If using Method A, the material should be at a water content that will result in closure of the groove in 25 to 35 blows of the liquid limit device to close the groove. For Method B, the number of blows should be between 20 and 30. Put, if necessary, the mixed material in the storage dish, cover to prevent loss of moisture, and allow to stand (cure) for at least 16 h. After the standing period, and immediately before starting the test, thoroughly remix the specimen.

(2) *Dry Preparation:*

(2.1) Dry the specimen at room temperature or in an oven at a temperature not exceeding 60°C until the soil clods will pulverize readily. Disaggregation is expedited if the material is not allowed to completely dry. However, the material should have a dry appearance when pulverized.

(2.2) Pulverize the material in a mortar with a rubber-tipped pestle or in some other way that does not cause breakdown of individual particles. When the coarse particles found during pulverization are concretions, shells, or other fragile particles, do not crush these particles to make them pass a 425-μm (No. 40) sieve, but remove by hand or other suitable means, such as washing. If a washing procedure is used, follow (1.2.1)–(1.2.4).

(2.3) Separate the sample on a 425-μm (No. 40) sieve, shaking the sieve by hand to assure thorough separation of the finer fraction. Return the material retained on the 425-μm (No. 40) sieve to the pulverizing apparatus and repeat the pulverizing and sieving operations. Stop this procedure when most of the fine material has been disaggregated and material retained on the 425-μm sieve consists of individual particles.

(2.4) Place material remaining on the 425-μm (No. 40) sieve after the final pulverizing operations in a dish and soak in a small amount of water. Stir the soil-water mixture and pour over the 425-μm (No. 40) sieve, catching the water and any suspended fines in the washing pan. Pour this suspension into a dish containing the dry soil previously sieved through the 425-μm sieve. Discard material retained on the 425-μm sieve.

(2.5) Proceed as described in (1.2.3) and (1.2.4).

ADJUSTMENT OF MECHANICAL DEVICE

The liquid limit device should be inspected prior to use to ensure that it is in good working order. In particular, the height to which the cup is raised prior to dropping it onto the base should be checked. At the raised position, the distance from the bottom of the cup (measured from the point where it touches the base when dropped) to the base should be 10 ± 0.2 mm (0.3937 ± 0.0079 in.) (see Figure 6–5). If the distance is incorrect, the device should be adjusted accordingly.

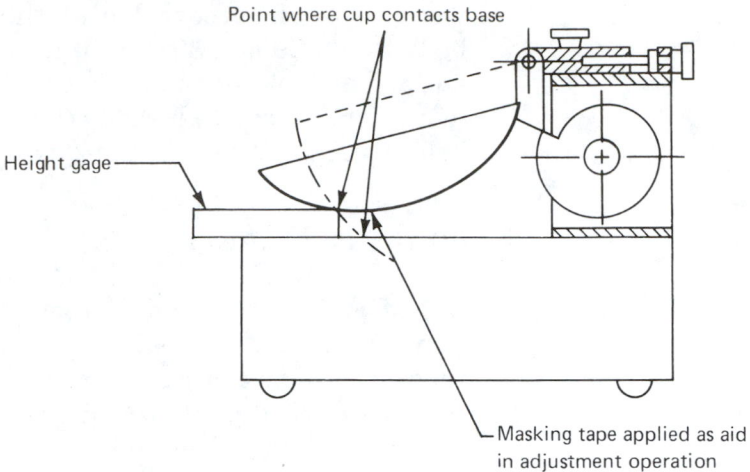

Point where cup contacts base

Height gage

Masking tape applied as aid
in adjustment operation

FIGURE 6–5   Calibration for Height of Drop [5]

PROCEDURE   As mentioned previously, the liquid limit is identified in the laboratory as that water content of the soil at which a groove of standard width is closed a distance of ½ in. when jarred in a standard manner by exactly 25 drops (or blows) from a height of 1 cm in a liquid limit device. Hence, the general procedure is to place a sample in the brass cup of the liquid limit device, cut the standard groove, and then count the number of drops of the device that are required to close the groove. If the number of drops turns out to be exactly 25, the water content of the sample could be determined, and it would be the liquid limit. This would happen only by chance, however; hence, the standard procedure is to carry out the foregoing operation at least three times for three different water contents. Each time the number of drops required to close the groove and the water content of the sample are determined. Then by plotting these data on semilogarithmic graph paper, the water content at which exactly 25 drops will presumably close the groove can be read. This water content is taken to be the liquid limit of the soil.

The generalized procedure presented in the preceding paragraph is denoted by ASTM as the "Multipoint Liquid Limit—Method A." An alternative ASTM procedure is known as the "One-Point Liquid Limit—Method B." In Method B, a soil sample is placed in the brass cup of the liquid limit device, the standard groove is cut, and the number of drops of the device that are required to close the groove is counted. If that number of drops is less than 20 or more than 30, the water content of the soil is adjusted and the procedure repeated until a test is done with a number of drops between 20 and 30, at which time the water content of the soil is determined. The liquid limit is computed using an equation involving the number of drops (between 20 and 30) required to close the groove and the associated water content of the soil. Two such determinations are done, and an average value is taken as the liquid limit.

Whether to use Method A or Method B should be specified by the requesting authority. If no method is specified, Method A is preferred, as it is generally more precise.

The actual step-by-step procedure for Multipoint Liquid Limit—Method A is as follows (ASTM D 4318-00 [5]):

(1) Thoroughly remix the specimen (soil) in its mixing cup, and, if necessary, adjust its water content until the consistency requires about 25 to 35 blows of the liquid limit device to close the groove. Using a spatula, place a portion(s) of the prepared soil in the cup of the liquid limit device at the point where the cup rests on the base, squeeze it down, and spread it into the cup to a depth of about 10 mm at its deepest point, tapering to form an approximately horizontal surface. Take care to eliminate air bubbles from the soil pat but form the pat with as few strokes as possible. Keep the unused soil in the storage dish. Cover the storage dish with a wet towel (or use other means) to retain the moisture in the soil.

(2) Form a groove in the soil pat by drawing the tool, beveled edge forward, through the soil on a line joining the highest point to the lowest point on the rim of the cup. When cutting the groove, hold the grooving tool against the surface of the cup and draw in an arc, maintaining the tool perpendicular to the surface of the cup throughout its movement. In soils where a groove cannot be made in one stroke without tearing the soil, cut the groove with several strokes of the grooving tool. Alternatively, cut the groove to slightly less than required dimensions with a spatula and use the grooving tool to bring the groove to final dimensions. Exercise extreme care to prevent sliding the soil pat relative to the surface of the cup.

(3) Verify that no crumbs of soil are present on the base or the underside of the cup. Lift and drop the cup by turning the crank at a rate of 1.9 to 2.1 drops per second until the two halves of the soil pat come in contact at the bottom of the groove along a distance of 13 mm (½ in.).

> *Note 1*—Use of a scale is recommended to verify that the groove has closed 13 mm (½ in.).

(4) Verify that an air bubble has not caused premature closing of the groove by observing that both sides of the groove have flowed together with approximately the same shape. If a bubble has caused premature closing of the groove, re-form the soil in the cup, adding a small amount of soil to make up for that lost in the grooving operation and repeat (1) to (3). If the soil slides on the surface of the cup, repeat (1) through (3) at a higher water content. If, after several trials at successively higher water contents, the soil pat continues to slide in the cup or if the number of blows required to close the groove is always less than 25, record that the liquid limit could not be determined, and report the soil as nonplastic without performing the plastic limit test.

(5) Record the number of drops N required to close the groove. Remove a slice of soil approximately the width of the spatula, extending from edge to edge of the soil cake at right angles to the groove and including that portion of the groove in which the soil flowed together, place in a weighed container, and cover.

(6) Return the soil remaining in the cup to the storage dish. Wash and dry the cup and grooving tool and reattach the cup to the carriage in preparation for the next trial.

(7) Remix the entire soil specimen in the storage dish adding distilled water to increase the water content of the soil and decrease the number of blows required to close the groove. Repeat (1) through (6) for at least two additional trials producing successively lower numbers of blows to close the groove. One of the trials shall be for a closure requiring 25 to 35 blows, one for closure between 20 and 30 blows, and one trial for a closure requiring 15 to 25 blows.

(8) Determine the water content of the soil specimen from each trial in accordance with ASTM Method D 2216 (Chapter 3). Initial weighings should be performed immediately after completion of the test. If the test is to be interrupted for more than about 15 minutes, the specimens already obtained should be weighed at the time of the interruption.

If using One-Point Liquid Limit—Method B, prepare the specimen in accordance with procedures described previously (page 56), except that at mixing, adjust the water content to a consistency requiring 20 to 30 drops of the liquid limit cup to close the groove. The step-by-step procedure is as follows (ASTM D 4318-95a [5]):

(1) Proceed as described in (1) through (5) for Method A except that the number of blows required to close the groove shall be 20 to 30. If less than 20 or more than 30 blows are required, adjust the water content of the soil and repeat the procedure.

(2) Immediately after removing a water content specimen as described in (5) for Method A, re-form the soil in the cup, adding a small amount of soil to make up for that lost in the grooving and water content sampling orientations. Repeat (2) through (5) for Method A, and, if the second closing of the groove requires the same number of drops or no more than two drops difference, secure another water content specimen. Otherwise, remix the entire specimen and repeat.

> *Note 2*—Excessive drying or inadequate mixing will cause the number of blows to vary.

(3) Determine water contents of specimens in accordance with (8) for Method A.

DATA   The data to be collected for each trial are the number of drops required to close the groove and the water content of the soil at that number of drops. Recall from Chapter 3 that, in order to determine water content, the mass of the container, mass of container plus wet soil, and mass of container plus oven-dried soil must be obtained. Because more than one water content must be determined, each container identification number should be recorded also. To summarize, the data collected for each trial in this test should include the following:

Number of drops, N

Container identification number

Mass of container, M_c

Mass of container plus moist soil, M_{cws}

Mass of container plus oven-dried soil, M_{cs}

CALCULATIONS   As indicated previously, the liquid limit is that water content of the soil at which the standard groove is closed a distance of ½ in. by exactly 25 drops. This is determined for Method A (Multipoint Liquid Limit) by plotting, on semilogarithmic graph paper, water content along the ordinate (arithmetical scale) versus number of drops along the abscissa (logarithmic scale) and drawing the best straight line through the plotted points. The resulting graph is referred to as a *flow curve*. From the flow curve, the liquid limit can be read as the water content at 25 drops.

For Method B (One-Point Liquid Limit), the liquid limit for each water content specimen can be determined using one of the following equations [5]:

$$LL = w\left(\frac{N}{25}\right)^{0.121} \tag{6--2}$$

or

$$LL = kw \tag{6--3}$$

where:
    $N =$ number of blows causing closure of the groove at water content
    $w =$ water content
    $k =$ factor given in Table 6–1

The liquid limit is the average of the two trial liquid limit values. If the difference between the two trial liquid limit values is greater than one percentage point, repeat the test.

Table 6–1 Factors for Obtaining Liquid Limit from Water Content and Number of Drops Causing Closure of Groove [5]

N (Number of Drops)	k (Factor for Liquid Limit)
20	0.974
21	0.979
22	0.985
23	0.990
24	0.995
25	1.000
26	1.005
27	1.009
28	1.014
29	1.018
30	1.022

NUMERICAL EXAMPLE

A laboratory test was conducted according to the previous procedure (Method A). The following data were obtained:

Determination No.	1	2	3
Number of drops	30	23	18
Can no.	A-1	A-2	A-3
Mass of can + moist soil, M_{cws}	34.06 g	32.47 g	37.46 g
Mass of can + dry soil, M_{cs}	27.15 g	25.80 g	29.00 g
Mass of can, M_c	11.80 g	11.61 g	11.69 g

The moisture content for each determination can be computed using Eq. (3–1). For the first determination,

$$M_w = M_{cws} - M_{cs}$$

$$M_w = 34.06 - 27.15 = 6.91 \text{ g}$$

$$M_s = M_{cs} - M_c$$

$$M_s = 27.15 - 11.80 = 15.35 \text{ g}$$

$$w = \frac{M_w}{M_s} \times 100 \tag{3–1}$$

$$w = \frac{6.91}{15.35} \times 100 = 45.0\%$$

Moisture contents for the second and third trials can be computed similarly. All results are shown on the form on page 71. (At the end of the chapter, two blank copies of this form are included for the reader's use.)

After the moisture contents have been determined, the flow curve can be obtained by plotting, on semilogarithmic graph paper, moisture

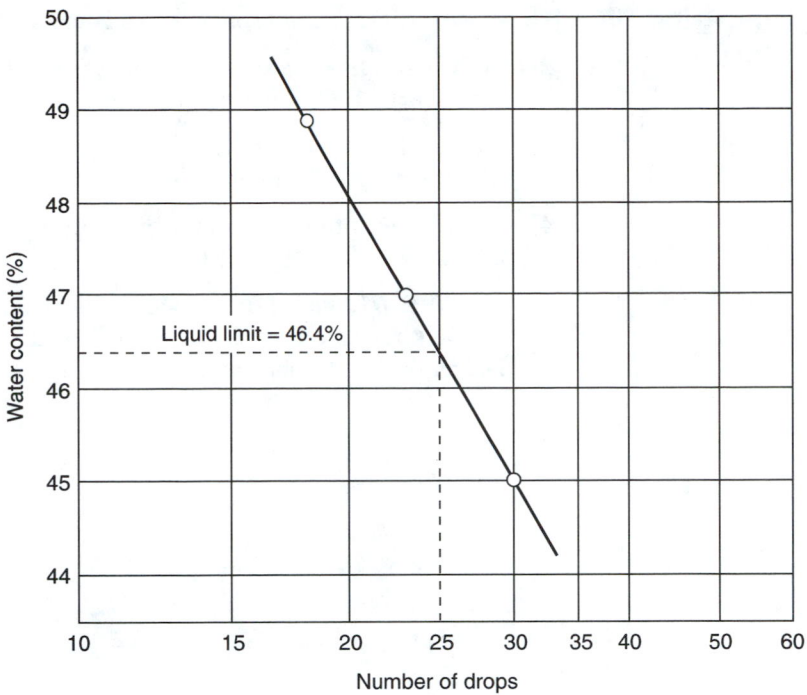

FIGURE 6–6   Flow Curve

content along the ordinate (arithmetical scale) versus number of drops along the abscissa (logarithmic scale) and drawing the best-fitting straight line through the plotted points. This is shown in Figure 6–6. (At the end of the text, a blank copy of semilogarithmic graph paper is included and may be photocopied as needed.) The liquid limit can be determined from the flow curve by noting the water content that corresponds to 25 drops. In this example, the liquid limit is observed to be 46.4%.

CONCLUSIONS   The primary result of the test performed is, of course, the value of the liquid limit. As pointed out earlier in this chapter, it can be used to approximate the compression index [Eq. (6–1)] for normally consolidated clays. In most cases, however, the liquid limit is determined and utilized in association with the other Atterberg limits and the plasticity index, all of which are used conjointly in soil identification and classification.

As mentioned previously, soil may be air dried prior to sieving (through a No. 40 sieve) to obtain the liquid limit sample, but it should not be oven dried. After sieving, care must be taken to ensure that the sample is well mixed with water prior to placing it in the brass cup of the liquid limit device. Entrapped air in the sample should be removed prior to the test, as it may decrease the accuracy of the results obtained.

The test shall always proceed from the dryer to the wetter condition of the soil. In no case shall dried soil be added to the seasoned soil being tested [5].

REFERENCES

[1] A. Atterberg, various papers published in the *Int. Mitt. Bodenkd.,* 1911, 1912.

[2] B. K. Hough, *Basic Soils Engineering,* 2d ed., The Ronald Press Company, New York, 1969. Copyright © 1969 by John Wiley & Sons, Inc.

[3] Ralph B. Peck, Walter E. Hansen, and Thomas H. Thornburn, *Foundation Engineering,* 2d ed., John Wiley & Sons, Inc., New York, 1974.

[4] A. W. Skempton, "Notes on the Compressibility of Clays," *Quart. J. Geol. Soc. Lond., C,* 119–135, 1944.

[5] ASTM, *2001 Annual Book of ASTM Standards,* West Conshohocken, PA, 2001. Copyright, American Society for Testing and Materials, 100 Barr Harbor Drive, West Conshohocken, PA 19428-2959. Reprinted with permission.

Soils Testing Laboratory
Liquid Limit Determination

Sample No. _____15_____   Project No. _____SR 2828_____

Boring No. _____B-21_____   Location _____Newell, N.C._____

Depth of Sample _____3 ft_____

Description of Sample _____Reddish brown silty clay_____

Tested by _____John Doe_____   Date _____1/26/02_____

Method used (Check one)   ☒ Method A
                                 ☐ Method B

Determination No.:	1	2	3
Number of drops	30	23	18
Can no.	A-1	A-2	A-3
Mass of can + moist soil, M_{cws} (g)	34.06	32.47	37.46
Mass of can + dry soil, M_{cs} (g)	27.15	25.80	29.00
Mass of can, M_c (g)	11.80	11.61	11.69
Mass of water, M_w (g)	6.91	6.67	8.46
Mass of dry soil, M_s (g)	15.35	14.19	17.31
Moisture content, w (%)	45.0	47.0	48.9

Method A: From the flow curve, the liquid limit = ___46.4___ %

Method B: From equation, the liquid limit for no. 1 determination = _____ %

From equation, the liquid limit for no. 2 determination = _____ %

The liquid limit (average of the two determinations) = _____ %

Soils Testing Laboratory
Liquid Limit Determination

Sample No. _____   Project No. _____

Boring No. _____   Location _____

Depth of Sample _____

Description of Sample _____

Tested by _____   Date _____

Method used (Check one)   ☐ Method A
                           ☐ Method B

Determination No.:	1	2	3
Number of drops			
Can no.			
Mass of can + moist soil, M_{cws} (g)			
Mass of can + dry soil, M_{cs} (g)			
Mass of can, M_c (g)			
Mass of water, M_w (g)			
Mass of dry soil, M_s (g)			
Moisture content, w (%)			

Method A: From the flow curve, the liquid limit = _____ %

Method B: From equation, the liquid limit for no. 1 determination = _____ %

From equation, the liquid limit for no. 2 determination = _____ %

The liquid limit (average of the two determinations) = _____ %

Soils Testing Laboratory
Liquid Limit Determination

Sample No. _____   *Project No.* _____

Boring No. _____   *Location* _____

Depth of Sample _____

Description of Sample _____

Tested by _____   *Date* _____

Method used (Check one)   ☐ Method A
                            ☐ Method B

Determination No.:	1	2	3
Number of drops			
Can no.			
Mass of can + moist soil, M_{cws} (g)			
Mass of can + dry soil, M_{cs} (g)			
Mass of can, M_c (g)			
Mass of water, M_w (g)			
Mass of dry soil, M_s (g)			
Moisture content, w (%)			

Method A: From the flow curve, the liquid limit = _____ %

Method B: From equation, the liquid limit for no. 1 determination = _____ %

From equation, the liquid limit for no. 2 determination = _____ %

The liquid limit (average of the two determinations) = _____ %

7

CHAPTER SEVEN

Determining the Plastic Limit and Plasticity Index of Soil

(Referenced Document: ASTM D 4318)

INTRODUCTION

As mentioned in Chapter 6, the *plastic limit* is the dividing line between the plastic and semisolid states. It is quantified for a given soil as a specific water content, and from a physical standpoint it is the water content at which the soil will begin to crumble when rolled into small threads. It is identified in the laboratory as the lowest water content at which the soil can be rolled into threads ⅛ in. (3.2 mm) in diameter without the threads breaking into pieces. The *plasticity index* is the difference between the liquid and plastic limits. Along with the other Atterberg limits, the plastic limit and plasticity index are valuable in identifying and classifying soils.

APPARATUS AND SUPPLIES

Evaporating dish (see Figure 7–1)

Spatula (see Figure 7–1)

Ground-glass plate—at least 30 cm (12 in.) square by 1 cm (⅜ in.) thick for mixing soil and rolling plastic limit threads (see Figure 7–1)

Balance (with accuracy to 0.01 g)

Containers (see Figure 7–1)

Oven

FIGURE 7–1   Apparatus for Determining Plastic Limit and Plasticity Index (Courtesy of Soiltest, Inc.)

PREPARATION OF TEST SPECIMEN [1]

Select a 20-g portion of soil from the material prepared for the liquid limit test, either after the second mixing before the test, or from the soil remaining after completion of the test. Reduce the water content of the soil to a consistency at which it can be rolled without sticking to the hands by spreading and mixing continuously on the glass plate or in the storage dish. The drying process may be accelerated by exposing the soil to the air current from an electric fan, or by blotting with paper that does not add any fiber to the soil. Paper such as hard surface paper toweling or high wet strength filter paper is adequate.

PROCEDURE

The procedure for determining the plastic limit is quite simple. The soil sample is rolled between the fingers and the rolling surface until a 3.2-mm (⅛-in.) diameter thread is obtained. This thread is broken into pieces, which are then squeezed together again. The resulting mass is rolled once more between the fingers and the rolling surface until another thread of the same diameter is obtained. Each time the soil is rolled out, it becomes drier by losing moisture. The entire process is repeated until the thread crumbles and the soil no longer can be rolled into a thread. At this point, the plastic limit is assumed to have been reached, and the water content (which is the plastic limit) is determined.

The actual step-by-step procedure is as follows (ASTM D 4318-00 [1]):

(1) From this plastic-limit specimen, select a 1.5 to 2.0 g portion. Form the selected portion into an ellipsoidal mass.

(2) Roll the soil mass by one of the following methods (hand or rolling device):

(2.1) *Hand Method*—Roll the mass between the palm or fingers and the ground-glass plate with just sufficient pressure to roll the mass into a thread of uniform diameter throughout its length (see Note 1). The thread shall be further deformed on each stroke so that its diameter reaches 3.2 mm (⅛ in.), taking no more than 2 min (see Note 2). The amount of hand or finger pressure required will vary greatly according to the soil being tested, that is, the required pressure typically increases with increasing plasticity. Fragile soils of low plasticity are best rolled under the outer edge of the palm or at the base of the thumb.

> *Note 1*—A normal rate of rolling for most soils should be 80 to 90 strokes per minute, counting a stroke as one complete motion of the hand forward and back to the starting position. This rate of rolling may have to be decreased for very fragile soils.

> *Note 2*—A 3.2-mm (⅛-in.) diameter rod or tube is useful for frequent comparison with the soil thread to ascertain when the thread has reached the proper diameter.

(2.2) *Rolling Device Method*—Attach smooth unglazed paper to both the top and bottom plates of the plastic limit-rolling device. Place the soil mass on the bottom plate at the midpoint between the slide rails. Place the top plate in contact with the soil mass(es). Simultaneously apply a slight downward force and back and forth motion to the top plate so that the top plate comes into contact with the side rails within 2 min (see Notes 1 and 3). During this rolling process, the end(s) the soil thread(s) shall not contact the side rail(s). If this occurs, roll a smaller mass of soil [even if it is less than that mentioned in (1)].

> *Note 3*—In most cases, two soil masses (threads) can be rolled simultaneously in the plastic limit-rolling device.

(3) When the diameter of the thread becomes 3.2 mm, break the thread into several pieces. Squeeze the pieces together, knead between the thumb and first finger of each hand, reform into an ellipsoidal mass, and reroll. Continue this alternate rolling to a thread 3.2 mm in diameter, gathering together, kneading and

rerolling, until the thread crumbles under the pressure required for rolling and the soil can no longer be rolled into a 3.2-mm diameter thread. It has no significance if the thread breaks into threads of shorter length. Roll each of these shorter threads to 3.2 mm in diameter. The only requirement for continuing the test is that they are able to be reformed into an ellipsoidal mass and rolled out again. The operator shall at no time attempt to produce failure at exactly 3.2 mm diameter by allowing the thread to reach 3.2 mm then reducing the rate of rolling or the hand pressure, or both, while continuing the rolling without further deformation until the thread falls apart. It is permissible, however, to reduce the total amount of deformation for feebly plastic soils by making the initial diameter of the ellipsoidal mass nearer to the required 3.2-mm final diameter. If crumbling occurs when the thread has a diameter greater than 3.2 mm, this shall be considered a satisfactory end point, provided the soil has been previously rolled into a thread 3.2 mm in diameter. Crumbling of the thread will manifest itself differently with the various types of soil. Some soils fall apart in numerous small aggregations of particles, others may form an outside tubular layer that starts splitting at both ends. The splitting progresses toward the middle, and finally, the thread falls apart in many small platy particles. Fat clay soils require much pressure to deform the thread, particularly as they approach the plastic limit. With these soils, the thread breaks into a series of barrel-shaped segments about 3.2 to 9.5 mm (⅛ to ⅜ in.) in length.

(4) Gather the portions of the crumbled thread together and place in a weighed container. Immediately cover the container.

(5) Select another 1.5- to 2.0-g portion of soil from the original 20 g specimen and repeat the operations described in (1) and (2) until the container has at least 6 g of soil.

(6) Repeat (1) through (5) to make another container holding at least 6 g of soil. Determine the water content of the soil contained in the containers in accordance with ASTM Method D 2216 (see Chapter 3).

DATA   As indicated in the preceding section, the plastic limit is the water content that exists when the thread crumbles and the soil no longer can be rolled into another thread. Hence, the only data collected and recorded are those required to determine the two water contents (see Chapter 3), namely:

Mass of container, M_c

Mass of container plus moist soil, M_{cws}

Mass of container plus oven-dried soil, M_{cs}

CALCULATIONS   After computing the water content of the soil contained in each container [Eq. (3–2)], the average of the two water contents is found. If the difference between the two water contents is greater than the acceptable range for two results listed in Table 7–1 for single-operator precision, the test must be repeated. The plastic limit is the average of the two water contents.

The plasticity index (PI) is determined simply by subtracting the plastic limit (PL) from the liquid limit (LL) (the latter to have been determined previously as explained in Chapter 6). In equation form,

$$PI = LL - PL \qquad\qquad (7\text{--}1)$$

NUMERICAL EXAMPLE A laboratory test was conducted according to the procedure previously described. The following data were obtained:

Can No.:	A-4	A-5
Mass of can plus moist soil, M_{cws}	15.47 g	16.27 g
Mass of can plus oven-dried soil, M_{cs}	14.79 g	15.46 g
Mass of can, M_c	11.56 g	11.68 g

These data are indicated on the form shown on page 83. (At the end of the chapter, two blank copies of this form are included for the reader's use.)

Inasmuch as sample water content calculations have been given in both Chapters 3 and 6, none is given here. The two computed water contents are 21.1% and 21.4% (see form on page 83). Because the difference between these two water contents (0.3%) is less than the acceptable range listed in Table 7–1 for single-operator precision, the plastic limit is the average of the two water contents, or 21.2%.

Table 7–1 Table of Precision Estimates[A] [1]

Material and Type Index	Standard Deviation[B]	Acceptable Range of Two Results[B]
Single-operator precison:		
Liquid Limit	0.8	2.4
Plastic Limit	0.9	2.6
Multilaboratory precision:		
Liquid Limit	3.5	9.9
Plastic Limit	3.7	10.6

[A]The figures given in Column 2 are the standard deviations that have been found to be appropriate for the test results described in Column 1. The figures given in Column 3 are the limits that should not be exceeded by the difference between the two properly conducted tests.
[B]These numbers represent, respectively, the (1S) and (D2S) limits as described in Practice C 670.

The plasticity index is determined by subtracting the plastic limit (21.2%, as just determined) from the liquid limit (46.4%, as determined in the section "Numerical Example" in Chapter 6). The value of the plasticity index in this example is therefore 46.4 − 21.2, or 25.2%.

CONCLUSIONS

If either the liquid or plastic limit cannot be determined, the plasticity index cannot be computed and should be reported as "NP" (indicating nonplastic). The plasticity index should also be reported as "NP" if the plastic limit turns out to be greater than or equal to the liquid limit. For very sandy soil, the plastic limit test should be performed before the liquid limit test; if it (the plastic limit) cannot be determined, both the liquid and the plastic limit should be reported as "NP."

The plastic limit is the lower boundary range of the plastic behavior of a given soil. It tends to increase in numerical value as grain size decreases. Its primary use is in association with the other Atterberg limits in soil identification and classification.

REFERENCE

[1]   ASTM, *2001 Annual Book of ASTM Standards,* West Conshohocken, PA, 2001. Copyright, American Society for Testing and Materials, 100 Barr Harbor Drive, West Conshohocken, PA 19428-2959. Reprinted with permission.

Soils Testing Laboratory
Plastic Limit Determination
and Plasticity Index

Sample No. _____15_____   Project No. _____SR 2828_____

Boring No. _____B-21_____   Location _____Newell, N.C._____

Depth of Sample _____3 ft_____

Description of Sample _____Reddish brown silty clay_____

Tested by _____John Doe_____   Date _____1/26/02_____

Determination No.:	1	2
Can no.	A-4	A-5
Mass of can + moist soil, M_{cws} (g)	15.47	16.27
Mass of can + dry soil, M_{cs} (g)	14.79	15.46
Mass of can, M_c (g)	11.56	11.68
Mass of water, M_w (g)	0.68	0.81
Mass of dry soil, M_s (g)	3.23	3.78
Water content, w (%)	21.1	21.4
Plastic limit (%)	21.2	

Liquid limit = __46.4__ %

Plastic limit = __21.2__ %

Plasticity index = liquid limit − plastic limit = __25.2__ %

Soils Testing Laboratory
Plastic Limit Determination
and Plasticity Index

Sample No. _____   Project No. _____

Boring No. _____   Location _____

Depth of Sample _____

Description of Sample _____

Tested by _____   Date _____

Determination No.:	1	2
Can no.		
Mass of can + moist soil, M_{cws} (g)		
Mass of can + dry soil, M_{cs} (g)		
Mass of can, M_c (g)		
Mass of water, M_w (g)		
Mass of dry soil, M_s (g)		
Water content, w (%)		
Plastic limit (%)		

Liquid limit = _____ %

Plastic limit = _____ %

Plasticity index = liquid limit − plastic limit = _____ %

85

Soils Testing Laboratory
Plastic Limit Determination
and Plasticity Index

Sample No. _____   Project No. _____

Boring No. _____   Location _____

Depth of Sample _____

Description of Sample _____

Tested by _____   Date _____

Determination No.:	1	2
Can no.		
Mass of can + moist soil, M_{cws} (g)		
Mass of can + dry soil, M_{cs} (g)		
Mass of can, M_c (g)		
Mass of water, M_w (g)		
Mass of dry soil, M_s (g)		
Water content, w (%)		
Plastic limit (%)		

Liquid limit = _____ %

Plastic limit = _____ %

Plasticity index = liquid limit − plastic limit = _____ %

CHAPTER EIGHT

Determining the Shrinkage Limit of Soil

(Referenced Document: ASTM D 427)

INTRODUCTION

As discussed in Chapter 6, the *shrinkage limit* is the dividing line between the semisolid and solid states. It is quantified for a given soil as a specific water content, and from a physical standpoint it is the water content that is just sufficient to fill the voids when the soil is at the minimum volume it will attain on drying. In other words, the smallest water content at which a soil can be completely saturated is the shrinkage limit. Below the shrinkage limit, any water content change will *not* result in volume change; above the shrinkage limit, any water content change will result in accompanying volume change (see Figure 8–1).

Another soil parameter that is often determined in conjunction with the shrinkage limit is the *shrinkage ratio,* which is an indicator of how much volume change may occur as changes in water content above the shrinkage limit take place. The shrinkage ratio is defined as the ratio of a given volume change, expressed as a percentage of the dry volume, to the corresponding change in water content above the shrinkage limit, expressed as a percentage of the mass of oven-dried soil. In equation form,

$$R = \frac{\Delta V / V_o}{\Delta w / M_o} \qquad (8\text{–}1)$$

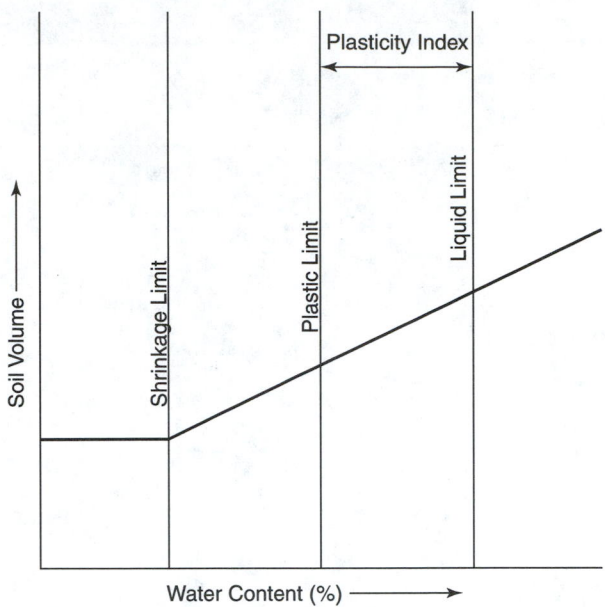

FIGURE 8-1   Definition of Shrinkage Limit

where:
R = shrinkage ratio
ΔV = soil volume change, cm^3
V_o = volume of oven-dried soil, cm^3
Δw = change in water content, g
M_o = mass of oven-dried soil, g

Because $\Delta w = (\Delta V)(\rho_w)$, where ρ_w is the unit mass of water in g/cm^3, Eq. (8–1) may be rewritten as

$$R = \frac{\Delta V/V_o}{(\Delta V)(\rho_w)/M_o}$$

or

$$R = \frac{M_o}{(V_o)(\rho_w)}$$

Because the unit mass of water, ρ_w, is 1 g/cm^3, the shrinkage ratio may be expressed simply as [1]

$$R = \frac{M_o}{V_o} \tag{8–2}$$

APPARATUS AND SUPPLIES

Evaporating dish [about 140 mm (5½ in.) in diameter; see Figures 8–2 and 8–3]

Spatula

Shrinkage dish (see Figures 8–2 and 8–3)

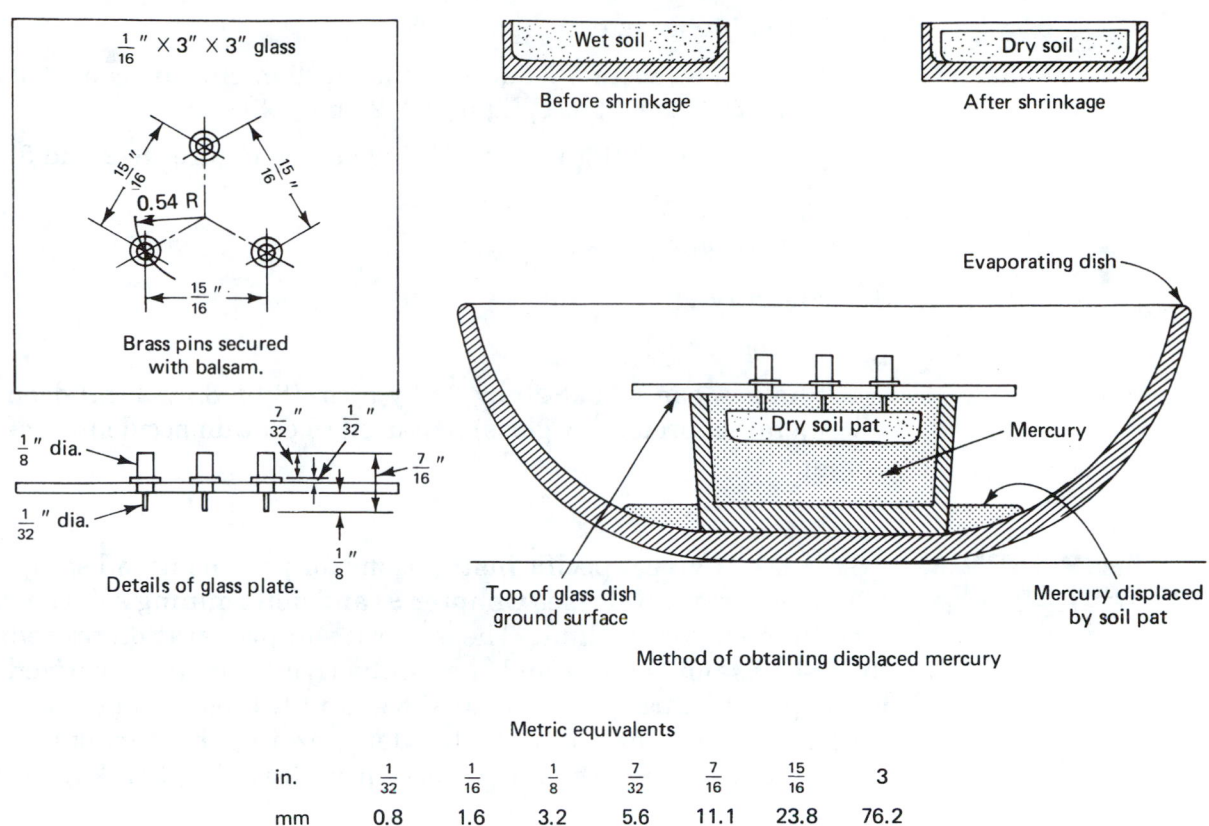

Method of obtaining displaced mercury

Metric equivalents

in.	$\frac{1}{32}$	$\frac{1}{16}$	$\frac{1}{8}$	$\frac{7}{32}$	$\frac{7}{16}$	$\frac{15}{16}$	3
mm	0.8	1.6	3.2	5.6	11.1	23.8	76.2

FIGURE 8–2 Apparatus for Determining the Volumetric Change of Subgrade Soils [1]

FIGURE 8–3 Apparatus for Determining Shrinkage Limit (Courtesy of Soiltest, Inc.)

Straightedge (steel)

Glass cup [about 57 mm (2½ in.) in diameter and about 31 mm (1¼ in.) in height; see Figures 8–2 and 8–3]

Glass plate (with three metal prongs; see Figures 8–2 and 8–3)

Graduate (glass) (see Figure 8–3)

Balance (with accuracy to 0.1 g)

Mercury

Petroleum jelly

Shallow pan [about 20 by 20 by 5 cm (8 by 8 by 2 in.) deep, non-metallic (preferably glass) pan used to contain accidental mercury spills]

PREPARATION OF TEST SAMPLE

ASTM D 421 gives specific instructions for preparing soil samples for particle-size analysis (see Chapter 9) and determining soil constants, including shrinkage limit. The general soil preparation procedure is to air-dry samples first and then pulverize them. A pulverized sample is passed through a No. 40 sieve, and the portion of the sample that passed through is used for determining the shrinkage limit. Approximately 80 g should be used in finding the shrinkage limit of a soil.

PROCEDURE

The general procedure for determining shrinkage limit is begun by placing the sample in an evaporating dish and mixing it with enough distilled water to fill the soil voids completely. After the shrinkage dish is coated with petroleum jelly, wet soil is taken from the evaporating dish with the spatula and placed in the shrinkage dish. This should be done in three parts, with steps taken each time to drive all air out of the soil. After the shrinkage dish and wet soil are weighed, the soil is set aside to dry in air. It is then oven-dried overnight, after which the shrinkage dish and dry soil are weighed again. After the oven-dried soil pat is removed from the shrinkage dish, its volume can be determined by mercury displacement. The weight and volume of the empty shrinkage dish must also be determined. The latter (i.e., the volume of the shrinkage dish) is also done by mercury displacement, and it is the same as the volume of the wet soil pat. With these data known, the shrinkage limit and shrinkage ratio can be determined by formulas.

The actual step-by-step procedure is as follows (ASTM D 427-98 [1]):

(1) Place the soil in the evaporating dish and thoroughly mix with distilled water. The amount of water added should produce a soil of the consistency somewhat above the liquid limit (Chapter 6) based on visual inspection. In physical terms, this is a consistency that is not a slurry but one that will flow sufficiently to expel air

bubbles when using gentle tapping action. It is desirable to use the minimum possible water content. This is of some importance with very plastic soils so that they do not crack during the drying process.

(2) Coat the inside of the shrinkage dish with a thin layer of petroleum jelly, silicone grease, or similar lubricant to prevent the adhesion of the soil to the dish. Determine and record the mass in grams of the empty dish, M_T.

(3) Place the shrinkage dish in the shallow pan in order to catch any mercury overflow. Fill the shrinkage dish to overflowing with mercury. Remove the excess mercury by pressing the glass plate firmly over the top of the shrinkage dish. Observe that there is no air trapped between the plate and mercury and, if there is, refill the dish and repeat the process. Determine the volume of mercury held in the shrinkage dish either by means of the glass graduate or by dividing the measured mass of mercury by the mass density of mercury (equal to 13.55 Mg/m^3). Record this volume in cubic centimeters of the wet soil pat, V.

> *Note 1—**Caution**—*Mercury is a hazardous substance which can cause serious health effects from prolonged inhalation of the vapor or contact with the skin.

> *Note 2—*It is not necessary to measure the volume of the shrinkage dish (wet soil pat) during each test. The value of a previous measurement may be used provided that it was obtained as specified in (3) and the shrinkage dish is properly identified and in good physical condition.

(4) Place an amount of the wetted soil equal to about one-third the volume of the dish in the center of the dish, and cause the soil to flow to the edges by tapping the dish on a firm surface cushioned by several layers of blotting paper or similar material. Add an amount of soil approximately equal to the first portion, and tap the dish until the soil is thoroughly compacted and all included air has been brought to the surface. Add more soil and continue the tapping until the dish is completely filled and excess soil stands out above its edge. Strike off the excess soil with a straightedge, and wipe off all soil adhering to the outside of the dish. Immediately after it is filled and struck off, determine and record the mass in grams of the dish and wet soil, M_w.

(5) Allow the soil pat to dry in air until the color of the soil turns from dark to light. Oven-dry the soil pat to constant mass at $110 \pm 5°C$ ($230 \pm 9°F$). If the soil pat is cracked or has broken in pieces, return to (1) and prepare another soil pat using a lower water content. Determine and record the mass in g of the dish and dry soil, M_D.

(6) Determine the volume of the dry soil pat by removing the pat from the shrinkage dish and immersing it in the glass cup full of mercury in the following manner.

(6.1) Place the glass cup in the shallow pan in order to catch any mercury overflow. Fill the glass cup to overflowing with mercury. Remove the excess mercury by pressing the glass plate with the three prongs (Figure 8–2) firmly over the top of the cup. Observe that there is no air trapped between the plate and mercury and, if there is, refill the dish and repeat the process. Carefully wipe off any mercury that may be adhering to the outside of the cup.

(6.2) Place the evaporating dish in the shallow pan in order to catch any mercury overflow. Place the cup filled with mercury in the evaporating dish and rest the soil pat on the surface of the mercury (it will float). Using the glass plate with the three prongs, gently press the pat under the mercury and press the plate firmly over the top of the cup to expel any excess mercury. Observe that no air is trapped between the plate and mercury and, if there is, repeat the process starting from (6.1). Measure the volume of the mercury displaced into the evaporating dish either by means of the glass graduate or by dividing the measured mass of mercury by the mass density of mercury. Record the volume in cubic centimeters (cubic feet) of the dry soil pat, V_o.

DATA   Data collected in this test should include the following:

Mass of dish coated with petroleum jelly, M_T

Mass of dish coated with petroleum jelly plus wet soil, M_w

Mass of dish coated with petroleum jelly plus oven-dried soil, M_D

Volume of dish, V (which is equal to the volume of the wet soil pat)

Volume of oven-dried soil pat, V_o

CALCULATIONS   From the known mass of the dish coated with petroleum jelly, M_T, mass of the dish coated with petroleum jelly plus wet soil, M_w, and mass of the dish coated with petroleum jelly plus oven-dried soil, M_D, the water content of the wet soil pat, w, can be computed using the following equations [1]:

$$\text{Initial wet soil mass } (M) = M_w - M_T \tag{8–3}$$

$$\text{Dry soil mass } (M_o) = M_D - M_T \tag{8–4}$$

$$w = \left(\frac{M - M_o}{M_o}\right) \times 100 \tag{8–5}$$

With the water content known, the shrinkage limit can be computed using equation [1]:

$$SL = w - \left[\frac{(V - V_o)\rho_w}{M_o} \right] \times 100 \qquad \text{(8–6)}$$

where:

SL = shrinkage limit (expressed as a percentage)

w = water content of wet soil in the shrinkage dish (expressed as a percentage)

V = volume of wet soil pat (same as volume of shrinkage dish), cm^3

V_o = volume of oven-dried soil pat, cm^3

ρ_w = approximate density of water equal to 1.0 g/cm^3

M_o = mass of oven-dried soil pat, g

The shrinkage ratio R can be computed using equation [1]:

$$R = \frac{M_o}{V_o \times \rho_w} \qquad \text{(8–7)}$$

NUMERICAL EXAMPLE

A laboratory test was conducted according to the procedure described previously. The following data were obtained:

Mass of dish coated with petroleum jelly, M_T = **11.30 g**

Mass of dish coated with petroleum jelly plus wet soil, M_w = **38.51 g**

Mass of dish coated with petroleum jelly plus oven-dried soil, M_D = **32.81 g**

Volume of wet soil pat (same as volume of shrinkage dish), V = **15.26 cm^3**

Volume of oven-dried soil pat, V_o = **12.83 cm^3**

The water content of the wet soil pat can be computed using Eqs. (8–3) through (8–5).

$$M = M_w - M_T \qquad \text{(8–3)}$$

$$M = 38.51 - 11.30 = 27.21 \text{ g}$$

$$M_o = M_D - M_T \qquad \text{(8–4)}$$

$$M_o = 32.81 - 11.30 = 21.51 \text{ g}$$

$$w = \left(\frac{M - M_o}{M_o} \right) \times 100 \qquad \text{(8–5)}$$

$$w = \left(\frac{27.21 - 21.51}{21.51} \right) \times 100 = 26.5\%$$

Equation (8–6) can then be used to compute the shrinkage limit:

$$SL = w - \left[\frac{(V - V_o)\rho_w}{M_o}\right] \times 100 \qquad (8\text{–}6)$$

$$SL = 26.5 - \left[\frac{(15.26 - 12.83)(1.0)}{21.51}\right] \times 100 = 15.2\%$$

Finally, Eq. (8–7) can be used to find the shrinkage ratio:

$$R = \frac{M_o}{V_o \times \rho_w} \qquad (8\text{–}7)$$

$$R = \frac{21.51}{12.83 \times 1.0} = 1.68$$

These results, together with the initial data, are summarized in the form on page 97. At the end of the chapter, two blank copies of this form are included for the reader's use.

CONCLUSIONS

The shrinkage limit and shrinkage ratio are particularly useful in analyzing soils that undergo large volume changes with changes in water content (such as clays). The shrinkage ratio gives an indication of how much volume change may occur as changes in water content above the shrinkage limit take place. Large changes in soil volume are important considerations for soils that are to be used as fill material for highways and railroads or for soils that are to support structural foundations. Unequal settlements resulting from such volume changes can result in cracks in structures or unevenness in roadbeds.

REFERENCE

[1]   ASTM, *2001 Annual Book of ASTM Standards,* West Conshohocken, PA, 2001. Copyright, American Society for Testing and Materials, 100 Barr Harbor Drive, West Conshohocken, PA 19428-2959. Reprinted with permission.

Soils Testing Laboratory
Shrinkage Limit Determination

Sample No. _____15_____   Project No. _____SR 2828_____

Boring No. _____B-21_____   Location _____Newell, N.C._____

Depth of Sample _____3 ft_____

Description of Sample _____Reddish brown silty clay_____

Tested by _____John Doe_____   Date _____1/26/02_____

Trial No.:	1	2
Mass of dish coated with petroleum jelly + wet soil, M_w (g)	38.51	
Mass of dish coated with petroleum jelly + oven-dried soil, M_D (g)	32.81	
Mass of dish coated with petroleum jelly (i.e., empty coated dish), M_T (g)	11.30	
Mass of wet soil pat, M (g)	27.21	
Mass of oven-dried soil pat, M_o (g)	21.51	
Mass of water in wet soil pat (g)	5.70	
Water content of soil when placed in dish, w (%)	26.5	
Volume of wet soil pat (i.e., volume of dish), V (cm^3)	15.26	
Volume of oven-dried soil pat, V_o (cm^3)	12.83	
Shrinkage limit, SL (%) $= w - \left[\dfrac{(V - V_o)\rho_w}{M_o} \right] \times 100$	15.2	
Shrinkage ratio, $R = \dfrac{M_o}{V_o \times \rho_w}$	1.68	

Note: $\rho_w = 1.0$ g/cm^3

Soils Testing Laboratory
Shrinkage Limit Determination

Sample No. _____   Project No. _____

Boring No. _____   Location _____

Depth of Sample _____

Description of Sample _____

Tested by _____   Date _____

Trial No.:	1	2
Mass of dish coated with petroleum jelly + wet soil, M_w (g)		
Mass of dish coated with petroleum jelly + oven-dried soil, M_D (g)		
Mass of dish coated with petroleum jelly (i.e., empty coated dish), M_T (g)		
Mass of wet soil pat, M (g)		
Mass of oven-dried soil pat, M_o (g)		
Mass of water in wet soil pat (g)		
Water content of soil when placed in dish, w (%)		
Volume of wet soil pat (i.e., volume of dish), V (cm^3)		
Volume of oven-dried soil pat, V_o (cm^3)		
Shrinkage limit, $SL\ (\%) = w - \left[\dfrac{(V - V_o)r_w}{M_o} \right] \times 100$		
Shrinkage ratio, $R = \dfrac{M_o}{V_o \times \rho_w}$		

Note: $\rho_w = 1.0$ g/cm^3

Soils Testing Laboratory
Shrinkage Limit Determination

Sample No. _____   *Project No.* _____

Boring No. _____   *Location* _____

Depth of Sample _____

Description of Sample _____

Tested by _____   *Date* _____

Trial No.:	1	2
Mass of dish coated with petroleum jelly + wet soil, M_w (g)		
Mass of dish coated with petroleum jelly + oven-dried soil, M_D (g)		
Mass of dish coated with petroleum jelly (i.e., empty coated dish), M_T (g)		
Mass of wet soil pat, M (g)		
Mass of oven-dried soil pat, M_o (g)		
Mass of water in wet soil pat (g)		
Water content of soil when placed in dish, w (%)		
Volume of wet soil pat (i.e., volume of dish), V (cm^3)		
Volume of oven-dried soil pat, V_o (cm^3)		
Shrinkage limit, $SL\ (\%) = w - \left[\dfrac{(V - V_o)r_w}{M_o} \right] \times 100$		
Shrinkage ratio, $R = \dfrac{M_o}{V_o \times \rho_w}$		

Note: $\rho_w = 1.0$ g/cm^3

CHAPTER NINE

Grain-Size Analysis of Soil (Including Both Mechanical and Hydrometer Analyses)

(Referenced Document: ASTM D 422)

INTRODUCTION Grain-size analysis, which is among the oldest of soil tests, is widely used in engineering classifications of soils. Grain-size analysis is also utilized in part of the specifications of soil for airfields, roads, earth dams, and other soil embankment construction. Additionally, frost susceptibility of soils can be fairly accurately predicted from the results of grain-size analysis. The standard grain-size analysis test determines the relative proportions of different grain sizes as they are distributed among certain size ranges.

Grain-size analysis of soils containing relatively large particles is accomplished using sieves. A sieve is similar to a cook's flour sifter. It is an apparatus having openings of equal size and shape through which grains smaller than the size of the opening will pass, while larger grains are retained. Obviously, a sieve can be used to separate soil grains in a sample into two groups: one containing grains smaller than the size of the sieve opening and the other containing larger grains. By passing the sample downward through a series of sieves, each of decreasing size openings, the grains can be separated into several groups, each of which contains grains in a particular size range. The various sieve sizes are usually specified and are standardized.

Soils with small grain sizes cannot generally be analyzed using sieves, because of the very small size of sieve opening that would be required and the difficulty of getting such small particles to pass through.

Grain-size analysis for these soils is done, therefore, by another method—hydrometer analysis. The hydrometer method is based on *Stokes' law,* which says that the larger the grain size, the greater its settling velocity in a fluid.

If a soil sample contains both large and small particles, which is often the case, its grain-size analysis can be performed using a combination of the two methods described.

APPARATUS AND SUPPLIES

Sieves (see Figure 9–1)

A full set of sieves includes the following sizes [1]:

      3 in. (75 mm)

      2 in. (50 mm)

      1½ in. (38.1 mm)

      1 in. (25.0 mm)

      ¾ in. (19.0 mm)

      ⅜ in. (9.5 mm)

      No. 4 (4.75 mm)

      No. 10 (2.00 mm)

      No. 20 (0.850 mm)

      No. 40 (0.425 mm)

      No. 60 (0.250 mm)

      No. 140 (0.140 mm)

      No. 200 (0.075 mm)

FIGURE 9–1　Set of Sieves (Courtesy of Soiltest, Inc.)

A set of sieves that gives uniform spacing of points for the grain-size distribution curve to be prepared may be used. It consists of the following sizes [1]:

3 in. (75 mm)

1½ in. (38.1 mm)

¾ in. (19.0 mm)

⅜ in. (9.5 mm)

No. 4 (4.75 mm)

No. 8 (2.36 mm)

No. 16 (1.18 mm)

No. 30 (0.600 mm)

No. 50 (0.300 mm)

No. 100 (0.150 mm)

No. 200 (0.075 mm)

Balance (with accuracy to 0.01 g)

Mechanically operated stirring device (such as a malt mixer) (see Figures 9–2 and 9–3)

Hydrometer (conforming to requirements for hydrometer 151H or 152H in ASTM Specification E 100) (see Figure 9–3)

Sedimentation cylinder (with a volume of 1,000 ml) (see Figure 9–3)

Thermometer [with accuracy to 1°F (0.5°C)]

Water bath or constant-temperature room (see Figure 9–3)

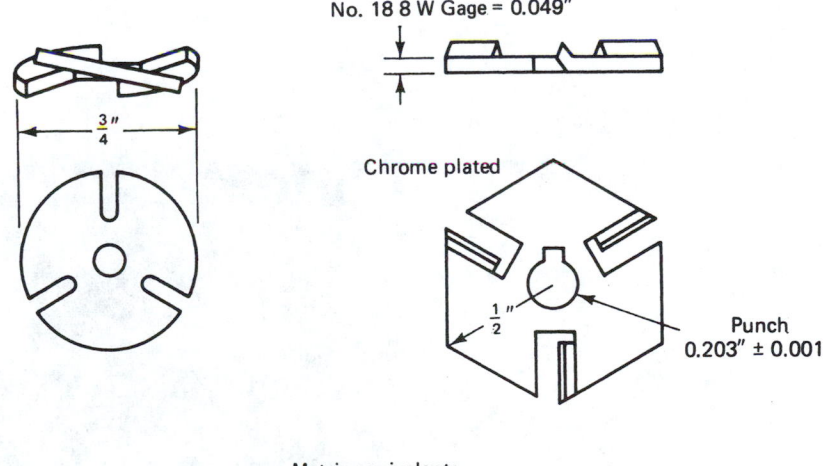

Metric equivalents					
in	0.001	0.049	0.203	½	¾
mm	0.03	1.24	5.16	12.7	19.0

(a) Detail of stirring paddles

FIGURE 9–2 Mechanically Operating Stirring Device [1]

(continued on next page)

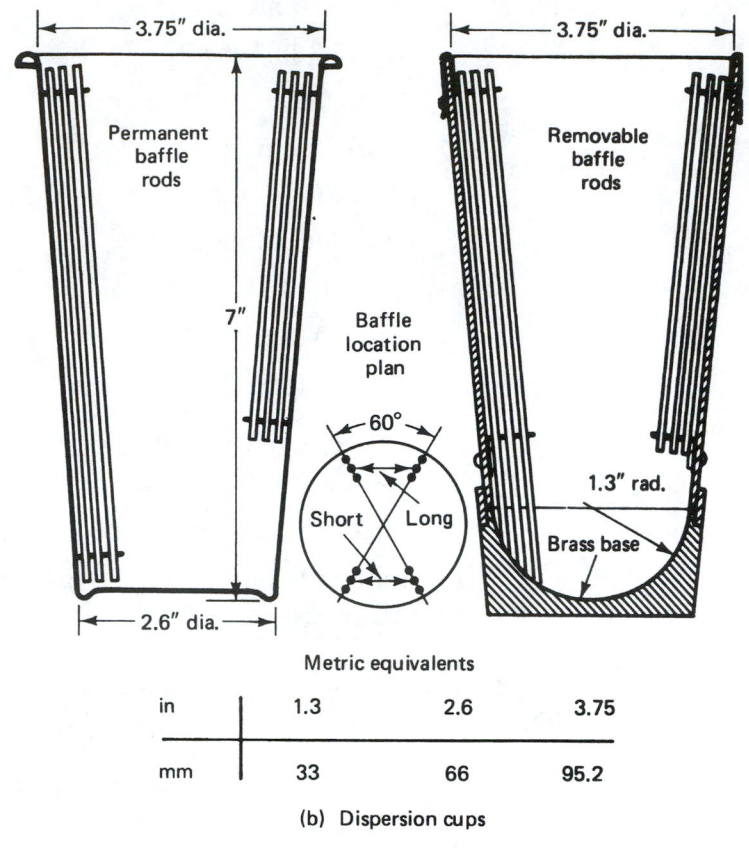

Metric equivalents

in	1.3	2.6	3.75
mm	33	66	95.2

(b) Dispersion cups

FIGURE 9–2   Mechanically Operating Stirring Device [1]
(continued from previous page)

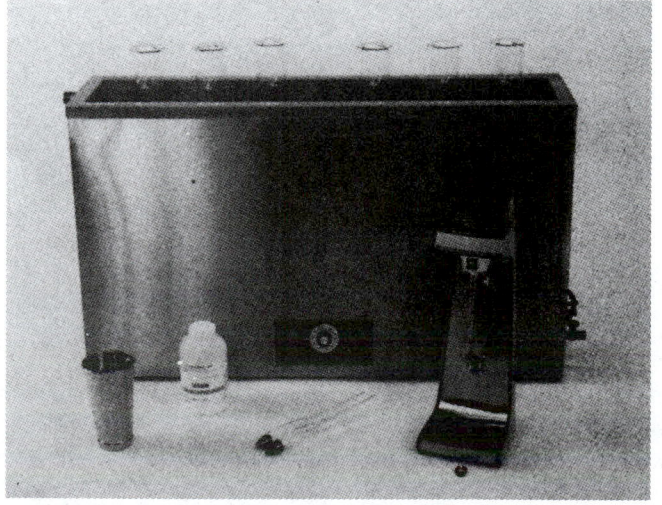

FIGURE 9–3   Apparatus Used in Grain-Size Analysis (Courtesy of Soiltest, Inc.)

Timing device

Beaker (150-mL capacity)

Containers

PREPARATION OF TEST SAMPLES

After air drying (if necessary), a representative sample must be weighed and then separated into two portions: one containing all particles retained on the No. 10 sieve and the other containing all particles passing through the No. 10 sieve. The amount of air-dried soil sample selected for this test should be sufficient to yield the following quantities for the sieve and hydrometer analyses:

1. The required minimum amount retained on the No. 10 sieve can be determined from Table 9–1.

2. The required amount passing through the No. 10 sieve should be approximately 115 g for sandy soils and approximately 65 g for silty and clayey soils.

The separation of soil on the No. 10 sieve should be accomplished by dry sieving and washing. The sample remaining on the No. 10 sieve after washing should be oven-dried and weighed. With the weight retained on the No. 10 sieve known, as well as the weight of the original sample, both the percentage retained on, and the percentage passing through, the No. 10 sieve can be calculated.

Table 9–1 Required Minimum Amount Retained on No. 10 Sieve [1]

Nominal Diameter of Largest Particles [in. (mm)]		Approximate Minimum Mass of Portion (g)
⅜	(9.5)	500
¾	(19.0)	1,000
1	(25.4)	2,000
1½	(38.1)	3,000
2	(50.8)	4,000
3	(76.2)	5,000

PROCEDURE FOR SIEVE ANALYSIS OF PORTION RETAINED ON NO. 10 SIEVE

The portion retained on the No. 10 sieve is tested for grain-size distribution by passing the sample through a number of sieves of different size openings. Normally, the sieves are stacked in order, with the sieve with the largest size opening at the top. The sieves should be agitated, preferably by mechanical means, over a period of time. When the sieving operation has been completed, the weight of the soil particles retained on each sieve is determined, from which the percentage passing each sieve can be computed.

The actual step-by-step procedure is as follows (ASTM D 422-63 (Reapproved 1990) [1]):

(1) Separate the portion retained on the No. 10 (2.00-mm) sieve into a series of fractions using the 3-in. (75-mm), 2-in. (50-mm), 1½-in. (38.1-mm), 1-in (25.0-mm), ¾-in. (19.0-mm), ⅜-in. (9.5-mm), No. 4 (4.75-mm), and No. 10 sieves, or as many as may be needed depending on the sample, or upon the specifications for the material under test.

(2) Conduct the sieving operation by means of a lateral and vertical motion of the sieve, accompanied by a jarring action in order to keep the sample moving continuously over the surface of the sieve. In no case turn or manipulate fragments in the sample through the sieve by hand. Continue sieving until not more than 1 mass percent of the residue on a sieve passes that sieve during 1 min of sieving. When mechanical sieving is used, test the thoroughness of sieving by using the hand method of sieving as described above.

(3) Determine the mass of each fraction on a balance. At the end of weighing, the sum of the masses retained on all sieves used should equal closely the original mass of the quantity sieved.

PROCEDURE FOR HYDROMETER ANALYSIS OF PORTION PASSING THROUGH NO. 10 SIEVE

The portion of the original sample that passed through the No. 10 sieve is tested for grain-size distribution using a hydrometer and later by sieve analysis.

Determination of Composite Correction for the Hydrometer Reading

Prior to performing a hydrometer test, a "composite correction" for hydrometer readings must be determined to correct for three items that tend to produce errors in hydrometer analysis. The first of these items needing correction results from the fact that a dispersing agent is used in the water, and this agent increases the specific gravity of the resulting liquid. The second is the effect of variation of temperature of the liquid from the hydrometer calibration temperature, and the third results from the fact that it is not possible to read the bottom of the meniscus when it is in a soil suspension. Composite corrections can be determined experimentally. Inasmuch as hydrometer analyses are performed often, it is convenient to prepare a graph or table giving composite correction as a function of temperature for a given hydrometer and dispersing agent. For example, composite corrections can be determined for two temperatures spanning the range of anticipated test temperatures, and a straight-line relationship may be assumed for determining composite corrections at any temperatures.

The procedure for determining a composite correction is as follows (ASTM D 422-63 (Reapproved 1998) [1]):

Prepare 1,000 mL of liquid composed of distilled or demineralized water and dispersing agent in the same proportion as will prevail in the sedimentation (hydrometer) test. Place the liquid in a sedimentation cylinder and the cylinder in the constant-temperature water bath, set for one of the two temperatures to be used. When the temperature of the liquid becomes constant, insert the hydrometer and, after a short interval to permit the hydrometer to come to the temperature of the liquid, read the hydrometer at the top of the meniscus formed on the stem. For hydrometer 151H the composite correction is the difference between this reading and one; for hydrometer 152H it is the difference between the reading and zero. Bring the liquid and the hydrometer to the other temperature to be used, and secure the composite correction as before.

Hygroscopic Moisture

Another preliminary task is to determine the hygroscopic moisture correction factor. This is required because in the subsequent hydrometer analysis, the oven-dried mass of the soil used in the hydrometer test is needed; however, the soil actually used in the hydrometer test is air-dried soil and is not available for determining oven-dried mass. To determine the correction factor, a portion of the air-dried sample (from 10 to 15 g) is placed in a container, weighed, oven-dried at 230 ± 9°F (110 ± 5°C), and then weighed again. The hygroscopic moisture correction factor is computed by dividing the oven-dried mass by the air-dried mass.

Dispersion of Soil Sample

The soil sample to be used in the hydrometer test must first be dispersed in order to eliminate particle coagulation. This is accomplished by mixing the soil with a sodium hexametaphosphate solution and stirring the mixture thoroughly. (For specific details, see the step-by-step procedure given shortly.)

Hydrometer Test

After dispersion, the soil-water slurry is transferred to a glass sedimentation cylinder and agitated manually. The cylinder is then placed in a convenient location, and hydrometer readings are taken at specific time intervals until 24 h have elapsed. (For specific details, see the step-by-step procedure, which is given subsequently.) The grain-size distribution is determined from computations using the hydrometer readings as a function of time. (This will be explained in the section "Calculations.")

Sieve Analysis

After the last hydrometer reading has been made, the suspension is transferred to a No. 200 sieve, washed, and oven-dried. The sample that is retained on the No. 200 sieve is subjected to a sieve analysis, normally using No. 10, 40, 100, and 200 sieves. This gives a grain-size analysis for that portion of the original sample that passed through the No. 10 sieve but was retained on the No. 200 sieve.

There may be some overlapping of grain-size distribution of the portion that passed the No. 10 sieve but was retained on the No. 200 sieve, as determined by hydrometer readings and by sieve analysis.

The actual step-by-step procedure for dispersion of the soil sample, the hydrometer test, and the sieve analysis of the portion passing the No. 10 sieve is as follows (ASTM D 422-63 (Reapproved 1990) [1]):

(1) When the soil is mostly of the clay and silt sizes, weigh out a sample of air-dried soil of approximately 50 g. When the soil is mostly sand, the sample should be approximately 100 g.

(2) Place the sample in a 250-mL beaker and cover with 125 mL of sodium hexametaphosphate solution (40 g/litre). Stir until the soil is thoroughly wetted. Allow to soak for at least 16 h.

(3) At the end of the soaking period, disperse the sample further, using a mechanically operated stirring apparatus. Transfer the soil-water slurry from the beaker into the special dispersion cup shown in Figure 9–2, washing any residue from the beaker into the cup with distilled or demineralized water (Note 1). Add distilled or demineralized water if necessary so that the cup is more than half full. Stir for a period of 1 min.

> Note 1—A large-size syringe is a convenient device for handling the water in the washing operation. Other devices include the wash-water bottle and a hose with nozzle connected to a pressurized distilled water tank.

(4) Immediately after dispersion, transfer the soil-water slurry to the glass sedimentation cylinder, and add distilled or demineralized water until the total volume is 1,000 mL.

(5) Using the palm of the hand over the open end of the cylinder (or a rubber stopper in the open end), turn the cylinder upside down and back for a period of 1 min to complete the agitation of the slurry (Note 2). At the end of 1 min set the cylinder in a convenient location and take hydrometer readings at the following intervals of time (measured from the beginning of sedimentation), or as many as may be needed, depending on the sample or the specification for the material under test: 2, 5, 15, 30, 60, 250, and 1,440 min. If the controlled water bath is used, the sedimentation cylinder should be placed in the bath between the 2- and 5-min readings.

> Note 2—The number of turns during this minute should be approximately 60, counting the turn upside down and back as two turns. Any soil remaining in the bottom of the cylinder during the first few turns should be loosened by vigorous shaking of the cylinder while it is in the inverted position.

(6) When it is desired to take a hydrometer reading, carefully insert the hydrometer about 20 to 25 s (seconds) before the reading is due to approximately the depth it will have when the reading is taken. As soon as the reading is taken, carefully remove the hydrometer and place it with a spinning motion in a graduate of clean distilled or demineralized water.

> Note 3—It is important to remove the hydrometer immediately after each reading. Readings shall be taken at the top of the meniscus formed by the suspension around the stem, since it is not possible to secure readings at the bottom of the meniscus.

(7) After each reading, take the temperature of the suspension by inserting the thermometer into the suspension.

(8) After taking the final hydrometer reading, transfer the suspension to a No. 200 sieve and wash with tap water until the wash water is clear. Transfer the material on the No. 200 sieve to a suitable container, dry in an oven at 230 ± 9°F (110 ± 5°C) and make a sieve analysis of the portion retained, using as many sieves as desired or required for the material, or upon the specification of the material under test.

DATA Data collected in this test should include the following:

Mass of total air-dried soil sample

Mass of fraction retained on No. 10 sieve (washed and oven-dried)

For sieve analysis of coarse aggregate:

Mass retained on each sieve

For hygroscopic moisture correction factor:

Mass of container plus air-dried soil

Mass of container plus oven-dried soil

Mass of container

For sieve analysis of fine aggregate:

Mass of container plus air-dried soil for hydrometer analysis

Mass of container

Mass retained on each sieve

For hydrometer analysis:

Hydrometer reading and temperature for each specified elapsed time

CALCULATIONS The calculations for this particular test are quite extensive:

1. Compute the percentage of sample retained on the No. 10 sieve by dividing mass of fraction retained by total mass of original sample and multiplying by 100.

2. Compute the percentage of sample passing through the No. 10 sieve by subtracting the previous value from 100.

3. For the sieve analysis of the coarse aggregates, compute the total percentage that passed each sieve by subtracting the summation of the mass retained on this and all previous sieves (beginning with the sieve with the largest screen opening) from the mass of the original sample, dividing this difference by the mass of the original sample, and multiplying the quotient by 100.

4. Compute the hygroscopic moisture correction factor by dividing the mass of the oven-dried soil by the mass of the air-dried soil.

5. In order to determine the sieve analysis for the fine aggregates and for use in computing the hydrometer analysis, the calculated mass of the total oven-dried hydrometer analysis sample must be determined by multiplying the mass of the air-dried sample for hydrometer analysis by the hygroscopic moisture correction factor, dividing the product by the percentage of sample passing the No. 10 sieve, and multiplying the quotient by 100. Denote this quantity M for later use.

6. For the sieve analysis of the fine aggregates, compute the total percentage that passed each sieve by subtracting the summation of the mass retained on this and all previous sieves from the mass of the oven-dried sample for hydrometer analysis, dividing this difference by the calculated mass of the total oven-dried hydrometer analysis sample (i.e., M from step 5), and multiplying the quotient by 100. The summation of the mass retained on this and all previous sieves begins with the No. 10 sieve, which retains no soil during the fine aggregate sieving (because the sample for hydrometer analysis consists only of soil that previously passed the No. 10 sieve); however, the mass of the oven-dried sample for hydrometer analysis is the mass passing the No. 10 sieve.

7. Recall that the data recorded in the hydrometer analysis consist of hydrometer readings and corresponding readings of temperature of the suspension for specified elapsed times. The first step in the computational procedure is to apply the composite correction to each hydrometer reading. (It will be recalled that for a given application, this correction will be a function of temperature.) Denote the corrected hydrometer reading R for later use. The percentage of soil remaining in suspension (i.e., percent of soil finer, or P) at each level at which the hydrometer measured the density of the suspension may be calculated using one of the following equations [1]

    For hydrometer 151H:

$$P = \left[\frac{(100,000/M) \times G}{G - G_1} \right] (R - G_1) \qquad (9\text{--}1)$$

For hydrometer 152H:

$$P = \frac{Ra}{M} \times 100 \qquad (9\text{–}2)$$

where:

P = percentage of soil remaining in suspension at level at which hydrometer measures density of suspension

M = mass of total oven-dried hydrometer analysis sample (see step 5)

G = specific gravity of soil particles

G_1 = specific gravity of liquid in which soil particles are suspended; use numerical value of one in both instances in equation (in the first instance, any possible variation produces no significant effect, and in the second, the composite correction for R is based on a value of one for G_1)

R = hydrometer reading with composite correction applied

a = correction factor to be applied to reading of hydrometer 152H. (Values shown on the scale are computed using a specific gravity of 2.65; correction factors can be obtained from Table 9–2.)

Table 9–2   Values of Correction Factor, *a,* for Different Specific Gravities of Soil Particles[a] [1]

Specific Gravity	Correction Factor
2.95	0.94
2.90	0.95
2.85	0.96
2.80	0.97
2.75	0.98
2.70	0.99
2.65	1.00
2.60	1.01
2.55	1.02
2.50	1.03
2.45	1.05

[a]For use in equation for percentage of soil remaining in suspension when using hydrometer 152H.

8. Compute the diameter of soil particle corresponding to the percentage indicated by a given hydrometer reading. This can be done using equation [1]

$$D = K\sqrt{\frac{L}{T}} \qquad (9\text{–}3)$$

Table 9–3   Values of *K* for Use in Equation for Computing Diameter of Particle in Hydrometer Analysis [1]

Temperature (°C)	Specific Gravity of Soil Particles								
	2.45	2.50	2.55	2.60	2.65	2.70	2.75	2.80	2.85
16	0.01510	0.01505	0.01481	0.01457	0.01435	0.01414	0.01394	0.01374	0.01356
17	0.01511	0.01486	0.01462	0.01439	0.01417	0.01396	0.01376	0.01356	0.01338
18	0.01492	0.01467	0.01443	0.01421	0.01399	0.01378	0.01359	0.01339	0.01321
19	0.01474	0.01449	0.01425	0.01403	0.01382	0.01361	0.01342	0.01323	0.01305
20	0.01456	0.01431	0.01408	0.01386	0.01365	0.01344	0.01325	0.01307	0.01289
21	0.01438	0.01414	0.01391	0.01369	0.01348	0.01328	0.01309	0.01291	0.01273
22	0.01421	0.01397	0.01374	0.01353	0.01332	0.01312	0.01294	0.01276	0.01258
23	0.01404	0.01381	0.01358	0.01337	0.01317	0.01297	0.01279	0.01261	0.01243
24	0.01388	0.01365	0.01342	0.01321	0.01301	0.01282	0.01264	0.01246	0.01229
25	0.01372	0.01349	0.01327	0.01306	0.01286	0.01267	0.01249	0.01232	0.01215
26	0.01357	0.01334	0.01312	0.01291	0.01272	0.01253	0.01235	0.01218	0.01201
27	0.01342	0.01319	0.01297	0.01277	0.01258	0.01239	0.01221	0.01204	0.01188
28	0.01327	0.01304	0.01283	0.01264	0.01244	0.01225	0.01208	0.01191	0.01175
29	0.01312	0.01290	0.01269	0.01249	0.01230	0.01212	0.01195	0.01178	0.01162
30	0.01298	0.01276	0.01256	0.01236	0.01217	0.01199	0.01182	0.01165	0.01149

where:

D = diameter of particle, mm

K = constant depending on temperature of suspension and specific gravity of soil particles; values of *K* can be obtained from Table 9–3

L = distance from surface of suspension to level at which density of suspension is being measured, cm; values of *L* can be obtained from Table 9–4

T = interval of time from beginning of sedimentation to taking of reading, min

9. Grain-size analysis results are normally presented in graphical form on semilogarithmic graph paper, with grain diameter along the abscissa on a logarithmic scale that increases from right to left and "percentage passing" along the ordinate on an arithmetic scale. Such a graph is referred to as the *grain-size distribution curve.*

NUMERICAL EXAMPLE

A laboratory test was conducted according to the procedure described previously. The following data were obtained:

Mass of total air-dried soil = **540.94 g**

Mass of fraction retained on No. 10 sieve (washed and oven-dried) = **2.20 g**

For sieve analysis of coarse aggregate:

Sieve Size	Mass Retained (g)
⅜ in.	**0**
No. 4	**0.97**
No. 10	**1.23**

Table 9–4 Values of Effective Depth Based on Hydrometer and Sedimentation Cylinder of Specified Sizes [1]

Hydrometer 151H		Hydrometer 152H			
Actual Hydrometer Reading	Effective Depth, L (cm)	Actual Hydrometer Reading	Effective Depth, L (cm)	Actual Hydrometer Reading	Effective Depth, L (cm)
1.000	16.3	0	16.3	31	11.2
1.001	16.0	1	16.1	32	11.1
1.002	15.8	2	16.0	33	10.9
1.003	15.5	3	15.8	34	10.7
1.004	15.2	4	15.6	35	10.6
1.005	15.0	5	15.5	36	10.4
1.006	14.7	6	15.3	37	10.2
1.007	14.4	7	15.2	38	10.1
1.008	14.2	8	15.0	39	9.9
1.009	13.9	9	14.8	40	9.7
1.010	13.7	10	14.7	41	9.6
1.011	13.4	11	14.5	42	9.4
1.012	13.1	12	14.3	43	9.2
1.013	12.9	13	14.2	44	9.1
1.014	12.6	14	14.0	45	8.9
1.015	12.3	15	13.8	46	8.8
1.016	12.1	16	13.7	47	8.6
1.017	11.8	17	13.5	48	8.4
1.018	11.5	18	13.3	49	8.3
1.019	11.3	19	13.2	50	8.1
1.020	11.0	20	13.0	51	7.9
1.021	10.7	21	12.9	52	7.8
1.022	10.5	22	12.7	53	7.6
1.023	10.2	23	12.5	54	7.4
1.024	10.0	24	12.4	55	7.3
1.025	9.7	25	12.2	56	7.1
1.026	9.4	26	12.0	57	7.0
1.027	9.2	27	11.9	58	6.8
1.028	8.9	28	11.7	59	6.6
1.029	8.6	29	11.5	60	6.5
1.030	8.4	30	11.4		
1.031	8.1				
1.032	7.8				
1.033	7.6				
1.034	7.3				
1.035	7.0				
1.036	6.8				
1.037	6.5				
1.038	6.2				

Values of effective depth are calculated from the equation:

$$L = L_1 + 1/2[L_2 - (V_B/A)]$$

where:

L = effective depth, cm
L_1 = distance along the stem of the hydrometer from the top of the bulb to the mark for a hydrometer reading, cm
L_2 = overall length of the hydrometer bulb, cm
V_B = volume of hydrometer bulb, cm^3
A = cross-sectional area of sedimentation cylinder, cm^2

Values used in calculating the values in Table 9–4 are as follows:

For both hydrometers, 151H and 152H:
L_2 = 14.0 cm
V_B = 67.0 cm^3
A = 27.8 cm^2

For hydrometer 151H:
L_1 = 10.5 cm for a reading of 1.000
    = 2.3 cm for a reading of 1.031

For hydrometer 152H:
L_1 = 10.5 cm for a reading of 0 g/litre
    = 2.3 cm for a reading of 50 g/litre

For hygroscopic moisture correction factor:

Mass of container plus air-dried soil = **109.57 g**

Mass of container plus oven-dried soil = **108.85 g**

Mass of container = **59.57 g**

For sieve analysis of fine aggregate:

Mass of container plus air-dried soil for hydrometer analysis = **170.49 g**

Mass of container = **110.21 g**

Sieve Size	Mass Retained (g)
No. 10	**0**
No. 40	**5.13**
No. 100	**5.31**
No. 200	**5.19**

For hydrometer analysis:

Date	Time	Hydrometer Reading	Temperature, (°C)	Composite Correction
4/10	8:30 A.M.			
	8:32	1.026	20°C	0.002
	8:35	1.024	20°C	0.002
	8:45	1.022	20°C	0.002
	9:00	1.020	20°C	0.002
	9:30	1.017	20°C	0.002
	12:40 P.M.	1.013	20°C	0.002
4/11	8:30 A.M.	1.009	20°C	0.002

The specific gravity of the soil particles G was determined by a previous test to be **2.70.**

The calculations for this example follow the steps outlined in the preceding section. A form prepared for recording both initial data and computed results is given on pages 117 to 119. (At the end of the chapter, two blank copies of this form are included for the reader's use.) The reader is referred to both the steps in the section "Calculations" and the form on pages 117 to 119 to help in understanding the calculations that follow.

1. Percentage of sample retained on No. 10 sieve = **(2.20/540.94)** × 100 = 0.4%.

2. Percentage of sample passing through No. 10 sieve = 100 − 0.4 = 99.6%.

Soils Testing Laboratory
Grain-Size Analysis

Sample No. _____ 20 _____   Project No. _____ SR 2828 _____

Boring No. _____ B-9 _____   Location _____ Newell, N.C. _____

Depth _____ 5 ft _____

Description of Sample _____ Brown silty clay _____

Tested by _____ John Doe _____   Date _____ 4/10/02 _____

(1) Mass of total air-dried sample _540.94_ g
(2) Mass of fraction retained on No. 10 sieve (washed and oven-dried) _2.20_ g
(3) Percentage of sample retained on No. 10 sieve _0.4_ %
(4) Percentage of sample passing No. 10 sieve _99.6_ %

Sieve Analysis of Coarse Aggregate

Sieve Size	Mass Retained (g)	Mass Passed (g)	Total Percent Passed
3 in.			
2 in.			
$1\frac{1}{2}$ in.			
1 in.			
3/4 in.			
3/8 in.	0	540.94	100.0
No. 4	0.97	539.97	99.8
No. 10	1.23	538.74	99.6

Hygroscopic Moisture Correction Factor

(5) Container no. _125A_
(6) Mass of container + air-dried soil _109.57_ g
(7) Mass of container + oven-dried soil _108.85_ g
(8) Mass of container _59.57_ g
(9) Mass of water [(6) − (7)] _0.72_ g
(10) Mass of oven-dried soil [(7) − (8)] _49.28_ g
(11) Mass of air-dried soil [(6) − (8)] _50.00_ g

(12) Hygroscopic moisture content $\left[\dfrac{(9)}{(10)} \times 100\right]$ _1.46_ %

(13) Hygroscopic moisture correction factor $\left[\dfrac{(10)}{(11)}\right]$ _0.9856_

Sieve Analysis of Fine Aggregate

(14) Container no. __126A__

(15) Mass of container + air-dried soil (for hydrometer analysis) __170.49__ g

(16) Mass of container __110.21__ g

(17) Mass of air-dried soil sample for hydrometer analysis __60.28__ g

(18) Hygroscopic moisture correction factor [(13)] __0.9856__

(19) Mass of oven-dried sample for hydrometer analysis [(17) × (18)] __59.41__ g

(20) Calculated mass of total hydrometer analysis sample,

$$M = \left[\frac{(19)}{(4)} \times 100 \right] \underline{59.65} \text{ g}$$

Sieve Size	Mass Retained (g)	Mass Passed (g)[a]	Total Percent Passed[b]
No. 10	0	59.41	99.6
No. 40	5.13	54.28	91.0
No. 100	5.31	48.97	82.1
No. 200	5.19	43.78	73.4

[a]Mass passed No. 10 sieve should be equal to item (19).

[b]Total percent passed $= \dfrac{mass\ passed}{(20)} \times 100$

Hydrometer Analysis

(21) Type of hydrometer used __151H__

(22) Specific gravity of soil (G) __2.70__

(23) Mass of oven-dried soil sample for hydrometer analysis [item (19)] __59.41__ g

(24) Calculated mass of total hydrometer analysis sample, M [item (20)] __59.65__ g

(25) Amount and type of dispersing agent used __125 mL of sodium__
__hexametaphosphate solution (40 g/litre)__

Date	Time	Elapsed Time, T (min)	Actual Hydrometer Reading	Composite Correction	Hydrometer Reading with Composite Correction Applied, R	Temperature, (°C)	Effective Depth of Hydrometer, L (cm) (from Table 9–4)	Value of K (from Table 9–3)	Diameter of Soil Particle, D (mm)	Soil in Suspension, P (i.e., % of Soil Finer) (%)
4/10	8:30 A.M.	0								
	8:32	2	1.026	0.002	1.024	20	9.4	0.01344	0.0291	63.9
	8:35	5	1.024	0.002	1.022	20	10.0	0.01344	0.0190	58.6
	8:45	15	1.022	0.002	1.020	20	10.5	0.01344	0.0112	53.3
	9:00	30	1.020	0.002	1.018	20	11.0	0.01344	0.0081	47.9
	9:30	60	1.017	0.002	1.015	20	11.8	0.01344	0.0060	39.9
	12:40 P.M.	250	1.013	0.002	1.011	20	12.9	0.01344	0.0031	29.3
4/11	8:30 A.M.	1440	1.009	0.002	1.007	20	13.9	0.01344	0.0013	18.6

3. For the ⅜-in. sieve, none of the sample is retained; hence, the mass passed is **540.94 g,** and the total percentage passed is 100. For the No. 4 sieve, **0.97 g** is retained; hence, the mass passed is **540.94** − **0.97,** or 539.97 g, and the total percentage passed is (539.97/**540.94**) × 100, or 99.8%. For the No. 10 sieve, **0.97** + **1.23,** or 2.20 g, is retained; hence the mass passed is **540.94** − 2.20, or 538.74 g, and the total percentage passed is (538.74/**540.94**) × 100, or 99.6%.

4. Hygroscopic moisture correction factor =

$$\frac{108.85 - 59.57}{109.57 - 59.57} = 0.9856$$

5. The mass of the air-dried sample for hydrometer analysis is **170.49** − **110.21,** or 60.28 g. Hence, the calculated mass of the total oven-dried hydrometer analysis sample can be computed as follows:

$$M = \frac{(60.28)(0.9856)}{99.6} \times 100 = 59.65 \text{ g}$$

6. For the No. 10 sieve, none of the sample is retained; the mass passed is (**170.49** − **110.21**) × 0.9856, or 59.41 g, and the total percentage passed is (59.41/59.65) × 100, or 99.6%. For the No. 40 sieve, **5.13 g** is retained; hence, the mass passed is 59.41 − **5.13,** or 54.28 g, and the total percentage passed is (54.28/59.65) × 100, or 91.0%. For the No. 100 sieve, **5.13** + **5.31,** or 10.44 g, is retained; hence, the mass passed is 59.41 − 10.44, or 48.97 g, and the total percentage passed is (48.97/59.65) × 100, or 82.1%. The computation for the No. 200 sieve is done in the same manner.

7. For this hydrometer analysis, a composite correction of **0.002** at **20°C** was determined. For the **8:32 A.M.** time, for which elapsed time is 2 min, the corrected hydrometer reading is **1.026** − **0.002,** or 1.024. The percentage of soil remaining in suspension can be computed using Eq. (9–1) (because the 151H hydrometer is being used):

$$P = \left[\frac{(100,000/M) \times G}{G - G_1}\right](R - G_1) \qquad \textbf{(9–1)}$$

In this application, M is 59.65 g (see step 5), G is **2.70** (known value from a previous specific gravity test), G_1 is taken to be 1, and R, the corrected hydrometer reading, is 1.024 (as determined previously). Substituting these values into Eq. (9–1) gives

$$P = \left(\frac{100,000}{59.65} \times \frac{2.70}{2.70 - 1}\right)(1.024 - 1) = 63.9\%$$

Computations of P for the remaining hydrometer readings are done in the same manner.

8. For the **8:32 A.M.** time, the diameter of soil particle can be computed using Eq. (9–3):

$$D = K\sqrt{\frac{L}{T}} \qquad (9\text{–}3)$$

In this application, K is 0.01344 (from Table 9–3 for a temperature of **20°C** and specific gravity of soil particles of **2.70**), L is 9.4 cm (from Table 9–4 for an actual hydrometer reading of **1.026**), and T is 2 min (elapsed time from **8:30 A.M.** to **8:32 A.M.**). Substituting these values into Eq. (9–3) gives

$$D = 0.01344\sqrt{\frac{9.4}{2}} = 0.0291 \text{ mm}$$

Computations of D for the remaining hydrometer readings are done in the same manner.

9. The grain-size distribution curve for the results of this test is plotted in Figure 9–4. (At the end of the text, a blank copy of the required graph paper is included for the reader's use.)

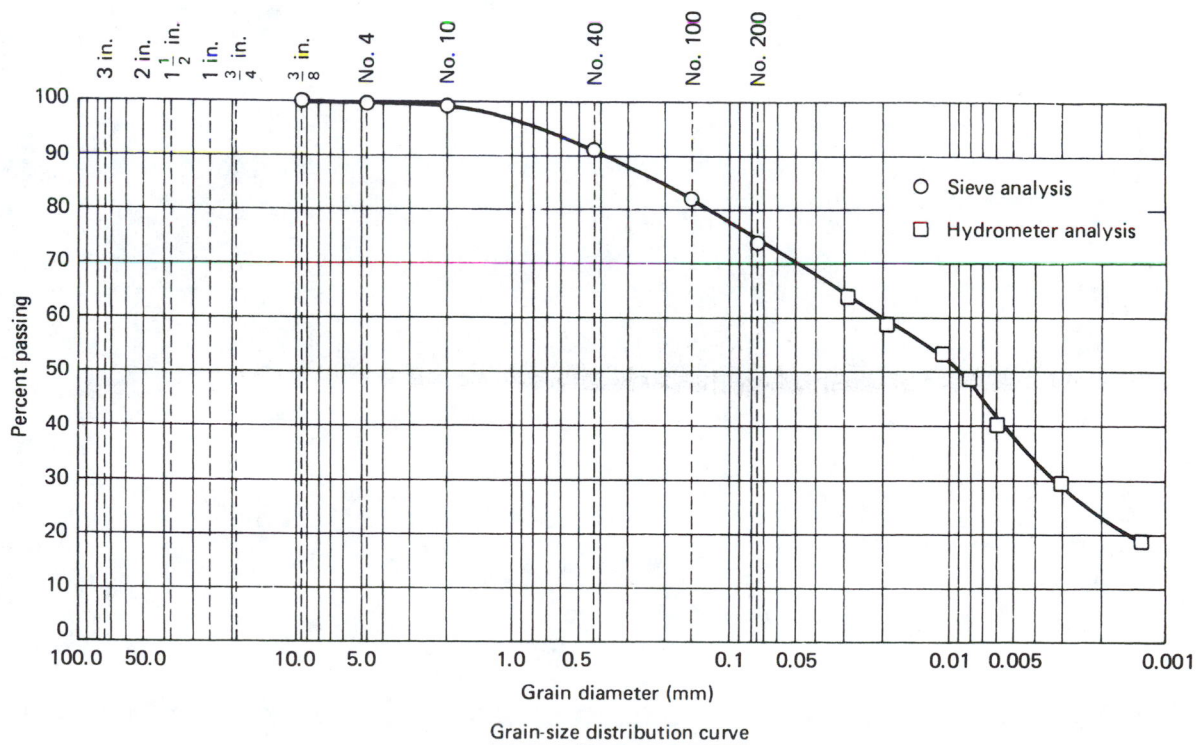

Grain-size distribution curve

FIGURE 9–4 Grain-Size Distribution Curve

CONCLUSIONS   Normally, the results of each grain-size analysis of soil are reported in the form of a grain-size distribution curve (Figure 9–4). As an alternative, the analysis may be reported in tabular form, giving percentages passing various sieve sizes or percentages found within various particle-size ranges.

As indicated previously, grain-size analysis is widely used in the identification and classification of soils. It is also utilized in part of the specifications of soil for airfields, roads, earth dams, and other soil embankment construction. Additionally, the frost susceptibility of soils can be fairly accurately predicted from the results of this analysis.

REFERENCE   [1]   ASTM, *2001 Annual Book of ASTM Standards,* West Conshohocken, PA, 2001. Copyright, American Society for Testing and Materials, 100 Barr Harbor Drive, West Conshohocken, PA 19428-2959. Reprinted with permission.

Soils Testing Laboratory
Grain-Size Analysis

Sample No. _____   Project No. _____

Boring No. _____   Location _____

Depth _____

Description of Sample _____

Tested by _____   Date _____

(1)   Mass of total air-dried sample _____ g
(2)   Mass of fraction retained on No. 10 sieve (washed and oven-dried) _____ g
(3)   Percentage of sample retained on No. 10 sieve _____%
(4)   Percentage of sample passing No. 10 sieve _____%

Sieve Analysis of Coarse Aggregate

Sieve Size	Mass Retained (g)	Mass Passed (g)	Total Percent Passed
3 in.			
2 in.			
$1\frac{1}{2}$ in.			
1 in.			
3/4 in.			
3/8 in.			
No. 4			
No. 10			

Hygroscopic Moisture Correction Factor

(5)   Container no. _____
(6)   Mass of container + air-dried soil _____ g
(7)   Mass of container + oven-dried soil _____ g
(8)   Mass of container _____ g
(9)   Mass of water [(6) − (7)] _____ g
(10) Mass of oven-dried soil [(7) − (8)] _____ g
(11) Mass of air-dried soil [(6) − (8)] _____ g

(12) Hygroscopic moisture content $\left[\dfrac{(9)}{(10)} \times 100\right]$ _____ %

(13) Hygroscopic moisture correction factor $\left[\dfrac{(10)}{(11)}\right]$ _____

Sieve Analysis of Fine Aggregate

(14) Container no. _____

(15) Mass of container + air-dried soil (for hydrometer analysis) _____ g

(16) Mass of container _____ g

(17) Mass of air-dried soil sample for hydrometer analysis _____ g

(18) Hygroscopic moisture correction factor [(13)] _____

(19) Mass of oven-dried sample for hydrometer analysis [(17) × (18)] _____ g

(20) Calculated mass of total hydrometer analysis sample,

$$M = \left[\frac{(19)}{(4)} \times 100 \right] \underline{\hspace{1cm}} \; g$$

Sieve Size	Mass Retained (g)	Mass Passed (g)[a]	Total Percent Passed[b]
No. 10			
No. 40			
No. 100			
No. 200			

[a]Mass passed No. 10 sieve should be equal to item (19).

[b]Total percent passed $= \dfrac{mass\ passed}{(20)} \times 100$.

Hydrometer Analysis

(21) Type of hydrometer used _____

(22) Specific gravity of soil (G) _____

(23) Mass of oven-dried soil sample for hydrometer analysis [item (19)] _____ g

(24) Calculated mass of total hydrometer analysis sample, M [item (20)] _____ g

(25) Amount and type of dispersing agent used _____

_____

Date	Time	Elapsed Time, T (min)	Actual Hydrometer Reading	Composite Correction	Hydrometer Reading with Composite Correction Applied, R	Temperature, (°C)	Effective Depth of Hydrometer, L (cm) (from Table 9–4)	Value of K (from Table 9–3)	Diameter of Soil Particle, D (mm)	Soil in Suspension, P (i.e., % of Soil Finer) (%)

Soils Testing Laboratory
Grain-Size Analysis

Sample No. _____   Project No. _____

Boring No. _____   Location _____

Depth _____

Description of Sample _____

Tested by _____   Date _____

    (1)   Mass of total air-dried sample _____ g

    (2)   Mass of fraction retained on No. 10 sieve (washed and oven-dried) _____ g

    (3)   Percentage of sample retained on No. 10 sieve _____%

    (4)   Percentage of sample passing No. 10 sieve _____%

Sieve Analysis of Coarse Aggregate

Sieve Size	Mass Retained (g)	Mass Passed (g)	Total Percent Passed
3 in.			
2 in.			
$1\frac{1}{2}$ in.			
1 in.			
3/4 in.			
3/8 in.			
No. 4			
No. 10			

Hygroscopic Moisture Correction Factor

    (5)   Container no. _____

    (6)   Mass of container + air-dried soil _____ g

    (7)   Mass of container + oven-dried soil _____ g

    (8)   Mass of container _____ g

    (9)   Mass of water [(6) − (7)] _____ g

    (10) Mass of oven-dried soil [(7) − (8)] _____ g

    (11) Mass of air-dried soil [(6) − (8)] _____ g

    (12) Hygroscopic moisture content $\left[\dfrac{(9)}{(10)} \times 100\right]$ _____ %

    (13) Hygroscopic moisture correction factor $\left[\dfrac{(10)}{(11)}\right]$ _____

Sieve Analysis of Fine Aggregate

(14) Container no. _____

(15) Mass of container + air-dried soil (for hydrometer analysis) _____ g

(16) Mass of container _____ g

(17) Mass of air-dried soil sample for hydrometer analysis _____ g

(18) Hygroscopic moisture correction factor [(13)] _____

(19) Mass of oven-dried sample for hydrometer analysis [(17) × (18)] _____ g

(20) Calculated mass of total hydrometer analysis sample,

$$M\left[\frac{(19)}{(4)} \times 100\right]$$ _____ g

Sieve Size	Mass Retained (g)	Mass Passed (g)[a]	Total Percent Passed[b]
No. 10			
No. 40			
No. 100			
No. 200			

[a]Mass passed No. 10 sieve should be equal to item (19).

[b]Total percent passed $= \dfrac{mass\ passed}{(20)} \times 100$.

Hydrometer Analysis

(21) Type of hydrometer used _____

(22) Specific gravity of soil (G) _____

(23) Mass of oven-dried soil sample for hydrometer analysis [item (19)] _____ g

(24) Calculated mass of total hydrometer analysis sample, M [item (20)] _____ g

(25) Amount and type of dispersing agent used _____

_____

Date	Time	Elapsed Time, T (min)	Actual Hydrometer Reading	Composite Correction	Hydrometer Reading with Composite Correction Applied, R	Temperature, (°C)	Effective Depth of Hydrometer, L (cm) (from Table 9-4)	Value of K (from Table 9-3)	Diameter of Soil Particle, D (mm)	Soil in Suspension, P (i.e., % of Soil Finer) (%)

CHAPTER TEN

Classification of Soils for Engineering Purposes

INTRODUCTION

To describe, in general, a specific soil without listing values of its parameters, it is convenient to have some kind of generalized classification system. In practice, a number of classification systems have been developed by different groups to meet their specific needs. Some examples are the AASHTO classification system, Unified Soil Classification System, and Federal Aviation Administration (FAA) classification system. All classifications of soil for engineering purposes use Atterberg limits (at least the liquid and plastic limits) and grain-size analysis as delimiting parameters.

The first two systems cited are covered in this chapter. In each case, one or more sample classifications are given to assist in understanding the system. Inasmuch as these classifications are based on Atterberg limits and grain-size analysis, laboratory procedures for which have been presented in Chapters 6 through 9, Chapter 10 presents no new laboratory procedure. Instead, the emphasis is on classification of soils once Atterberg limits and grain-size analyses are known.

AASHTO CLASSIFICATION SYSTEM

AASHTO stands for the American Association of State Highway and Transportation Officials; therefore, this classification system is widely used in highway work. Required parameters for each classification by this system are grain-size analysis, liquid limit, and plasticity index. With values of these parameters known for a given soil, one consults the second column of Table 10–1 and determines whether or not the known parameters meet the limiting values in that column. If they do, the soil classification is that listed at the top of the column (A-1-a, if the known parameters meet the limiting values in the second column). If they do not, one enters the next column (to the right) and determines whether or not the known parameters meet the limiting values in that column. This procedure is repeated until the first column is reached in which the known parameters meet the limiting values in that column. The given soil's classification is the one listed at the top of that particular column.

Once soils have been classified using Table 10–1 (and Figure 10–1), they can be further described with a "group index." The group index utilizes the percentage of soil passing a No. 200 sieve, the liquid limit, and the plasticity index and is computed from equation [1]

$$\text{group index} = (F - 35)\,[0.2 + 0.005\,(LL - 40)] \qquad \textbf{(10–1)}$$
$$+ \, 0.01\,(F - 15)\,(PI - 10)$$

where:

F = percentage of soil passing a No. 200 sieve
LL = liquid limit
PI = plasticity index

The group index computed from Eq. (10–1) is rounded off to the nearest whole number and appended in parentheses to the group designation determined from Table 10–1. If a computed group index is either zero or negative, the number zero is used as the group index and should be appended to the group designation. Also, if soil is nonplastic and the liquid limit cannot be determined, report the group index as zero. If preferred, Figure 10–2 may be used instead of Eq. (10–1) to obtain the group index.

As a general rule, the value of soil as a subgrade material is in inverse ratio to its group index. Table 10–2 gives some general descriptions of the various classification groups according to the AASHTO system. The AASHTO system is published as AASHTO M 145-87 (1990) and ASTM D 3282-93.

Example 10–1

Given:

A sample of soil was tested in a laboratory with the following results:

1. Liquid limit = **46.2%**

2. Plastic limit = **21.9%**

3. Sieve analysis data:

U.S. Sieve Size	Percentage Passing
No. 4	**100**
No. 10	**85.6**
No. 40	**72.3**
No. 200	**58.8**

Table 10–1 AASHTO Classification of Soils and Soil-Aggregate Mixtures [1]

General Classification:	Granular Materials (35% or less passing 0.075 mm)							Silt-Clay Materials (more than 35% passing 0.075 mm)			
	A-1		A-3	A-2				A-4	A-5	A-6	A-7
Group Classification:	A-1-a	A-1-b		A-2-4	A-2-5	A-2-6	A-2-7				A-7-5, A-7-6
Sieve analysis: percent passing											
2.00 mm (No. 10)	50 max.	—	—								
0.425 mm (No. 40)	30 max.	50 max.	51 min.								
0.075 mm (No. 200)	15 max.	25 max.	10 max.	35 max.	35 max.	35 max.	35 max.	36 min.	36 min.	36 min.	36 min.
Characteristics of fraction passing 0.425 mm (No. 40)											
Liquid limit				40 max.	41 min.	40 max.	41 min.	40 max.	41 min.	40 max.	41 min.
Plasticity index	6 max.		NP[a]	10 max.	10 max.	11 min.	11 min.	10 max.	10 max.	11 min.	11 min.[b]
Usual types of significant constituent materials	Stone fragments, gravel, and sand		Fine sand	Silty or clayey gravel and sand				Silty soils		Clayey soils	
General rating as subgrade	Excellent to good							Fair to poor			

[a]NP, nonplastic.
[b]Plasticity index of A-7-5 subgroup is equal to or less than LL minus 30. Plasticity index of A-7-6 subgroup is greater than LL minus 30 (see Figure 10–1).

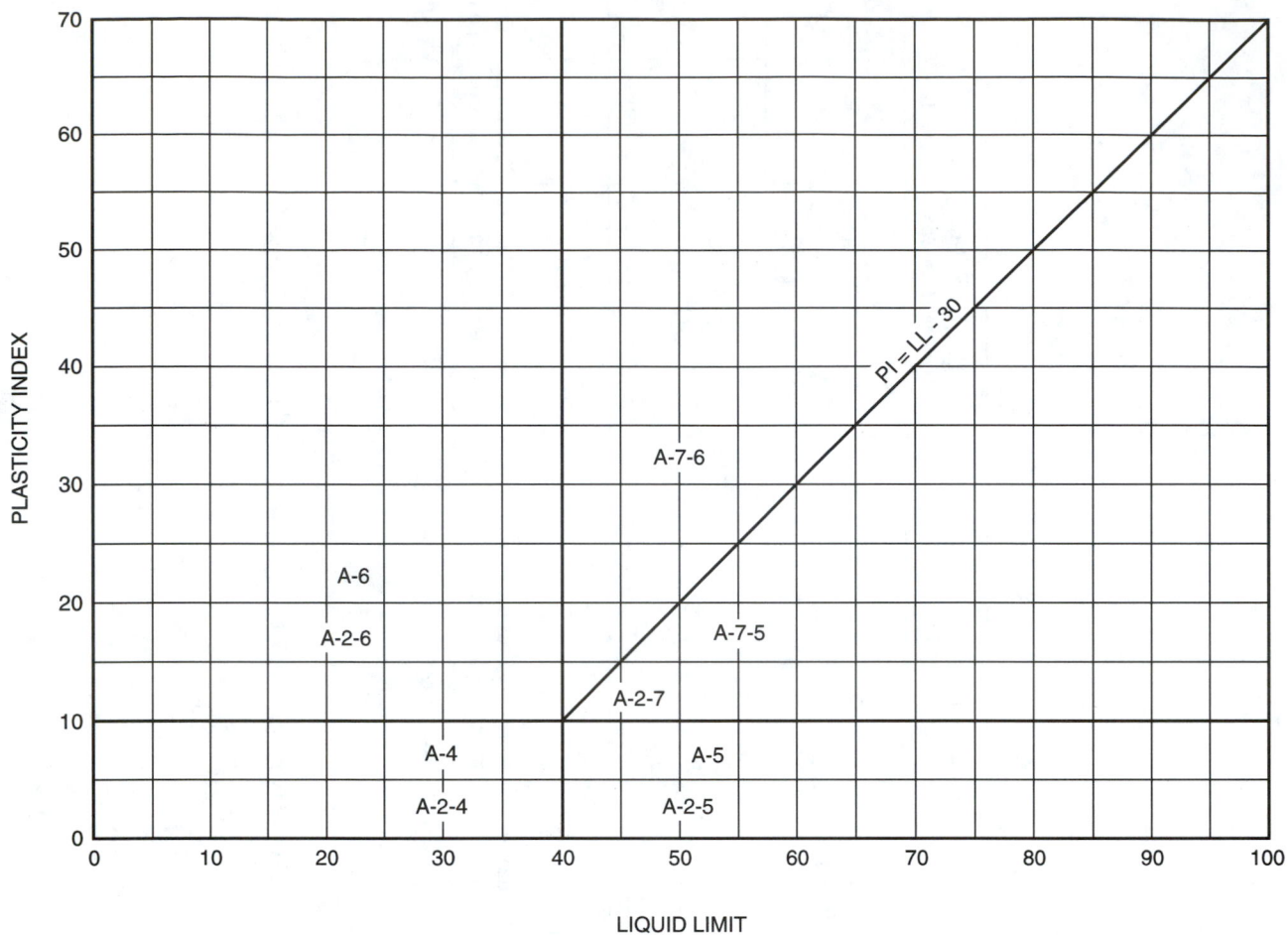

FIGURE 10–1   Liquid Limit and Plasticity Index Ranges for Silt-Clay Materials [2]

Required:
Classify the soil by the AASHTO system.

Solution:

$$\text{Plasticity index } (PI) = \text{liquid limit } (LL) - \text{plastic limit } (PL)$$

$$PI = \mathbf{46.2} - \mathbf{21.9} = 24.3\%$$

With **85.6%** passing the No. 10 sieve, **72.3%** passing the No. 40 sieve, **58.8%** passing the No. 200 sieve, a liquid limit of **46.2%,** and a plasticity index of 24.3%, one proceeds across Table 10–1 from left to right until the first column is reached in which these parameters meet the limiting values in that column. This turns out to be A-7. According to the AASHTO classification system, the plasticity index of the A-7-5 subgroup is equal to or less than the liquid limit minus 30, whereas the plasticity index of the A-7-6 subgroup is greater than the liquid limit minus 30 (see footnote to Table 10–1). For this soil, the liquid limit minus

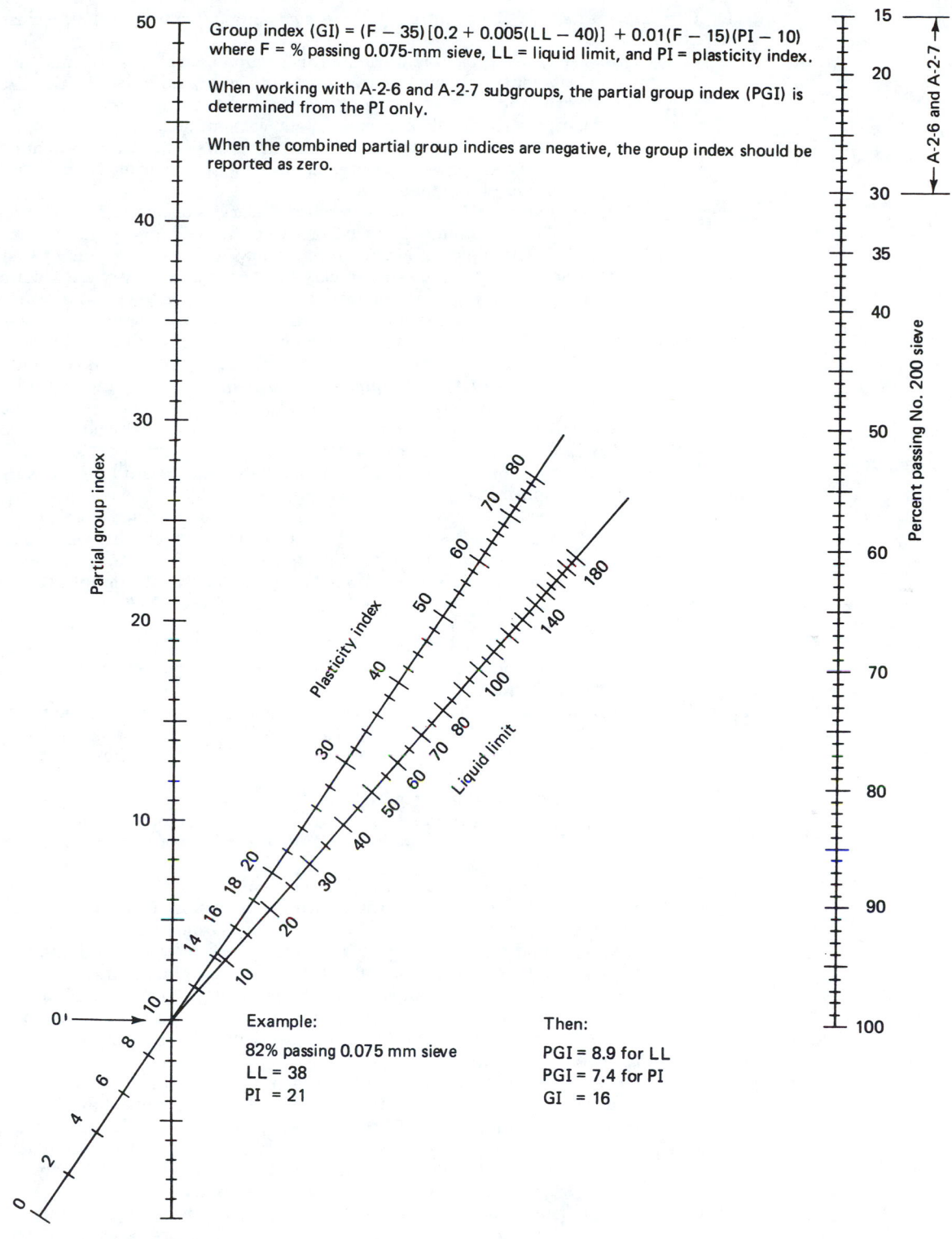

Group index (GI) = (F − 35) [0.2 + 0.005(LL − 40)] + 0.01(F − 15)(PI − 10) where F = % passing 0.075-mm sieve, LL = liquid limit, and PI = plasticity index.

When working with A-2-6 and A-2-7 subgroups, the partial group index (PGI) is determined from the PI only.

When the combined partial group indices are negative, the group index should be reported as zero.

Example:

82% passing 0.075 mm sieve
LL = 38
PI = 21

Then:

PGI = 8.9 for LL
PGI = 7.4 for PI
GI = 16

FIGURE 10–2 Group Index Chart [1]

Table 10–2 Descriptions of AASHTO Classification Groups [1]

(1) *Granular Materials.* Containing 35% or less passing 0.075 mm (No. 200) sieve, Note 1.
    (1.1) *Group A-1:* The typical material of this group is a well-graded mixture of stone fragments or gravel, coarse sand, fine sand and a nonplastic or feebly plastic soil binder. However, this group includes also stone fragments, gravel, coarse sand, volcanic cinders, etc. without soil binder.
        (1.1.1) Subgroup A-1-a includes those materials consisting predominantly of stone fragments or gravel, either with or without a well-graded binder of fine material.
        (1.1.2) Subgroup A-1-b includes those materials consisting predominantly of coarse sand either with or without a well-graded soil binder.
    (1.2) *Group A-3:* The typical material of this group is fine beach sand or fine desert blow sand without silty or clay fines or with a very small amount of nonplastic silt. The group includes also stream-deposited mixtures of poorly-graded fine sand and limited amounts of coarse sand and gravel.
    (1.3) *Group A-2:* This group includes a wide variety of "granular" materials which are borderline between the materials falling in Groups A-1 and A-3 and silt-clay materials of Groups A-4, A-5, A-6, and A-7. It includes all materials containing 35% or less passing the 0.075-mm sieve which cannot be classified as A-1 or A-3, due to fines content or plasticity or both, in excess of the limitations for those groups.
        (1.3.1) Subgroups A-2-4 and A-2-5 include various granular materials containing 35% or less passing the 0.075-mm sieve and with a minus 0.425 mm (No. 40) portion having the characteristics of the A-4 and A-5 groups. These groups include such materials as gravel and coarse sand with silt contents or plasticity indexes in excess of the limitations of Group A-1, and fine sand with nonplastic silt content in excess of the limitations of Group A-3.
        (1.3.2) Subgroups A-2-6 and A-2-7 include materials similar to those described under Subgroups A-2-4 and A-2-5 except that the fine portion contains plastic clay having the characteristics of the A-6 or A-7 group.
Note 1: Classification of materials in the various groups applies only to the fraction passing the 75-mm sieve. Therefore, any specification regarding the use of A-1, A-2, or A-3 materials in construction should state whether boulders (retained on 3-in. sieve) are permitted.

(2) *Silt-Clay Materials.* Containing more than 35% passing the 0.075-mm sieve.
    (2.1) *Group A-4:* The typical material of this group is a nonplastic or moderately plastic silty soil usually having the 75% or more passing the 0.075-mm sieve. The group includes also mixtures of fine silty soil and up to 64% of sand and gravel retained on 0.075-mm sieve.
    (2.2) *Group A-5:* The typical material of this group is similar to that described under Group A-4, except that it is usually of diatomaceous or micaceous character and may be highly elastic as indicated by the high liquid limit.
    (2.3) *Group A-6:* The typical material of this group is a plastic clay soil usually having 75% or more passing the 0.075-mm sieve. The group includes also mixtures of fine clayey soil and up to 64% of sand and gravel retained on the 0.075-mm sieve. Materials of this group usually have high volume change between wet and dry states.
    (2.4) *Group A-7:* The typical material of this group is similar to that described under Group A-6, except that it has the high liquid limits characteristic of Group A-5 and may be elastic as well as subject to high volume change.
        (2.4.1) Subgroup A-7-5 includes those materials with moderate plasticity indexes in relation to liquid limit and which may be highly elastic as well as subject to considerable volume change.
        (2.4.2) Subgroup A-7-6 includes those materials with high plasticity indexes in relation to liquid limit and which are subject to extremely high volume change.
Note 2: Highly organic soils (peat or muck) may be classified as an A-8 group. Classification of these materials is based on visual inspection, and is not dependent on percentage passing the 0.075-mm (No. 200) sieve, liquid limit or plasticity index. The material is composed primarily of partially decayed organic matter, generally has a fibrous texture, dark brown or black color, and odor of decay. These organic materials are unsuitable for use in embankments and subgrades. They are highly compressible and have low strength.

30 is **46.2** − 30, or 16.2%. Because the plasticity index of 24.3% is greater than 16.2%, this soil is classified as A-7-6. (Also, see Figure 10–1.)

Next, the group index must be determined, using either Eq. (10–1) or Figure 10–2. By Eq. (10–1),

$$\text{Group index} = (F - 35)\,[0.2 + 0.005\,(LL - 40)] \tag{10-1}$$
$$+ 0.01\,(F - 15)\,(PI - 10)$$

$$\text{Group index} = (\mathbf{58.8} - 35)\,[0.2 + 0.005\,(\mathbf{46.2} - 40)]$$
$$+ 0.01\,(\mathbf{58.8} - 15)\,(24.3 - 10) = 11.8$$

By Figure 10–2, with a liquid limit of **46.2%** and with **58.8%** passing the No. 200 sieve, the partial group index for the liquid limit is 5.5. With a plasticity index of 24.3% and with **58.8%** passing the No. 200 sieve, the partial group index for the plasticity index is 6.3. The (total) group index is the sum of these (5.5 + 6.3), or 11.8.

Therefore, this soil is classified as A-7-6(12), according to the AASHTO system.

UNIFIED SOIL CLASSIFICATION SYSTEM [2, 3, 4]

The Unified Soil Classification System was developed by Casagrande [3] and is utilized by the U.S. Army Corps of Engineers. In this system, soils fall within one of three major categories: coarse grained, fine grained, and highly organic soils. These categories are further subdivided into 15 basic soil groups. The following symbols are used in the Unified System:

G	Gravel
S	Sand
M	Silt
C	Clay
O	Organic
PT	Peat
W	Well graded
P	Poorly graded

Normally, two group symbols are used to classify soils. For example, SW indicates well-graded sand. Table 10–3 lists the 15 soil groups, including each one's name and symbol as well as giving specific details for classifying soils by this system.

In order to classify a given soil by the Unified System, its grain-size distribution, liquid limit, and plasticity index must first be determined. With these values known, the soil can be classified using Table 10–3 and Figure 10–3. The soil can also be classified using the flowcharts of Figures 10–4, 10–5, and 10–6. The Unified Soil Classification System is published as ASTM D 2487-00.

Table 10-3 Soil Classification Chart [2]

Criteria for Assigning Group Symbols and Group Names Using Laboratory Tests[A]

				Soil Classification	
				Group Symbol	Group Name[B]
Coarse-Grained Soils More than 50% retained on No. 200 sieve	Gravels More than 50% of coarse fraction retained on No. 4 sieve	Clean Gravels Less than 5% fines[E]	$C_u \geq 4$ and $1 \leq C_c \leq 3^C$	GW	Well-graded gravel[D]
			$C_u < 4$ and/or $1 > C_c > 3^C$	GP	Poorly graded gravel[D]
		Gravels with Fines More than 12% fines[E]	Fines classify as ML or MH	GM	Silty gravel[D,F,G]
			Fines classify as CL or CH	GC	Clayey gravel[D,F,G]
	Sands 50% or more of coarse fraction passes No. 4 sieve	Clean Sands Less than 5% fines[I]	$C_u \geq 6$ and $1 \leq C_c \leq 3^C$	SW	Well-graded sand[H]
			$C_u < 6$ and/or $1 > C_c > 3^C$	SP	Poorly graded sand[H]
		Sands with Fines More than 12% fines[I]	Fines classify as ML or MH	SM	Silty sand[F,G,H]
			Fines classify as CL or CH	SC	Clayey sand[F,G,H]
Fine-Grained Soils 50% or more passes the No. 200 sieve	Silts and Clays Liquid limit less than 50	Inorganic	PI > 7 and plots on or above "A" line[J]	CL	Lean clay[K,L,M]
			PI < 4 or plots below "A" line[J]	ML	Silt[K,L,M]
		Organic	$\dfrac{\text{Liquid limit—oven dried}}{\text{Liquid limit—not dried}} < 0.75$	OL	Organic clay[K,L,M,N] / Organic silt[K,L,M,O]
	Silts and Clays Liquid limit 50 or more	Inorganic	PI plots on or above "A" line	CH	Fat clay[K,L,M]
			PI plots below "A" line	MH	Elastic silt[K,L,M]
		Organic	$\dfrac{\text{Liquid limit—oven dried}}{\text{Liquid limit—not dried}} < 0.75$	OH	Organic clay[K,L,M,P] / Organic silt[K,L,M,Q]
Highly organic soils			Primarily organic matter, dark in color, and organic odor	PT	Peat

[A] Based on the material passing the 3-in. (75-mm) sieve.

[B] If field sample contained cobbles or boulders, or both, add "with cobbles or boulders, or both" to group name.

[C] $C_u = D_{60}/D_{10}$, $C_c = \dfrac{(D_{30})^2}{D_{10} \times D_{60}}$

[D] If soil contains ≥ 15% sand, add "with sand" to group name.

[E] Gravels with 5 to 12% fines require dual symbols:
GW-GM, well-graded gravel with silt
GW-GC, well-graded gravel with clay
GP-GM, poorly graded gravel with silt
GP-GC, poorly graded gravel with clay

[F] If fines classify as CL-ML, use dual symbol GC-GM or SC-SM.

[G] If fines are organic, add "with organic fines" to group name.

[H] If soil contains ≥ 15% gravel, add "with gravel" to group name.

[I] Sands with 5 to 12% fines require dual symbols:
SW-SM, well-graded sand with silt
SW-SC, well-graded sand with clay
SP-SM, poorly graded sand with silt
SP-SC, poorly graded sand with clay

[J] If Atterberg limits plot in hatched area, soil is a CL-ML silty clay.

[K] If soil contains 15 to 29% plus No. 200, add "with sand" or "with gravel," whichever is predominant.

[L] If soil contains ≥ 30% plus No. 200, predominantly sand, add "sandy" to group name.

[M] If soil contains ≥ 30% plus No. 200, predominantly gravel, add "gravelly" to group name.

[N] PI ≥ 4 and plots on or above "A" line.

[O] PI < 4 or plots below "A" line.

[P] PI plots on or above "A" line.

[Q] PI plots below "A" line.

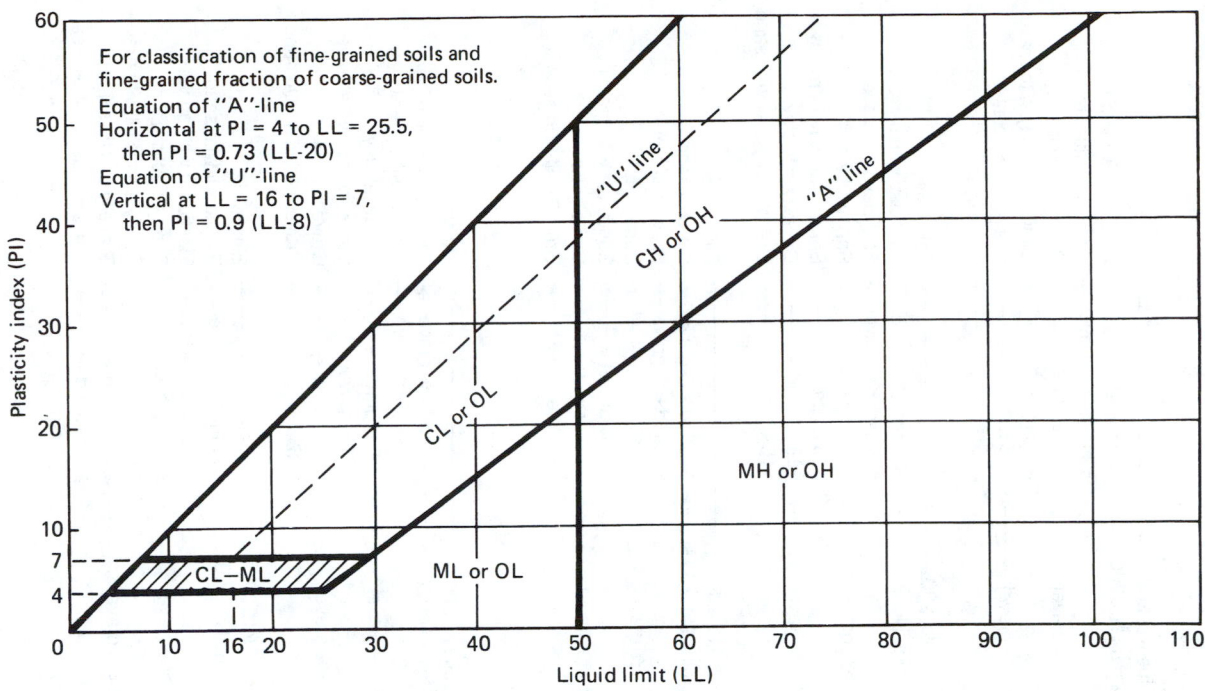

For classification of fine-grained soils and
fine-grained fraction of coarse-grained soils.
Equation of "A"-line
Horizontal at PI = 4 to LL = 25.5,
   then PI = 0.73 (LL-20)
Equation of "U"-line
Vertical at LL = 16 to PI = 7,
   then PI = 0.9 (LL-8)

FIGURE 10–3 Plasticity Chart [2]

Example 10–2

Given:

A sample of soil was tested in the laboratory with the following results:

1. Liquid limit = **30.0%**

2. Plastic limit = **12.0%**

3. Sieve analysis data:

U.S. Sieve Size	Percentage Passing
⅜ in.	100
No. 4	76.5
No. 10	60.0
No. 40	39.7
No. 200	15.2

Required:

Classify the soil by the Unified Soil Classification System.

Solution:

Because the percentage retained on the No. 200 sieve (100 − **15.2,** or 84.8%) is more than 50%, go to the block labeled "Coarse-Grained Soils" in Table 10–3. The sample consists of 100 − **15.2,** or 84.8%,

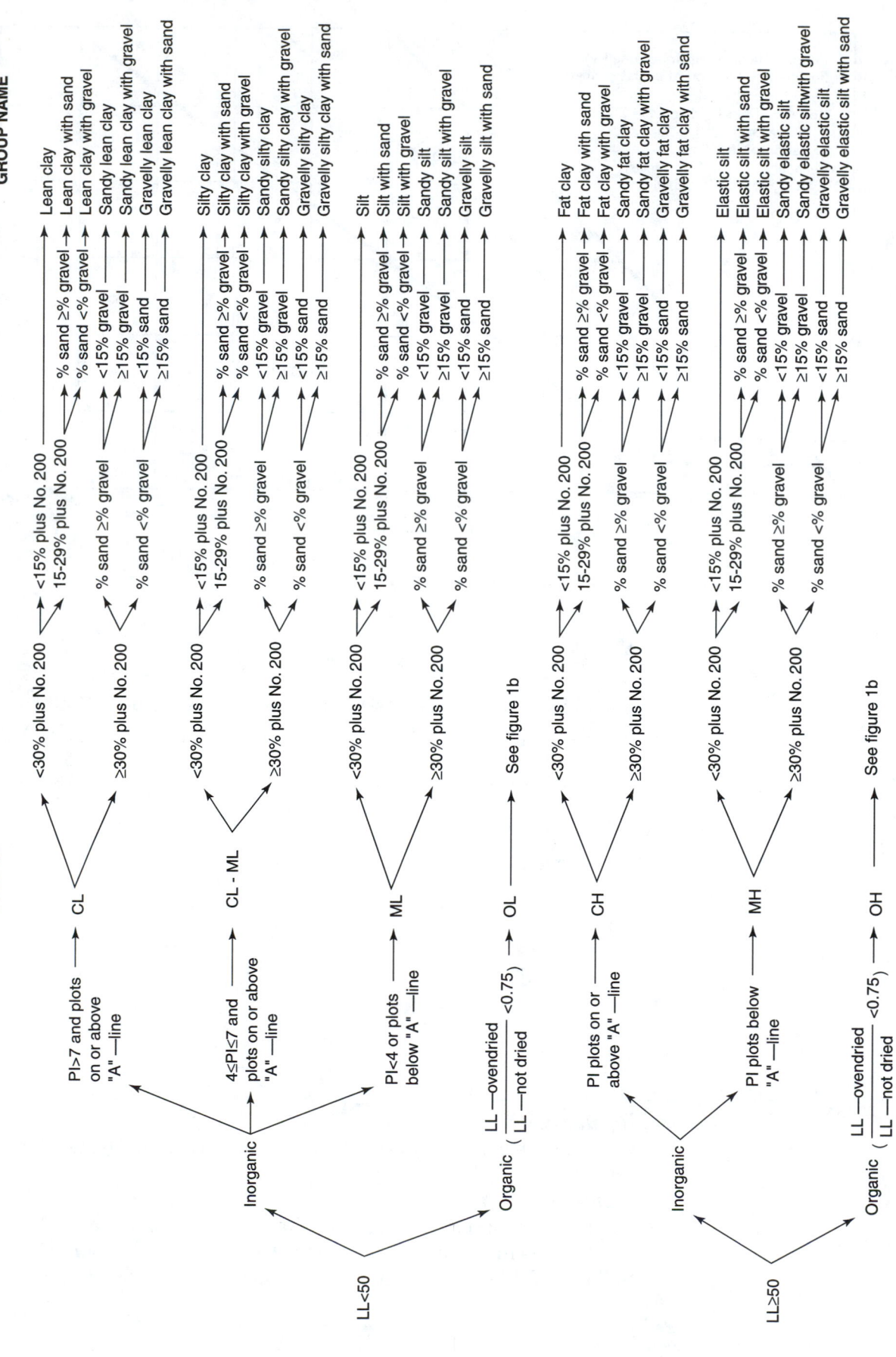

FIGURE 10–4 Flow Chart for Classifying Fine-Grained Soil (50% or More Passes No. 200 Sieve) [2]

140

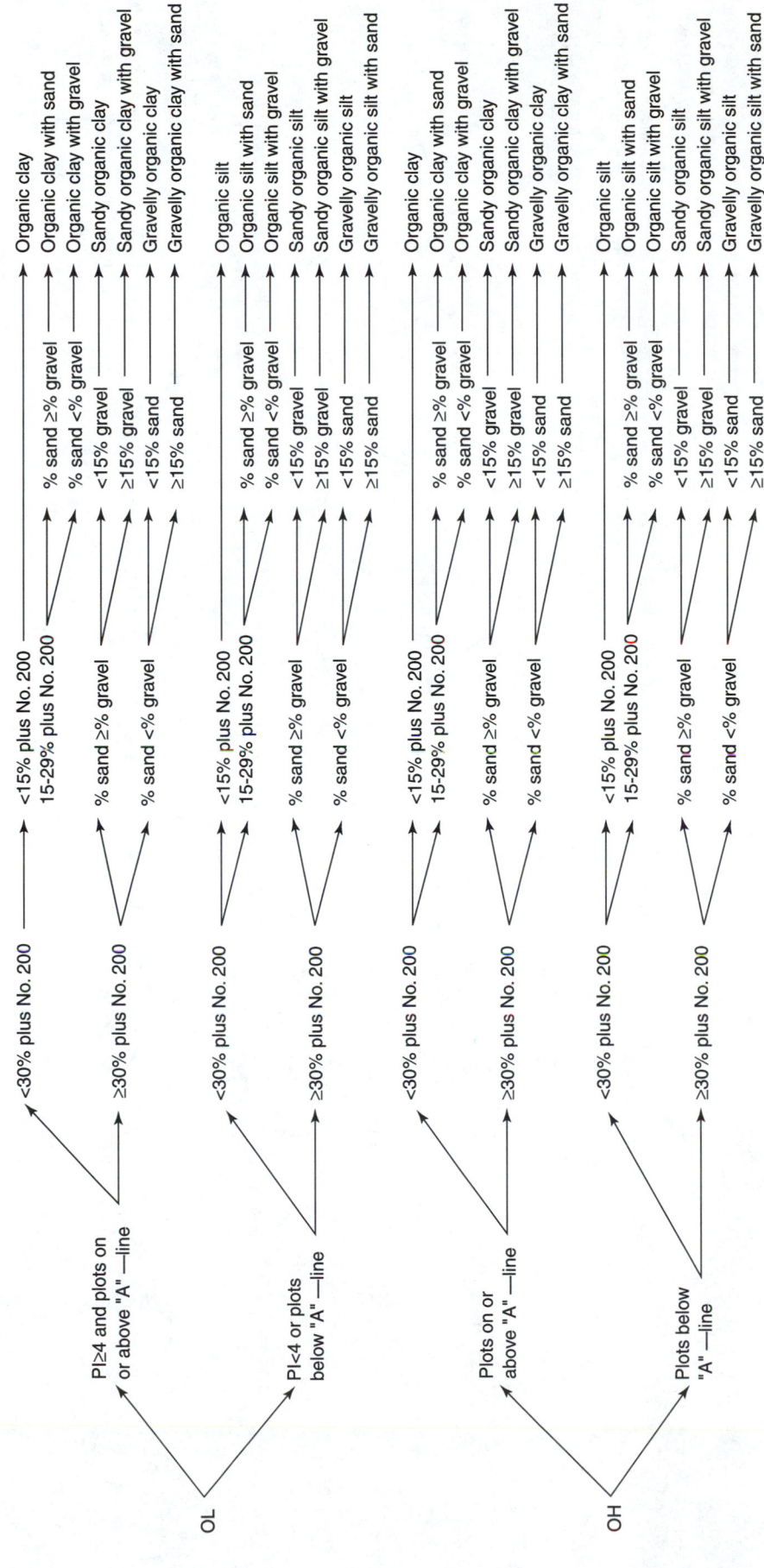

FIGURE 10–5 Flow Chart for Classifying Organic Fine-Grained Soil (50% or More Passes No. 200 Sieve)[2]

141

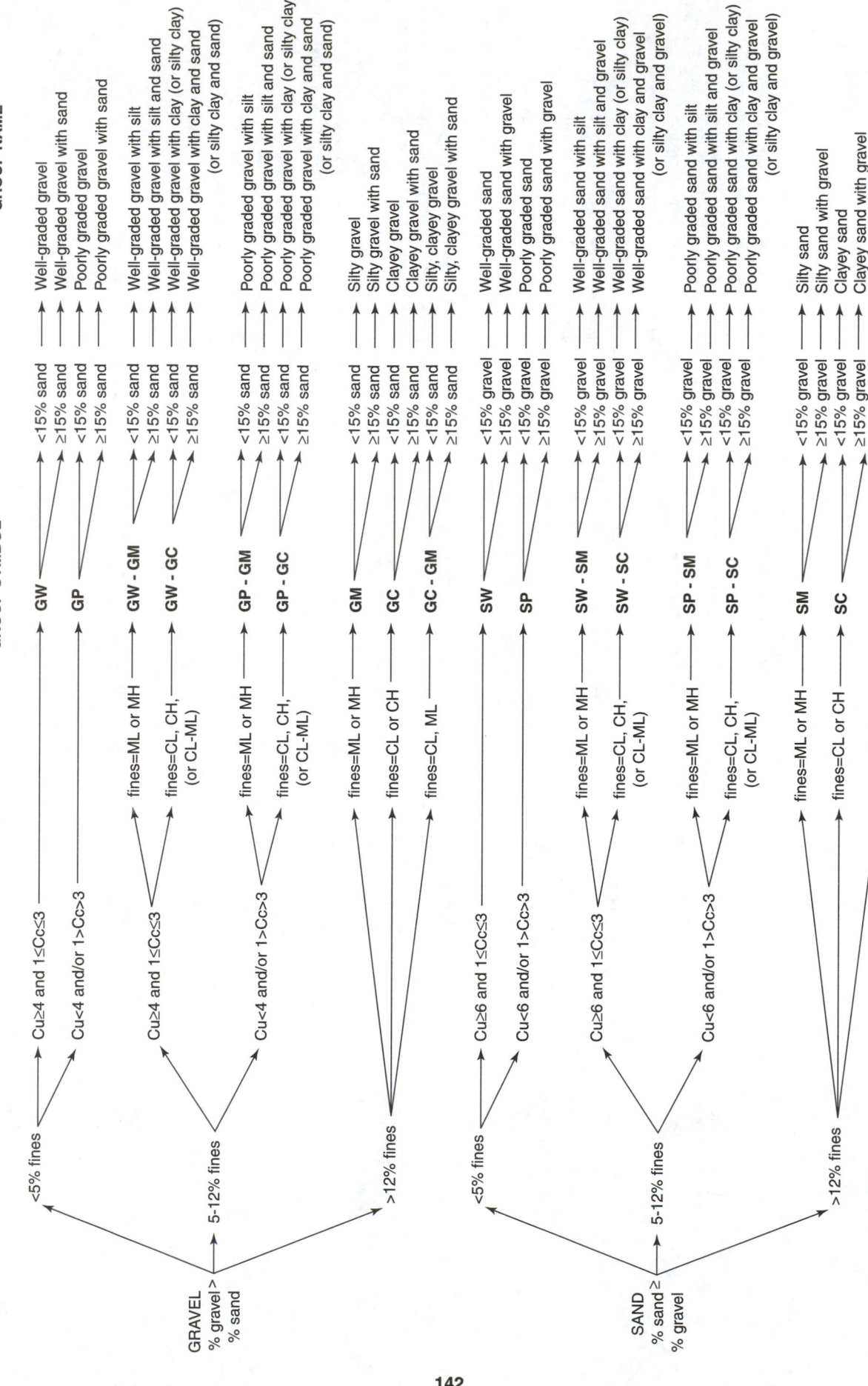

FIGURE 10–6 Flow Chart for Classifying Coarse-Grained Soil (More Than 50% Retained on No. 200 Sieve)[2]

142

coarse-grain sizes, and 100 − **76.5,** or 23.5%, was retained on the No. 4 sieve. Thus the percentage of coarse fraction retained on the No. 4 sieve is (23.5/84.8) (100), or 27.7%, and the percentage of coarse fraction that passed the No. 4 sieve is 72.3%. Because 72.3% is greater than 50%, go to the block labeled "Sands" in Table 10–3. The soil is evidently a sand. Because the sample contains **15.2%** passing the No. 200 sieve, which is greater than 12% fines, go to the block labeled "Sands with Fines—More than 12% fines." Refer next to the plasticity chart (Figure 10–3). With a liquid limit of **30.0%** and plasticity index of 18.0% (recall that the plasticity index is the difference between the liquid and plastic limits, or **30.0 − 12.0**), the sample is located above the A-line, and the fines are classified as CL. Return to Table 10–3, and go to the block labeled "SC." Thus, this soil is classified SC, according to the Unified Soil Classification System.

Example 10–3

Given:

A sample of soil was tested in the laboratory with the following results:

1. Liquid limit = NP (nonplastic)

2. Plastic limit = NP (nonplastic)

3. Sieve analysis data:

U.S. Sieve Size	Percentage Passing
1 in.	100
¾ in.	85
½ in.	70
⅜ in.	60
No. 4	48
No. 10	30
No. 40	16
No. 100	10
No. 200	2

Required:

Classify the soil by the Unified Soil Classification System.

Solution:

Because the percentage retained on the No. 200 sieve (100 − **2,** or 98%) is more than 50%, go to the block labeled "Coarse Grained Soils" in Table 10–3. The sample consists of 100 − **2,** or 98%, coarse-grained sizes, and 100 − **48,** or 52%, was retained on the No. 4 sieve. Thus the percentage of coarse fraction retained on the No. 4 sieve is 52/98, or 53.1%. Because 53.1% is greater than 50%, go to the block labeled "Gravels" in Table 10–3. The soil is evidently a gravel. Because the sample contains **2%** passing the No. 200 sieve, which is less than 5% fines, go to the block labeled "Clean Gravels— Less than 5% fines." The next block indicates that examination of the grain-size curve is necessary at this point. In Table 10–3, the two equations used to examine the grain-size curve are

$$C_u = \frac{D_{60}}{D_{10}} \qquad \textbf{(10–2)}$$

and

$$C_c = \frac{(D_{30})^2}{D_{10} \times D_{60}} \qquad \textbf{(10–3)}$$

where D_n represents the diameter of soil particles at which $n\%$ of the soil sample passes this diameter. In other words, D indicates particle size, and the subscript indicates percentage of soil sample that is smaller than that particular particle size. As indicated in Table 10–3, C_u must be greater than or equal to 4, and C_c must be between 1 and 3 in order for the sample to be classified as "GW." Otherwise, the sample would be classified "GP." Values of D_{60}, D_{30}, and D_{10} can be obtained from the grain-size distribution curve (see Figure 10–7). (In this particular example, the values of D_{60}, D_{30}, and D_{10} are available coincidentally from the information given.)

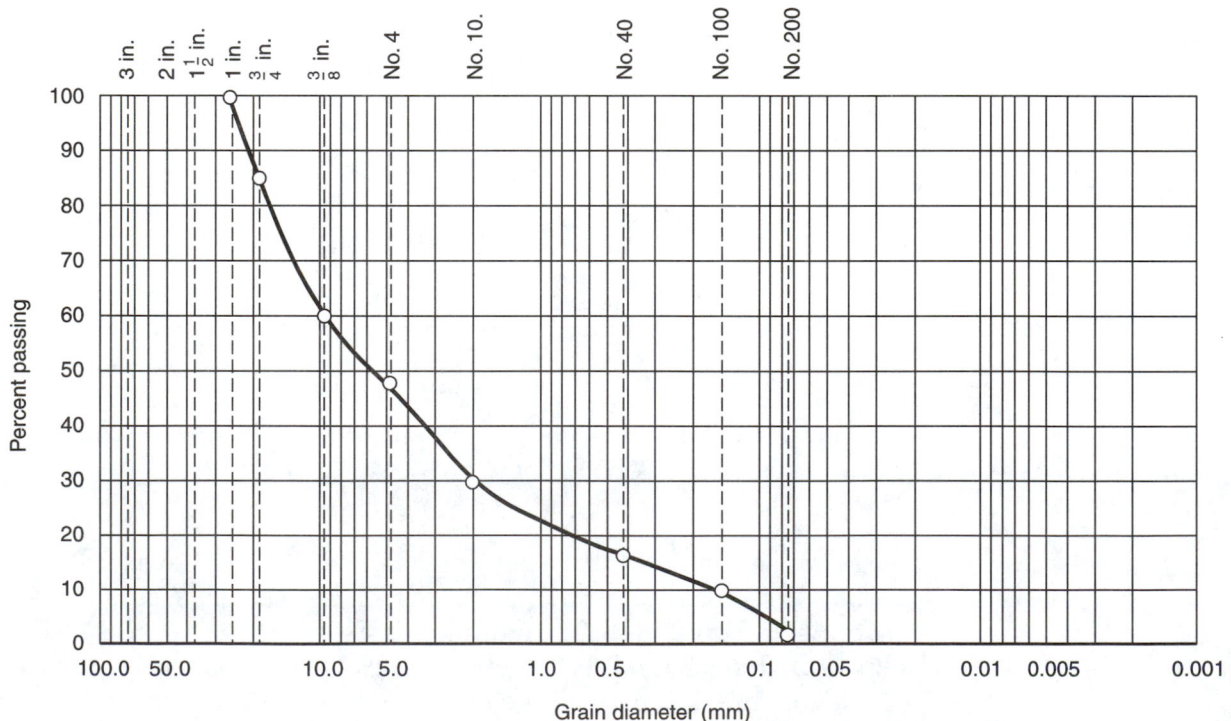

FIGURE 10–7   Grain-Size Distribution Curve

D_{60} corresponds to a ⅜-in. (9.5-mm) sieve, D_{30} corresponds to a No. 10 sieve (2.00 mm), and D_{10} corresponds to a No. 100 sieve (0.150 mm). Substituting these data into Eqs. (10–2) and (10–3) gives

$$C_u = \frac{9.5}{0.150} = 63.3$$

$$C_c = \frac{(2.00)^2}{(0.150)(9.5)} = 2.8$$

Because C_u (63.3) is greater than 4 and C_c (2.8) is between 1 and 3, this sample meets both criteria for a well-graded gravel. Hence, from Table 10–3, the soil is classified GW (i.e., well-graded gravel), according to the Unified Soil Classification System.

Example 10–4

Given:
A sample of inorganic soil was tested in the laboratory with the following results:

1. Liquid limit = **42.3%**

2. Plastic limit = **15.8%**

3. Sieve analysis data:

U.S. Sieve Size	Percentage Passing
No. 4	100
No. 10	93.2
No. 40	81.0
No. 200	60.2

Required:
Classify the soil sample by the Unified Soil Classification System.

Solution:
Because the percentage passing the No. 200 sieve is **60.2%,** which is greater than 50%, go to the lower block (labeled "Fine-Grained Soils") in Table 10–3. The liquid limit is **42.3%,** which is less than 50%, so go to the block labeled "Silts and Clays, Liquid limit less than 50." Now, because the sample is an inorganic soil and the plasticity index is **42.3** − **15.8**, or 26.5%, which is greater than 7, refer next to the Plasticity Chart (Figure 10–3). With a liquid limit of **42.3%** and plasticity index of 26.5%, the sample is located above the A-line. Return to Table 10–3 and go to the block labeled "CL." Thus

the soil is classified CL according to the Unified Soil Classification System.

Figure 10–8 lists a BASIC microcomputer program, prepared by Stevens [5], that classifies soils by the Unified Soil Classification System. Input to this program includes the following laboratory results that are supplied in response to prompts: (a) percent passing through standard sieves #200, #40, #4, ½ in., ¾ in., 1 in., 2 in., and 3 in.; (b) Atterberg limits (liquid and plastic limits); and (c) diameter at which 10%, 30%, and 60% of the particles are smaller. Output from the program is the soil classification of the sample in accordance with ASTM D 2487, based on the laboratory results. The program accepts the input, displays it for error checking, and then classifies the sample. In addition, it displays various calculated values used during the classification process. A value of 999 indicates that the associated parameter was not used. Figures 10–9, 10–10, and 10–11 demonstrate the use of the program to classify the soil samples considered in Examples 10–2, 10–3, and 10–4, respectively.

REFERENCES

[1]   AASHTO, *Standard Specifications for Transportation Materials and Methods of Sampling and Testing, Part I, Specifications,* 15th ed., Washington, D.C., 1990.

[2]   ASTM, *2002 Annual Book of ASTM Standards,* West Conshohocken, PA, 2002. Copyright, American Society for Testing and Materials, 100 Barr Harbor Drive, West Conshohocken, PA 19428-2959. Reprinted with permission.

[3]   A. Casagrande, "Classification and Identification of Soils," *Trans. ASCE, 113,* 901 (1948).

[4]   U.S. Army Corps of Engineers, *The Unified Soil Classification System,* Waterways Exp. Sta. Tech. Mem. 3-357, Vicksburg, Miss., 1953.

[5]   Jim Stevens, "Unified Soil Classification System," *Civil Engineering,* December 1982, pp. 61–2.

```
100 PAGE
110 PRINT @ 401: "UNIFIED SYSTEM ASTM D-2487"
120 PRINT @ 40:
130 PRINT @ 40: "_____"
140 PRINT @ 40: "INPUT LAB RESULTS"
150 PRINT @ 40: "_____"
160 REM -- INITIALIZE VARIABLES --
170 L15999
180 P15L1
190 P25L1
200 D15L1
210 D35L1
220 D65L1
230 C15L1
240 C25L1
250 REM -- ACCEPT SIEVE VALUES--
260 PRINT @40: " % PASSING ;NS200 SIEVE ? ";
270 INPUT @40: S2
280 PRINT @40: " % PASSING ;NS40 SIEVE ? ";
290 INPUT @40: S4
300 PRINT @40: " % PASSING ;NS10 SIEVE ? ";
310 INPUT @40: S1
320 PRINT @40: " % PASSING ;NS4 SIEVE ? ";
330 INPUT @40: S3
340 PRINT @40: " % PASSING 1/2 IN. ? ";
350 INPUT @40: S3
360 PRINT @40: " % PASSING 3/4 IN. ? ";
370 INPUT @40: S6
380 PRINT @40: " % PASSING 1.0 IN. ? ";
390 INPUT @40: S7
400 PRINT @40: " % PASSING 2.0 IN. ? ";
410 INPUT @40: S8
420 PRINT @40: " % PASSING 3.0 IN. ? ";
430 INPUT @40: S9
440 PRINT @40:
450 REM -- CALCULATE % RETAINED ON EACH SIEVE --
460 R1-S4-S2
470 R25S1-S4
480 R35S3-S1
490 R45S5-S3
500 R55S6-S5
510 R65S7-S6
520 R75S8-S7
530 R85S9-S8
540 R95100-S9
550 IF SW,512 THEN 610
560 REM -- GET LIQUID LIMIT, PLASTIC LIMIT --
570 PRINT @40: " LIQUID LIMIT ? ";
580 INPUT @40: L1
590 PRINT @40: " PLASTIC LIMIT ? ";
600 INPUT @40: P1
610 IF S2.12 THEN 720
620 REM -- INPUT @40: D6, D3, D1
630 PRINT @40: " DIAMETER AT WHICH 10% IS FINER ? ";
640 INPUT @40: D1
650 PRINT @40: " DIAMETER AT WHICH 30% IS FINER ? ";
660 INPUT @40: D3
670 PRINT @40: " DIAMETER AT WHICH 60% IS FINER ? ";
680 INPUT @40: D6
690 REM -- CALCULATE C1 AND C2 --
700 C15D6/D1
710 C25D3 2/ (D1*D6)
720 PRINT @40:
730 PRINT @40: "_____"
740 PRINT @40: " INPUT ERROR CHECK "
750 PRINT @40: "_____"
760 REM - PRINT INPUTS AND CALCULATED RETAINED VALUES FOR CONFIRMATION -
770 PRINT @40: " PASSING ;NS200 SIEVE: ";S2;" % RETAINED: ";R1;"%"
780 PRINT @40: " PASSING ;NS40 SIEVE: ";S4;" % RETAINED: ";R2;"%"
790 PRINT @40: " PASSING ;NS10 SIEVE: ";S1;" % RETAINED: ";R3;"%"
800 PRINT @40: " PASSING ;NS4 SIEVE: ";S3;" % RETAINED: ";R4;"%"
810 PRINT @40:
820 PRINT @40: "PASSING 1/2 IN. ? ";S5;" % RETAINED: ";R5;"%"
830 PRINT @40: " PASSING 3/4 IN. SIEVE: ";S6;" % RETAINED: "R6;"%"
840 PRINT @40: " PASSING 1.0 IN. SIEVE: ";S7;" % RETAINED: ";R7;"%"
850 PRINT @40: " PASSING 2.0 IN. SIEVE: ";S8;" % RETAINED: ";R8;"%"
860 PRINT @40: " PASSING 3.0 IN. SIEVE: ";S9;" % RETAINED: ";R9;"%"
870 PRINT @40:
880 PRINT @40: "** TO CONTINUE PRESS 'RETURN' ** ";
890 INPUT @40: R$
IAL ? ";
900 PAGE
910 PRINT @40:
920 PRINT @40:
930 PRINT @40: " D10 IS "; D1
940 PRINT @40: " D30 IS "; D3
950 PRINT @40: " D60 IS "; D6
960 PRINT @40:
970 PRINT @40: " LL IS "; L1
980 PRINT @40: " PL IS "; P1
990 P25L1-P1
1000 PRINT @40: " PI IS "; P2
1010 PRINT @40:
IAL ? ";
1020 PRINT @40: " CALCULATED VALUES OF CU AND CZ"
1030 PRINT @40: " CU IS "; C1
1040 PRINT @40: " CZ IS "; C2
1050 PRINT @40:
106- PRINT @40: "_____"
1070 PRINT @40: " YOUR SOIL CLASSIFICATION IS:"
1080 PRINT @40: "_____"
1090 REM -- SOLVE FOR 'PI' AND 'A LINE' EQUATION --

1100 A150
1110 A25A1
1120 A35A1
1130 A450, 73* (L1-20)-P2
1140 IF A4,0 THEN 1170
1150 A251
1160 GO TO 1240
1170 IF P2.7 THEN 1230
1180 IF P2,4 THEN 1230
1190 IF L1.26 THEN 1230
1200 IF L1,10 THEN 1230
1210 A351
1220 GO TO 1240
1230 A151
1240 REM -- BEGIN CLASSIFICATION --
1250 IF S25.50 THEN 1840
1260 PRINT @ 40: "--- COARSE-GRAINED ---"
1270 PRINT @40:
1280 F45(100-S3)/(100-S2)
1290 IF F4,0.5 THEN 1590
1300 A450, 73* (L1-20) -P2
1310 PRINT @40: "-- GRAVEL --"
1320 PRINT @40:
1330 IF S2,5 THEN 1360
1340 IF S2.12 THEN 1480
1350 GO TO 1440
1360 REM -- LESS THAN 5% PASS ;NS 200 SIEVE --
1370 IF C1,54 THEN 1420
1380 IF C2,1 THEN 1420
1390 IF C2.3 THEN 1420
1400 PRINT @40: " WELL-GRADED GRAVEL (GW)"
1410 END
1420 PRINT @40: " POORLY GRADED GRAVEL (GP)"
1430 END
1440 REM -- BETWEEN 5% AND 12% PASS ;NS 200 SIEVE --
1450 PRINT @40: " ** BORDERLINE ** "
1460 PRINT @40:
1470 CO TO 1360
1480 REM -- MORE THAN 12% PASS ;NS200 SIEVE --
1490 REM -- SOLVE FOR L1 AND P1 --
1500 IF A1,.1 THEN 1520
1510 IF P2.7 THEN 1570
1520 IF A351 THEN 1550
1530 PRINT @40: " SILTY GRAVEL (GM)"
1540 END
1550 PRINT @40: " SILTY,CLAYEY GRAVEL (GM-GC)"
1560 END
1570 PRINT @40: " CLAYEY GRAVEL (GC)"
1580 END
1590 PRINT @40: " -- SAND --"
1600 PRINT @40:
1610 IF S2,5 THEN 1640
1620 IF S2.12 THEN 1740
1630 GO TO 1710
1640 IF C1,56 THEN 1690
1650 IF C2.3 THEN 1690
1660 IF C2,51 THEN 1690
1670 PRINT @40: " WELL-GRADED SAND (SW)"
1680 END
1690 PRINT @40: " POORLY-GRADED SAND (SP)"
1700 END
1710 PRINT @40: " ** BORDERLINE ** "
1720 PRINT @40:
1730 GO TO 1640
1740 REM -- SOLVE FOR L1 AND P1 --
1750 IF A1,.1 THEN 1770
1760 IF P2.7 THEN 1820
1770 IF A351 THEN 1800
1780 PRINT @40: " SILTY SAND (SM)"
1790 END
1800 PRINT @40: " SILTY,CLAYEY SAND (SM-SC)"
1810 END
1820 PRINT @40: " CLAYEY SAND (SC)"
1830 END
1840 PRINT @40: " --- FINE GRAINED ---"
1850 PRINT @40:
1860 IF L1.50 THEN 2000
1870 IF A151 THEN 1980
1880 IF A351 THEN 1960
1890 PRINT @40: " DO COLOR AND ODOR INDICATE ORGANIC MATER-
1900 INPUT @40: R$
1910 IF R$5"NO" THEN 1940
1920 PRINT @40: "ORGANIC SILTS (OL)"
1930 END
1940 PRINT @40: " INORGANIC SILTS (ML)"
1950 END
1960 PRINT @40: " INORGANIC SILTS AND CLAYS (ML-CL)"
1970 END
1980 PRINT @40: " INORGANIC CLAY (CL)"
1990 END
2000 IF A151 THEN 2080
2010 PRINT @40: " DO COLOR AND ODOR INDICATE ORGANIC MATER-
2020 INPUT @40: R$
2030 IF R$5"YES" THEN 2060
2040 PRINT @40: " INORGANIC SILT (MH)"
2050 END
2060 PRINT @40: " ORGANIC SILT (OH)"
2070 END
2080 PRINT @40: " INORGANIC CLAY (CH)"
2090 END
```

FIGURE 10-8 Program to Classify Soils by Unified Soil Classification System [5]

```
UNIFIED SYSTEM ASTM D-2487

----------------------------------------
INPUT LAB RESULTS
----------------------------------------
%  PASSING #200 SIEVE ?      ? 15.2
%  PASSING #40 SIEVE ?       ? 39.7
%  PASSING #10 SIEVE ?       ? 60.0
%  PASSING #4  SIEVE ?       ? 76.5
%  PASSING 1/2 IN . ?        ? 100
%  PASSING 3/4 IN . ?        ? 100
%  PASSING 1.0 IN . ?        ? 100
%  PASSING 2.0 IN . ?        ? 100
%  PASSING 3.0 IN . ?        ? 100


LIQUID LIMIT ?          ? 30
PLASTIC LIMIT ?         ? 12

----------------------------------------
INPUT ERROR CHECK
----------------------------------------
PASSING # 200 SIEVE 15.2   % RETAINED: 24.5 %
PASSING # 40 SIEVE: 39.7   % RETAINED: 20.3 %
PASSING # 10 SIEVE: 60     % RETAINED: 16.5 %
PASSING #  4 SIEVE: 76.5   % RETAINED: 23.5 %

PASSING 1/2 IN. SIEVE: 100   % RETAINED: 0 %
PASSING 3/4 IN. SIEVE: 100   % RETAINED: 0 %
PASSING 1.0 IN. SIEVE: 100   % RETAINED: 0 %
PASSING 2.0 IN. SIEVE: 100   % RETAINED: 0 %
PASSING 3.0 IN. SIEVE: 100   % RETAINED: 0 %

****TO CONTINUE PRESS 'RETURN'****     ?
NOTE:  VALUE OF 999 INDICATES DATA NOT APPLICABLE

D-sub-10 IS  999
D-sub-30 IS  999
D-sub-60 IS  999


LIQUID LIMIT -- LL IS  30
PLASTIC LIMIT - PL IS  12
PLASTICITY INDEX -- Ip IS  18


CALCULATED VALUES OF Cu AND Cc:
COEFFICIENT OF UNIFORMITY -- Cu IS  999
COEFFICIENT OF CONCAVITY --- Cc IS  999

----------------------------------------
YOUR   SOIL CLASSIFICATION IS:
----------------------------------------
----- COARSE GRAINED -----

---- SAND ----

CLAYEY SAND (SC)
```

FIGURE 10–9 Application of Program of Figure 10–8 to Solve Example 10–2

```
UNIFIED SYSTEM ASTM D-2487

----------------------------------------
INPUT LAB RESULTS
----------------------------------------
%  PASSING #200 SIEVE ?      ? 2.0
%  PASSING #40 SIEVE ?       ? 16.0
%  PASSING #10 SIEVE ?       ? 30.0
%  PASSING #4  SIEVE ?       ? 48.0
%  PASSING 1/2 IN . ?        ? 70.0
%  PASSING 3/4 IN . ?        ? 85.0
%  PASSING 1.0 IN . ?        ? 100
%  PASSING 2.0 IN . ?        ? 100
%  PASSING 3.0 IN . ?        ? 100


DIAMETER AT WHICH 10% IS FINER ?     ? .15
DIAMETER AT WHICH 30% IS FINER ?     ? 2.0
DIAMETER AT WHICH 60% IS FINER ?     ? 9.5

----------------------------------------
INPUT ERROR CHECK
----------------------------------------
PASSING # 200 SIEVE 2      % RETAINED: 14 %
PASSING # 40 SIEVE: 16     % RETAINED: 14 %
PASSING # 10 SIEVE: 30     % RETAINED: 18 %
PASSING #  4 SIEVE: 48     % RETAINED: 22 %

PASSING 1/2 IN. SIEVE: 70    % RETAINED: 15 %
PASSING 3/4 IN. SIEVE: 85    % RETAINED: 15 %
PASSING 1.0 IN. SIEVE: 100   % RETAINED: 0 %
PASSING 2.0 IN. SIEVE: 100   % RETAINED: 0 %
PASSING 3.0 IN. SIEVE: 100   % RETAINED: 0 %

****TO CONTINUE PRESS 'RETURN'****     ?
NOTE:  VALUE OF 999 INDICATES DATA NOT APPLICABLE

D-sub-10 IS  .15
D-sub-30 IS  2
D-sub-60 IS  9.5


LIQUID LIMIT -- LL IS  999
PLASTIC LIMIT - PL IS  999
PLASTICITY INDEX -- Ip IS  0


CALCULATED VALUES OF Cu AND Cc:
COEFFICIENT OF UNIFORMITY -- Cu IS  63.33333
COEFFICIENT OF CONCAVITY --- Cc IS  2.807017

----------------------------------------
YOUR   SOIL CLASSIFICATION IS:
----------------------------------------
----- COARSE GRAINED -----

------ GRAVEL ------

WELL-GRADED GRAVEL (GW)
```

FIGURE 10–10 Application of Program of Figure 10–8 to Solve Example 10–3

```
UNIFIED SYSTEM ASTM D-2487

---------------------------------------
INPUT LAB RESULTS
---------------------------------------
% PASSING #200 SIEVE ?      ? 60.2
% PASSING #40 SIEVE ?       ? 81.0
% PASSING #10 SIEVE ?       ? 93.2
% PASSING #4  SIEVE ?       ? 100
% PASSING 1/2 IN . ?        ? 100
% PASSING 3/4 IN . ?        ? 100
% PASSING 1.0 IN . ?        ? 100
% PASSING 2.0 IN . ?        ? 100
% PASSING 3.0 IN . ?        ? 100


LIQUID LIMIT ?        ? 42.3
PLASTIC LIMIT ?       ? 15.8

---------------------------------------------
INPUT ERROR CHECK
---------------------------------------------
PASSING # 200 SIEVE 60.2   % RETAINED: 20.8 %
PASSING # 40 SIEVE: 81     % RETAINED: 12.2 %
PASSING # 10 SIEVE: 93.2   % RETAINED: 6.800003 %
PASSING #  4 SIEVE: 100    % RETAINED: 0 %

PASSING 1/2 IN. SIEVE: 100  % RETAINED: 0 %
PASSING 3/4 IN. SIEVE: 100  % RETAINED: 0 %
PASSING 1.0 IN. SIEVE: 100  % RETAINED: 0 %
PASSING 2.0 IN. SIEVE: 100  % RETAINED: 0 %
PASSING 3.0 IN. SIEVE: 100  % RETAINED: 0 %

****TO CONTINUE PRESS 'RETURN'****      ?
NOTE:  VALUE OF 999 INDICATES DATA NOT APPLICABLE

D-sub-10 IS  999
D-sub-30 IS  999
D-sub-60 IS  999

LIQUID LIMIT -- LL IS  42.3
PLASTIC LIMIT - PL IS  15.8
PLASTICITY INDEX -- Ip IS  26.5

CALCULATED VALUES OF Cu AND Cc:
COEFFICIENT OF UNIFORMITY -- Cu IS  999
COEFFICIENT OF CONCAVITY --- Cc IS  999

---------------------------------------
YOUR  SOIL CLASSIFICATION IS:
---------------------------------------
----- FINE GRAINED -----

INORGANIC CLAY (CL)
```

FIGURE 10–11 Application of Program of Figure 10–8 to Solve Example 10–4

CHAPTER ELEVEN

Determining Moisture-Unit Weight Relations of Soil (Compaction Test)

(Referenced Document: ASTM D 698)

INTRODUCTION   The general meaning of the verb "compact" is "to press closely together." In soil mechanics, it means to press soil particles tightly together by expelling air from void spaces between the particles. Compaction is normally done deliberately, often by heavy compaction rollers, and proceeds rapidly during construction. Compaction increases soil unit weight, thereby producing three important effects: (1) an increase in sheer strength, (2) a decrease in future settlement, and (3) a decrease in permeability [1]. These three changes in soil characteristics are beneficial for some types of earth construction, such as highways, airfields, and earth dams; as a general rule, the greater the compaction, the greater the benefits will be. Compaction is actually a rather cheap and effective way to improve the properties of a soil.

   The amount of compaction is quantified in terms of the dry unit weight of the soil. Usually, dry soils can be compacted best (and thus a greater unit weight achieved) if for each soil, a certain amount of water is added. In effect, water acts as a lubricant, allowing soil particles to be packed together better. However, if too much water is added, a lower unit weight will result. Thus, for a given compactive effort, there is a particular moisture content at which dry unit weight is greatest and compaction is best. This moisture content is known as the *optimum moisture content,* and the associated dry unit weight is called the *maximum dry unit weight.*

The usual practice in a construction project is to perform laboratory compaction tests on representative samples from the construction site to determine the optimum moisture content and maximum dry unit weight. The maximum dry unit weight is used by designers in specifying design sheer strength, resistance to future settlement, and permeability characteristics. The soil is then compacted by field compaction methods to achieve the laboratory maximum dry unit weight (or a percentage of it). In-place soil unit weight tests (discussed in Chapters 13, 14, and 15) are used to determine if and when the laboratory maximum dry unit weight (or an acceptable percentage thereof) has been achieved.

A common compaction test is known as the *Standard Proctor test.* The exact procedure for conducting a Standard Proctor test is described later in this chapter, but the basic premise of the test is that a soil sample is compacted in a 4- or 6-in. (101.6- or 152.5-mm) diameter mold by dropping a 5.5-lb (24.4-N) hammer onto the sample from a height of 12 in. (305 mm), producing a compactive effort of 12,400 ft-lb/ft^3 (600 kN-m/m^3). An alternative test, known as the *Modified Proctor test,* uses a 10-lb (44.5-N) hammer that is dropped 18 in. (457 mm). The latter produces greater compaction and, hence, greater soil unit weight (because the hammer is heavier, drops farther, and therefore exerts greater compaction effort on the soil sample). Therefore, the Modified Proctor test may be used when greater soil unit weight is required. Only the Standard Proctor test is described in detail in this book. For details regarding the Modified Proctor test, the reader is referred to ASTM D 1557 [2].

Three alternative methods are provided for carrying out a Standard Proctor test [2]:

(1) *Method A:*

(1.1) *Mold*—4-in. (101.6-mm) diameter.

(1.2) *Material*—Passing No. 4 (4.75-mm) sieve.

(1.3) *Layers*—Three.

(1.4) *Blows per layer*—25.

(1.5) *Use*—May be used if 20% or less by mass of the material is retained on the No. 4 (4.75-mm) sieve.

(1.6) *Other Use*—If this method is not specified, materials that meet these gradation requirements may be tested using Methods B or C.

(2) *Method B:*

(2.1) *Mold*—4-in. (101.6-mm) diameter.

(2.2) *Material*—Passing ⅜-in. (9.5-mm) sieve.

(2.3) *Layers*—Three.

(2.4) *Blows per layer*—25.

(2.5) *Use*—Shall be used if more than 20% by mass of the material is retained on the No. 4 (4.75-mm) sieve and 20% or less by mass of the material is retained on the ⅜-in. (9.5-mm) sieve.

(2.6) *Other Use*—If this method is not specified, materials that meet these gradation requirements may be tested using Method C.

(3) *Method C:*

(3.1) *Mold*—6-in. (152.4-mm) diameter.

(3.2) *Material*—Passing ¾-in. (19.0-mm) sieve.

(3.3) *Layers*—Three.

(3.4) *Blows per layer*—56.

(3.5) *Use*—Shall be used if more than 20% by mass of the material is retained on the ⅜-in. (9.5-mm) sieve and less than 30% by mass of the material is retained on the ¾-in. (19.0-mm) sieve.

(4) The 6-in. (152.4-mm) diameter mold shall not be used with Method A or B.

> *Note 1*—Results have been found to vary slightly when a material is tested at the same compactive effort in different size molds.

This test method applies only to soils that have 30% or less by mass of particles retained on the the ¾-inch (19.0-mm) sieve.

> *Note 2*—For relationships between unit weights and water contents of soils with 30% or less by weight of material retained on the ¾-in. (19.0-mm) sieve to unit weights and water contents of the fraction passing ¾-in. (19.0-mm) sieve, see ASTM Practice D 4718.

If the specimen contains more than 5% by weight oversize fraction (coarse fraction) and the material will not be included in the test, corrections must be made to the unit weight and water content of the specimen or to the appropriate field in-place density test specimen using ASTM Practice D 4718.

APPARATUS AND SUPPLIES

Compaction test equipment (see Figures 11–1 through 11–5)

> Mold, 4 in.—A mold having a 4.000 ± 0.016-in. (101.6 ± 0.4-mm) average inside diameter, a height of 4.584 ± 0.018-in. (116.4 ± 0.5-mm) and a volume of 0.0333 ± 0.0005 ft³ (944 ± 14 cm³). A mold assembly having the minimum required features is shown in Figure 11–1.

> Mold, 6 in.—A mold having a 6.000 ± 0.026-in. (152.4 ± 0.7- mm) average inside diameter, a height of 4.584 ± 0.018-in.

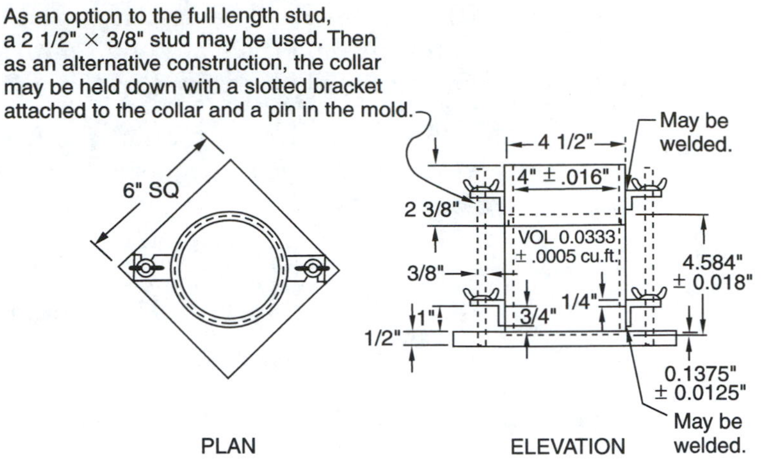

As an option to the full length stud,
a 2 1/2" × 3/8" stud may be used. Then
as an alternative construction, the collar
may be held down with a slotted bracket
attached to the collar and a pin in the mold.

6" SQ

May be welded.

4 1/2"
4" ± .016"
2 3/8"
VOL 0.0333
± .0005 cu.ft.
3/8"
4.584"
± 0.018"
1/4"
1"
3/4"
1/2"
0.1375"
± 0.0125"
May be welded.

PLAN

ELEVATION

FIGURE 11–1 4-in. Cylindrical Mold [2]

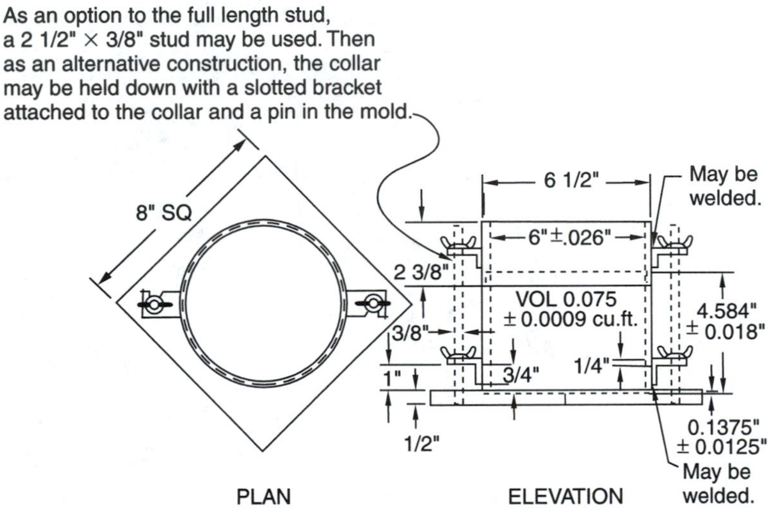

As an option to the full length stud,
a 2 1/2" × 3/8" stud may be used. Then
as an alternative construction, the collar
may be held down with a slotted bracket
attached to the collar and a pin in the mold.

8" SQ

May be welded.

6 1/2"
6" ± .026"
2 3/8"
VOL 0.075
± 0.0009 cu.ft.
3/8"
4.584"
± 0.018"
1/4"
1"
3/4"
1/2"
0.1375"
± 0.0125"
May be welded.

PLAN

ELEVATION

FIGURE 11–2 6-in. Cylindrical Mold [2]

$(116.4 \pm 0.5$-mm), and a volume of $0.075 \pm 0.0009 \text{ft}^3 (2124 \pm 25 \text{ cm}^3)$. A mold assembly having the minimum required features is shown in Figure 11–2.

5.5-lb (24.4-N) hammer, to be operated manually or mechanically to drop a free fall of 12 in. (305 mm) onto the soil (see Figure 11–3)

Sample extruder

Balances

One with a 20-kilogram (kg) capacity and accuracy to 1 g

One with a 1,000-g capacity and accuracy to 0.01 g

Drying oven

Straightedge

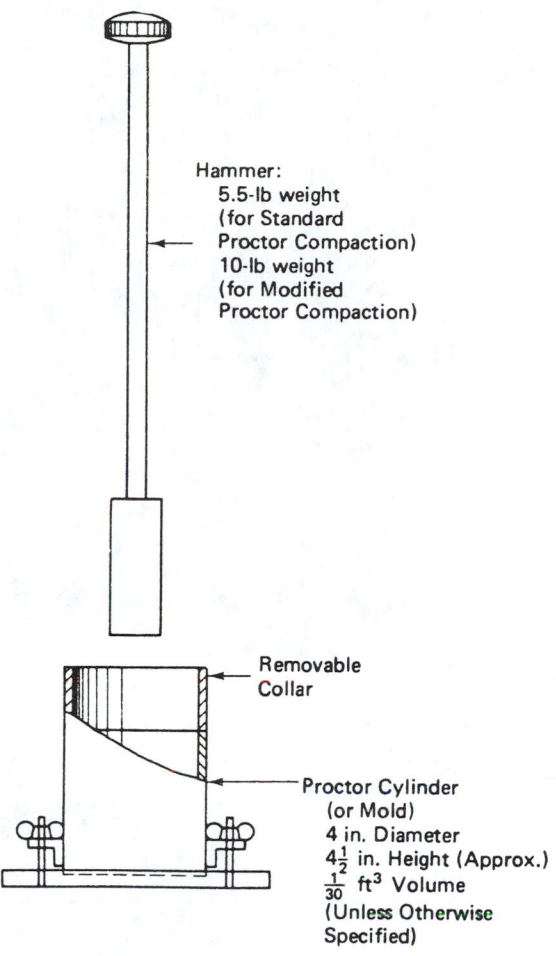

Hammer:
5.5-lb weight
(for Standard
Proctor Compaction)
10-lb weight
(for Modified
Proctor Compaction)

Removable
Collar

Proctor Cylinder
(or Mold)
4 in. Diameter
$4\frac{1}{2}$ in. Height (Approx.)
$\frac{1}{30}$ ft^3 Volume
(Unless Otherwise
Specified)

FIGURE 11–3　Compaction Test Equipment [3]

Sieves (3 in., ¾ in., ⅜ in., and No. 4)

Soil mixer (see Figure 11–5)

Miscellaneous mixing tools, such as a mixing pan, spoon, trowel, etc. (see Figure 11–4).

CALIBRATION [2]　Perform calibrations before initial use, after repairs or other occurrences that might affect the test results, at intervals not exceeding 1,000 test specimens, or annually, whichever occurs first, for the following apparatus:

Balance—Evaluate in accordance with ASTM Specification D 4753.

Molds—Determine the volume as described in the next section.

Manual Rammer—Verify the free-fall distance, rammer mass, rammer face, and guide sleeve requirements.

Mechanical Rammer—Calibrate and adjust the mechanical rammer in accordance with ASTM Test Methods D 2168. In addition,

FIGURE 11–4 Apparatus Used in Compaction Test (Courtesy of Soiltest, Inc.)

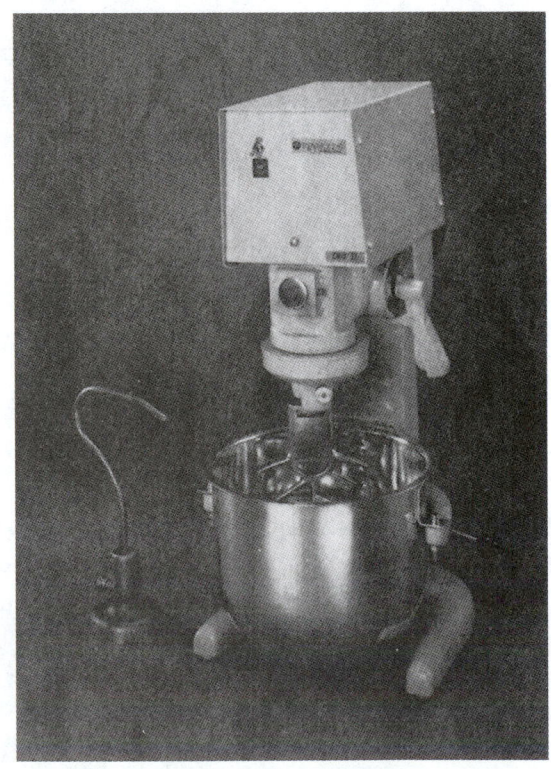

FIGURE 11–5 Soil Mixer (Courtesy of Soiltest, Inc.)

the clearance between the rammer and the inside surface of the mold shall be verified.

VOLUME OF COMPACTION MOLD [2]

(1) *Scope*

(1.1) This describes the procedure for determining the volume of a compaction mold.

(1.2) The volume is determined by a water-filled method and checked by a linear-measurement method.

(2) *Apparatus*

(2.1) In addition to the apparatus listed previously, the following items are required:

(2.1.1) *Vernier or Dial Caliper*—having a measuring range of at least 0 to 6 in. (0 to 150 mm) and readable to at least 0.001 in. (0.02 mm).

(2.1.2) *Inside Micrometer*—having a measuring range of at least 2 to 12 in. (50 to 300 mm) and readable to at least 0.001 in. (0.02 mm).

(2.1.3) *Plastic or Glass Plates*—Two plastic or glass plates approximately 8 in. square by ¼ in. thick (200 by 200 mm by 6 mm).

(2.1.4) *Thermometer*—0 to 50°C range, 0.5°C graduations.

(2.1.5) *Stopcock Grease* or similar sealant.

(2.1.6) *Miscellaneous Equipment*—Bulb syringe, towels, etc.

(3) *Precautions*

(3.1) Perform this procedure in an area isolated from drafts or extreme temperature fluctuations.

(4) *Procedure*

(4.1) *Water-Filling Method:*

(4.1.1) Lightly grease the bottom of the compaction mold and place it on one of the plastic or glass plates. Lightly grease the top of the mold. Be careful not to get grease on the inside of the mold. If it is necessary to use the base plate, place the greased mold onto the base plate and secure with the locking studs.

(4.1.2) Determine the mass of the greased mold and both plastic or glass plates to the nearest 0.01 lb (1 g) and record. When the base plate is being used in lieu of the bottom plastic or glass plate, determine the mass of the mold, base plate, and a single plastic or glass plate to be used on top of the mold to the nearest 0.01 lb (1 g) and record.

(4.1.3) Place the mold and the bottom plastic or glass plate on a firm, level surface and fill the mold with water to slightly above its rim.

(4.1.4) Slide the second plate over the top surface of the mold so that the mold remains completely filled with water and air bubbles are not entrapped. Add or remove water as necessary with a bulb syringe.

(4.1.5) Completely dry any excess water from the outside of the mold and plates.

(4.1.6) Determine the mass of the mold, plates, and water and record to the nearest 0.01 lb (1 g).

(4.1.7) Determine the temperature of the water in the mold to the nearest 1°C and record. Determine and record the absolute density of water from Table 11–1.

(4.1.8) Calculate the mass of water in the mold by subtracting the mass determined in (4.1.2) from the mass determined in (4.1.6).

(4.1.9) Calculate the volume of water by dividing the mass of water by the density of water and record to the nearest 0.0001 ft^3 (1 cm^3).

(4.1.10) When the base plate is used for the calibration of the mold volume, repeat (4.1.3) through (4.1.9).

(4.2) *Linear-Measurement Method:*

(4.2.1) Using either the vernier caliper or the inside micrometer, measure the diameter of the mold six times at the top of the mold and six times at the bottom of the mold, spacing each of the six top and bottom measurements equally around the circumference of the mold. Record the values to the nearest 0.001 in. (0.02 mm).

(4.2.2) Using the vernier caliper, measure the inside height of the mold by making three measurements equally spaced around the circumference of the mold. Record values to the nearest 0.001 in. (0.02 mm).

(4.2.3) Calculate the average top diameter, average bottom diameter, and average height.

(4.2.4) Calculate the volume of the mold and record to the nearest 0.0001 ft^3 (1 cm^3) as follows:

$$V = \frac{(\pi)(h)(d_\text{t} + d_\text{b})^2}{(16)(1728)} \text{ (English Gravitational System)}$$

$$V = \frac{(\pi)(h)(d_\text{t} + d_\text{b})^2}{(16)(10^3)} \text{ (International System)}$$

where:

V = volume of mold, ft^3 (cm^3)

h = average height, in. (mm)

d_b = average top diameter, in. (mm)

d_t = average bottom diameter, in. (mm)

$\frac{1}{1728}$ = constant to convert in.^{3} to ft^3

$\frac{1}{10^3}$ = constant to convert mm^3 to cm^3

(5) *Comparison of Results*

(5.1) The volume obtained by either method should be within the volume tolerance requirements of the "Apparatus and Supplies" section.

(5.2) The difference between the two methods should not exceed 0.5% of the nominal volume of the mold.

(5.3) Repeat the determination of volume if these criteria are not met.

(5.4) Failure to obtain satisfactory agreement between the two methods, even after several trials, is an indication that the mold is badly deformed and should be replaced.

(5.5) Use the volume of the mold determined using the water-filling method as the assigned volume value for calculating the moist and dry density.

Table 11–1   Density of Water[a] [2]

Temperature, °C (°F)	Density of Water, g/ml
18 (64.4)	0.99862
19 (66.2)	0.99843
20 (68.0)	0.99823
21 (69.8)	0.99802
22 (71.6)	0.99779
23 (73.4)	0.99756
24 (75.2)	0.99733
25 (77.0)	0.99707
26 (78.8)	0.99681

[A]Values other than shown may be obtained by referring to the *Handbook of Chemistry and Physics*, Chemical Rubber Publishing Co., Cleveland, Ohio.

TEST SAMPLE [2]

(1) The required sample mass for Methods A and B is approximately 35 lb (16 kg), and for Method C is approximately 65 lb (29 kg) of dry soil. Therefore, the field sample should have a moist mass of at least 50 lb (23 kg) and 100 lb (45 kg), respectively.

(2) Determine the percentage of material by mass retained on the No. 4 (4.75-mm), ⅜-in. (9.5-mm), or ¾-in. (19.0-mm) sieve as appropriate for choosing Method A, B, or C. Make this determination by separating out a representative portion from the total sample and

determining the percentages passing the sieves of interest by ASTM Test Methods D 422 or C 136. It is only necessary to calculate percentages for the sieve or sieves for which information is desired.

PREPARATION OF APPARATUS [2]

(1) Select the proper compaction mold in accordance with the method (A, B, or C) being used. Determine and record its mass to the nearest gram. Assemble the mold, base and extension collar. Check the alignment of the inner wall of the mold and mold extension collar. Adjust if necessary.

(2) Check that the rammer assembly is in good working condition and that parts are not loose or worn. Make any necessary adjustments or repairs. If adjustments or repairs are made, the rammer must be recalibrated.

PROCEDURE [2]

To carry out a laboratory compaction test, a soil at a selected water content is placed in three layers into a mold of given dimensions, with each layer compacted by 25 (Methods A and B) or 56 (Method C) blows of a 5.5-lb (24.4-N) rammer dropped from a distance of 12 in. (305 mm), subjecting the soil to a total compactive effort of about 12,400 ft-lb/ft^3 (600 kN-m/m^3). The resulting dry unit weight is determined. The procedure is repeated for a sufficient number of water contents to establish a relationship between the dry unit weight and the water content for the soil. These data, when plotted, represent a curvilinear relationship known as the *compaction curve*. The values of optimum water content and standard maximum dry unit weight are determined from the compaction curve.

The actual step-by-step procedure is as follows (ASTM D 698-00 [2]):

(1) *Soils:*

(1.1) Do not reuse soil that has been previously laboratory compacted.

(1.2) When using this test method for soils containing hydrated halloysite, or where past experience with a particular soil indicates that results will be altered by air drying, use the moist preparation method [see (2)].

(1.3) Prepare the soil specimens for testing in accordance with (2) (preferred) or with (3).

(2) *Moist Preparation Method (preferred)*—Without previously drying the sample, pass it through a No. 4 (4.75-mm), ⅜-in. (9.5-mm), or ¾-in. (19.0-mm) sieve, depending on the method (A, B, or C) being used. Determine the water content of the processed soil.

(2.1) Prepare at least four (preferably five) specimens having water contents such that they bracket the estimated optimum water content. A specimen having a water content close to optimum should be prepared first by trial additions of water and mixing (see Note 1). Select water contents for the rest of the specimens to provide at least two specimens wet and two specimens dry of optimum, and water contents varying by about 2%. At least two water contents are necessary on the wet and dry side of optimum to accurately define the dry unit weight compaction curve [see (5)]. Some soils with very high optimum water content or a relatively flat compaction curve may require larger water content increments to obtain a well-defined maximum dry unit weight. Water content increments should not exceed 4%.

> *Note 1*—With practice it is usually possible to visually judge a point near optimum water content. Typically, soil at optimum water content can be squeezed into a lump that sticks together when hand pressure is released, but will break cleanly into two sections when "bent." At water contents dry of optimum, soils tend to crumble; wet of optimum, soils tend to stick together in a sticky cohesive mass. Optimum water content is typically slightly less than the plastic limit.

(2.2) Use approximately 5 lbm (2.3 kg) of the sieved soil for each specimen to be compacted using Method A or B, or 13 lbm (5.9 kg) using Method C. To obtain the specimen water contents selected in (2.1), add or remove the required amounts of water as follows: to add water, spray it into the soil during mixing; to remove water, allow the soil to dry in air at ambient temperature or in a drying apparatus such that the temperature of the sample does not exceed 140°F (60°C). Mix the soil frequently during drying to maintain an even water content distribution. Thoroughly mix each specimen to ensure even distribution of water throughout and then place in a separate covered container and allow to stand in accordance with Table 11–2 prior to compaction. For the purpose of selecting a standing time, the soil may be classified using ASTM Test Method D 2487, Practice D 2488, or data on other samples from the same material source. For referee testing, classification shall be by ASTM Test Method D 2487.

Table 11–2 Required Standing Times of Moisturized Specimens

Classification	Minimum Standing Time, h
GW, GP, SW, SP	No Requirement
GM, SM	3
All other soils	16

(3) *Dry Preparation Method*—If the sample is too damp to be friable, reduce the water content by air drying until the material is friable. Drying may be in air or by the use of drying apparatus such that the temperature of the sample does not exceed 140°F (60°C). Thoroughly break up the aggregations in such a manner as to avoid breaking individual particles. Pass the material through the appropriate sieve: No. 4 (4.75-mm), ⅜-in. (9.5-mm), or ¾-in. (19.0-mm). When preparing the material by passing over the ¾-in. sieve for

compaction in the 6-in. mold, break up aggregations sufficiently to at least pass the ⅜-in. sieve in order to facilitate the distribution of water throughout the soil in later mixing.

(3.1) Prepare at least four (preferably five) specimens in accordance with (2.1).

(3.2) Use approximately 5 lbm (2.3 kg) of the sieved soil for each specimen to be compacted using Method A or B, or 13 lbm (5.9 kg) using Method C. Add the required amounts of water to bring the water contents of the specimens to the values selected in (3.1). Follow the specimen preparation procedure specified in (2.2) for drying the soil or adding water into the soil and curing each test specimen.

(4) *Compaction*—After curing, if required, each specimen shall be compacted as follows:

(4.1) Determine and record the mass of the mold or mold and base plate.

(4.2) Assemble and secure the mold and collar to the base plate. The mold shall rest on a uniform rigid foundation, such as provided by a cylinder or cube of concrete with a mass of not less than 200 lbm (91 kg). Secure the base plate to the rigid foundation. The method of attachment to the rigid foundation shall allow easy removal of the assembled mold, collar, and base plate after compaction is completed.

(4.3) Compact the specimen in three layers. After compaction, each layer should be approximately equal in thickness. Prior to compaction, place the loose soil into the mold and spread into a layer of uniform thickness. Lightly tamp the soil prior to compaction until it is not in a fluffy or loose state, using either the manual compaction rammer or a 2-in. (5-mm) diameter cylinder. Following compaction of each of the first two layers, any soil adjacent to the mold walls that has not been compacted or extends above the compacted surface shall be trimmed. The trimmed soil may be included with the additional soil for the next layer. A knife or other suitable device may be used. The total amount of soil used shall be such that the third compacted layer slightly extends into the collar, but does not exceed ¼ in. (6 mm) above the top of the mold. If the third layer does extend above the top of the mold by more than ¼ in. (6 mm), the specimen shall be discarded. The specimen shall be discarded when the last blow of the rammer for the third layer results in the bottom of the rammer extending below the top of the compaction mold.

(4.4) Compact each layer with 25 blows for the 4-in. (101.6-mm) mold or with 56 blows for the 6-in. (152.4-mm) mold.

> *Note 2*—When compacting specimens wetter than optimum water content, uneven compacted surfaces can occur and operator judgment is required as to the average height of the specimen.

(4.5) In operating the manual rammer, take care to avoid lifting the guide sleeve during the rammer upstroke. Hold the guide sleeve steady and within 5° of vertical. Apply the blows at a uniform rate of approximately 25 blows/min and in such a manner as to provide complete, uniform coverage of the specimen surface.

(4.6) Following compaction of the last layer, remove the collar and base plate from the mold, except as noted in (4.7). A knife may be used to trim the soil adjacent to the collar to loosen the soil from the collar before removal to avoid disrupting the soil below the top of the mold.

(4.7) Carefully trim the compacted specimen even with the top of the mold by means of the straightedge scraped across the top of the mold to form a plane surface even with the top of the mold. Initial trimming of the specimen above the top of the mold with a knife may prevent the soil from tearing below the top of the mold. Fill any holes in the top surface with unused or trimmed soil from the specimen, press in with the fingers, and again scrape the straightedge across the top of the mold. Repeat the appropriate preceding operations on the bottom of the specimen when the mold volume was determined without the base plate. For very wet or dry soils, soil or water may be lost if the base plate is removed. For these situations, leave the base plate attached to the mold. When the base plate is left attached, the volume of the mold must be calibrated with the base plate attached to the mold rather than a plastic or glass plate as noted in the "Volume of Compaction Mold" section.

(4.8) Determine and record the mass of the specimen and mold to the nearest gram. When the base plate is left attached, determine and record the mass of the specimen, mold, and base plate to the nearest gram.

(4.9) Remove the material from the mold. Obtain a specimen for water content by using either the whole specimen (preferred method) or a representative portion. When the entire specimen is used, break it up to facilitate drying. Otherwise, obtain a portion by slicing the compacted specimen axially through the center and removing about 500 g of material from the cut faces. Obtain the water content in accordance with ASTM Test Method D 2216.

(5) Following compaction of the last specimen, compare the wet unit weights to ensure that a desired pattern of obtaining data on each side of the optimum water content will be attained for the dry unit weight compaction curve. Plotting the wet unit weight and water content of each compacted specimen can be an aid in making the above evaluation. If the desired pattern is not obtained, additional compacted specimens will be required. Generally, one water content value greater than the water content defining the maximum wet unit weight is sufficient to ensure data on the wet side of optimum water content for the maximum dry unit weight.

DATA Data collected in this test should include the following for each repetition of the compaction procedure:

[A] Density Data

Mass of mold, M_m

Mass of compacted soil plus mold, M_{sm}

[B] Moisture Content Data

Mass of container, M_c

Mass of container plus moist soil, M_{cws}

Mass of container plus oven-dried soil, M_{cs}

CALCULATIONS For each repetition of the compaction procedure, moisture content and both wet and dry unit weights must be determined. The moisture contents can be computed using Eq. (3–2). The unit weights can be determined as follows:

$$\rho_m = \frac{(M_t - M_{md})}{1000\ V} \qquad (11\text{--}1)$$

where:

ρ_m = moist density of compacted specimen, Mg/m^3,
M_t = mass of moist specimen and mold, kg,
M_{md} = mass of compaction mold, kg, and
V = volume of compaction mold, m^3

$$\rho_d = \frac{\rho_m}{1 + \dfrac{w}{100}} \qquad (11\text{--}2)$$

where:

ρ_d = dry density of compacted specimen, Mg/m^3, and
w = water content, %.

$$\gamma_d = 62.43\ \rho_d \text{ in lb/ft}^3 \qquad (11\text{--}3)$$

or

$$\gamma_d = 9.807\ \rho_d \text{ in kN/m}^3 \qquad (11\text{--}4)$$

where:

γ_d = dry unit weight of compacted specimen

The moisture-unit weight relationship for the soil sample being tested can be analyzed by plotting a graph with moisture contents along the abscissa and corresponding dry unit weights along the ordinate. The moisture content and dry unit weight corresponding to the peak of the plotted curve are termed "optimum moisture content" and "maximum dry unit weight," respectively.

NUMERICAL EXAMPLE

A laboratory test was conducted according to the procedure for Method A described previously. The following data were obtained for the first trial specimen of the compaction test:

[A] Unit Weight Data

Mass of mold, M_{md} = **1,990.0 g**, or **1.990 kg**

Mass of moist specimen and mold, M_t = **3,718.2g**, or **3.718 kg**

Volume of compaction mold, V = **0.03333 ft³**, or **0.0009438 m³**

[B] Moisture Content Data

Mass of container, M_c = **45.20 g**

Mass of container plus moist soil, M_{cws} = **235.65 g**

Mass of container plus oven-dried soil, M_{cs} = **210.38 g**

These data for the first trial specimen of the compaction test, together with the data for all succeeding trial specimens, are presented on the form on the following page. At the end of the chapter, two blank copies of this form are included for the reader's use.

The moisture content, computed using Eq. (3–2), was determined to be 15.3%. The unit weights can be computed as follows:

$$\rho_m = \frac{M_t - M_{md}}{1000\,V} \tag{11–1}$$

$$\rho_m = \frac{3.718 - 1.990}{(1000)(0.0009438)} = 1.831 \text{ Mg/m}^3$$

$$\rho_d = \frac{\rho_m}{1 + \dfrac{w}{100}} \tag{11–2}$$

$$\rho_d = \frac{1.831}{1 + 15.3/100} = 1.588 \text{ Mg/m}$$

$$\gamma_{wet} = (62.43)(1.831) = 114.3 \text{ lb/ft}^3$$

$$\gamma_d = (62.43)(1.588) = 99.1 \text{ lb/ft}^3$$

These values are presented on the first line of the form on page 166. Similar calculations for the four remaining trial specimens were made, and the results are also given on the form.

Soils Testing Laboratory
Compaction Test

Method Used: ASTM D 698 Method A

Project No.: SR 2828

Location: Newell, N.C.

Description of Sample: Brown silty clay

Tested by: John Doe

Date: 4/24/02

Sample No.: 20

Boring No.: B-9

Depth: 5 ft

Volume of Mold (V): 0.0009438 m³

Moisture Content Determination

Trial No.	Mass of Moist Specimen + Mold, M_t (kg)	Mass of Mold, M_{md} (kg)	Mass of Moist Specimen (kg)	Moist Density of Compacted Specimen ρ_m (Mg/m³)	Can No.	Mass of Wet Soil + Can, M_{cws} (g)	Mass of Dry Soil + Can, M_{cs} (g)	Mass of Water, M_w (g)	Mass of Can, M_c (g)	Mass of Dry Soil, M_s (g)	Moisture Content, w (%)	Dry Density of Compacted Specimen ρ_d (Mg/m³)	Dry Unit Weight γ_d (lb/ft³)
A	B	C	D	E	F	G	H	I	J	K	L	M	N
1	3.718	1.990	1.728	1.831	A-1	235.65	210.38	25.27	45.20	165.18	15.3	1.588	99.1
2	3.802	1.990	1.812	1.920	A-2	231.50	203.55	27.95	43.83	159.72	17.5	1.634	102.0
3	3.903	1.990	1.913	2.027	A-3	236.48	203.27	33.21	43.14	160.13	20.7	1.679	104.8
4	3.921	1.990	1.931	2.046	A-4	256.63	217.16	39.47	44.80	172.36	22.9	1.665	103.9
5	3.901	1.990	1.911	2.025	A-5	243.99	204.07	39.92	42.89	161.18	24.8	1.623	101.3

Notes: $D = B - C$; $E = \dfrac{D}{1000\ V}$; $I = G - H$; $K = H - J$; $L = \dfrac{I}{K} \times 100$; $M = \dfrac{E}{100 + L} \times 100$; $N = M \times 62.43$

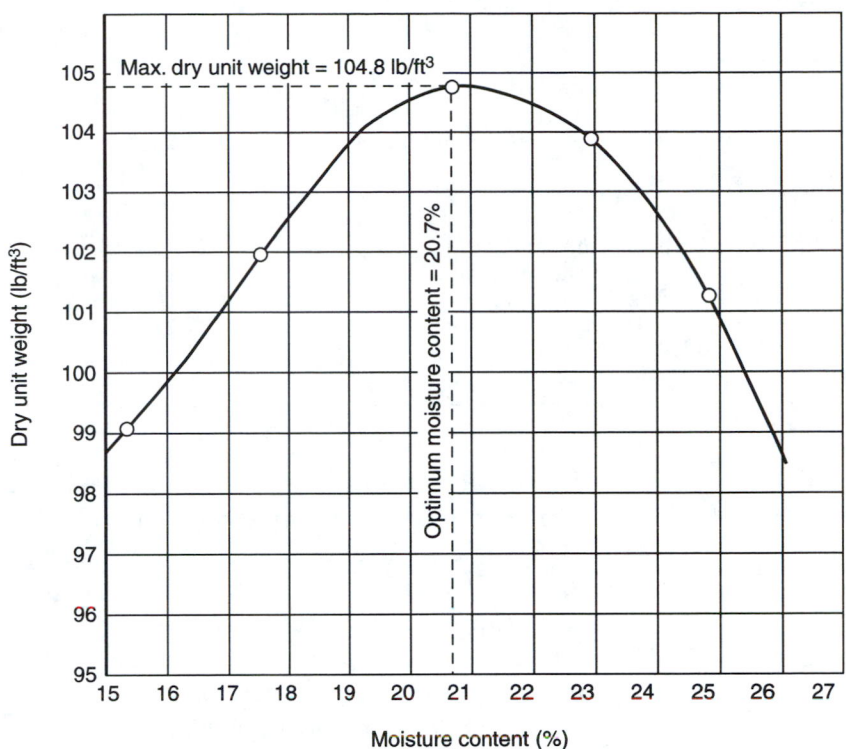

Optimum moisture content = 20.7%
Maximum dry unit weight = 104.8 lb/ft³

FIGURE 11–6 Compaction Curve

(In using the form on page 166 the reader will note that columns B, C, F, G, H, and J contain data that were obtained during the test. Other columns contain values computed from test data. A key for facilitating computation of these values is given at the bottom of the form.)

The moisture-unit weight relationship (compaction curve) is obtained by plotting a graph of dry unit weight versus moisture content, as shown in Figure 11–6. (At the end of the text a copy of the blank graph form is included and may be photocopied as needed.) From this graph, the optimum moisture content is determined to be 20.7% and the maximum dry unit weight is 104.8 lb/ft³.

CONCLUSIONS

The primary values determined in a compaction test are, of course, the optimum moisture content and maximum dry unit weight; however, the written report would normally also include the compaction curve and the data form. In addition, the origin of the material tested, as well as a description of it, would normally be included, together with an indication of the method used (A, B, or C) and the preparation (moist or dry).

Type of soil is the primary factor affecting maximum dry unit weight and optimum moisture content for a given compactive effort and compaction method. Maximum dry unit weights may range from around 60 lb/ft³ for organic soils to about 145 lb/ft³ for well-graded, granular

Table 11-3   General Guide to Selection of Soils on Basis of Anticipated Embankment Performance [4, 5]

HBR Classification	Visual Description	Maximum Dry Unit Weight Range lb/ft³	Optimum Moisture Range (%)	Anticipated Embankment Performance
A-1-a	Granular material	115–142	7–15	Good to excellent
A-1-b				
A-2-4	Granular material with soil	110–135	9–18	Fair to excellent
A-2-5				
A-2-6				
A-2-7				
A-3	Fine sand and sand	110–115	9–15	Fair to good
A-4	Sandy silts and silts	95–130	10–20	Poor to good
A-5	Elastic silts and clays	85–100	20–35	Unsatisfactory
A-6	Silt-clay	95–120	10–30	Poor to good
A-7-5	Elastic silty clay	85–100	20–35	Unsatisfactory
A-7-6	Clay	90–115	15–30	Poor to fair

material containing just enough fines to fill small voids. Optimum moisture contents may range from around 5% for granular material to about 35% for elastic silts and clays. Higher optimum moisture contents are generally associated with lower dry unit weights. Higher dry unit weights are associated with well-graded granular materials. Uniformly graded sand, clays of high plasticity, and organic silts and clays typically respond poorly to compaction [4].

Tables 11–3 and 11–4 give some general compaction characteristics of various types of soil. Table 11–3 gives ranges for both maximum dry unit weight and optimum moisture content, along with anticipated embankment performance for soils classified according to the AASHTO system. Table 11–4 gives ranges for maximum dry unit weight together with values as embankment, subgrade, and base material for soils classified according to the Unified Soil Classification System.

REFERENCES

[1]   T. William Lambe, *Soil Testing for Engineers,* John Wiley & Sons, Inc., New York, 1951.

[2]   ASTM, *2001 Annual Book of ASTM Standards,* West Conshohocken, PA, 2001. Copyright, American Society for Testing and Materials, 100 Barr Harbor Drive, West Conshohocken, PA 19428-2959. Reprinted with permission.

[3]   B. K. Hough, *Basic Soils Engineering,* 2d ed., The Ronald Press Company, New York, 1969. Copyright © 1969 by John Wiley & Sons, Inc.

[4]   Robert D. Krebs and Richard D. Walker, *Highway Materials,* McGraw-Hill Book Company, New York, 1971.

[5]   L. E. Gregg, "Earthwork," in K. B. Woods, ed., *Highway Engineering Handbook,* McGraw-Hill Book Company, New York, 1960.

[6]   U.S. Army Corps of Engineers, *The Unified Soil Classification System,* Waterways Exp. Sta. Tech. Mem. 3-357 (including Appendix A, 1953, and Appendix B, 1957), Vicksburg, Miss., 1953.

Table 11–4 Compaction Characteristics and Ratings of Unified Soil Classification Classes for Soil Construction [4, 6]

Class	Compaction Characteristics	Maximum Dry Unit Weight Standard AASHTO (lb/ft^3)	Compressibility and Expansion	Value as Embankment Material	Value as Subgrade Material	Value as Base Course
GW	Good: tractor, rubber-tired, steel wheel, or vibratory roller	125–135	Almost none	Very stable	Excellent	Good
GP	Good: tractor, rubber-tired, steel wheel, or vibratory roller	115–125	Almost none	Reasonably stable	Excellent to good	Poor to fair
GM	Good: rubber-tired or light sheepsfoot roller	120–135	Slight	Reasonably stable	Excellent to good	Fair to poor
GC	Good to fair: rubber-tired or sheepsfoot roller	115–130	Slight	Reasonably stable	Good	Good to fair
SW	Good: tractor, rubber-tired, or vibratory roller	110–130	Almost none	Very stable	Good	Fair to poor
SP	Good: tractor, rubber-tired, or vibratory roller	100–120	Almost none	Reasonably stable when dense	Good to fair	Poor
SM	Good: rubber-tired or sheepsfoot roller	110–125	Slight	Reasonably stable when dense	Good to fair	Poor
SC	Good to fair: rubber-tired or sheepsfoot roller	105–125	Slight to medium	Reasonably stable	Good to fair	Fair to poor
ML	Good to poor: rubber-tired or sheepsfoot roller	95–120	Slight to medium	Poor stability, high density required	Fair to poor	Not suitable
CL	Good to fair: sheepsfoot or rubber-tired roller	95–120	Medium	Good stability	Fair to poor	Not suitable
OL	Fair to poor: sheepsfoot or rubber-tired roller	80–100	Medium to high	Unstable, should not be used	Poor	Not suitable
MH	Fair to poor: sheepsfoot or rubber-tired roller	70–95	High	Poor stability, should not be used	Poor	Not suitable
CH	Fair to poor: sheepsfoot roller	80–105	Very high	Fair stability, may soften on expansion	Poor to very poor	Not suitable
OH	Fair to poor: sheepsfoot roller	65–100	High	Unstable, should not be used	Very poor	Not suitable
PT	Not suitable	—	Very high	Should not be used	Not suitable	Not suitable

Soils Testing Laboratory
Compaction Test

Method Used _____   Tested by _____

Sample No. _____   Project No. _____

Boring No. _____   Location _____   Date _____

Depth _____   Description of Sample _____

Volume of Mold (V) _____ m^3

Trial No.	Mass of Moist Specimen + Mold, M_t (kg)	Mass of Mold, M_{md} (kg)	Mass of Moist Specimen (kg)	Moist Density of Compacted Specimen ρ_m (Mg/m³)	Moisture Content Determination								Dry Density of Compacted Specimen ρ_d (Mg/m³)	Dry Unit Weight γ_d (lb/ft³)
					Can No.	Mass of Wet Soil + Can, M_{cws} (g)	Mass of Dry Soil + Can, M_{cs} (g)	Mass of Water, M_w (g)	Mass of Can, M_c (g)	Mass of Dry Soil, M_s (g)	Moisture Content, w (%)			
A	B	C	D	E	F	G	H	I	J	K	L		M	N

Notes: $D = B - C$; $E = \dfrac{D}{1000\,V}$; $I = G - H$; $K = H - J$; $L = \dfrac{I}{K} \times 100$; $M = \dfrac{E}{100 + L} \times 100$; $N = M \times 62.43$

Soils Testing Laboratory
Compaction Test

Method Used _____

Sample No. _____   Tested by _____

Boring No. _____   Project No. _____

Depth _____   Location _____   Date _____

Volume of Mold (V) _____ m³ Description of Sample _____

Trial No.	Mass of Moist Specimen + Mold, M_t (kg)	Mass of Mold, M_{md} (kg)	Mass of Moist Specimen (kg)	Moist Density of Compacted Specimen ρ_m (Mg/m³)	Moisture Content Determination							Dry Density of Compacted Specimen ρ_d (Mg/m³)	Dry Unit Weight γ_d (lb/ft³)
					Can No.	Mass of Wet Soil + Can, M_{cws} (g)	Mass of Dry Soil + Can, M_{cs} (g)	Mass of Water, M_w (g)	Mass of Can, M_c (g)	Mass of Dry Soil, M_s (g)	Moisture Content, w (%)		
A	B	C	D	E	F	G	H	I	J	K	L	M	N

Notes: $D = B - C$; $E = \dfrac{D}{1000\,V}$; $I = G - H$; $K = H - J$; $L = \dfrac{I}{K} \times 100$; $M = \dfrac{E}{100 + L} \times 100$; $N = M \times 62.43$

CHAPTER TWELVE

Determining the Moisture Content of Soil (Calcium Carbide Gas Moisture Tester)

(Referenced Document: ASTM D 4944)

INTRODUCTION
In Chapter 11, the laboratory compaction test was described. Results of this test give the *maximum dry unit weight* and *optimum moisture content* for a given soil. The maximum dry unit weight is used by designers in specifying design shear strength, resistance to future settlement, and permeability characteristics of the soil for a given construction site. The soil is then compacted by field compaction methods to achieve the laboratory maximum dry unit weight (or a percentage of it as specified in design).

Normally, soil is compacted in the field in layers. After a fill layer has been compacted by the contractor, it is important that the in-place dry unit weight of the compacted soil be determined in order to tell whether the maximum laboratory dry unit weight has been achieved. There are several methods for determining in-place unit weight, three of which will be described in the next three chapters. The in-place unit weight determined by these methods is, however, the wet unit weight of the soil. To obtain in-place dry unit weight, it is also necessary to find the moisture content of the soil [for use in Eq. (11–2)].

Moisture content can, of course, be determined by oven drying (see Chapters 3 and 4); however, such methods may be too time consuming or inconvenient for field compaction testing, where test results are commonly needed quickly. Drying of soil can be accomplished by putting the sample in a skillet and placing it over the open flame of a camp stove. (If the soil contains significant organic material, this method should not be used.) A calcium carbide gas pressure moisture tester can also be used to determine moisture content quickly with fairly good results. The latter method is described in this chapter.

This test method (using a calcium carbide gas pressure moisture tester) determines the water (moisture) content of soil by chemical reaction using calcium carbide as a reagent to react with the available water in the soil, producing a gas. A measurement is made of the gas pressure produced when a specified mass of wet or moist soil is placed in a testing device with an appropriate volume of reagent and mixed. [1]

This method is applicable for most soils. Calcium carbide, used as a reagent, reacts with water as it is mixed with the soil by shaking and agitating with the aid of steel balls in the apparatus. To produce accurate results, the reagent must react with all the water that is not chemically hydrated with soil minerals or compounds in the soil. Some highly plastic clay soils or other soils not friable enough to break up may not produce representative results because some of the water may be trapped inside soil clods or clumps that cannot come in contact with the reagent. There may be some soils containing certain compounds or chemicals that will react unpredictably with the reagent and give erroneous results. Any such problem will become evident as calibration or check tests with ASTM Test Method D 2216 (Chapter 3) are made. Some soils containing compounds or minerals that dehydrate with heat (such as gypsum) that are to have special temperature control with Test Method D 2216 may not be affected (dehydrated) in this test method. [1]

This test method is limited to using calcium carbide moisture test equipment made for 20 g or larger soil specimens and to testing soil that contains particles no larger than the No. 4 Standard sieve size. [1]

APPARATUS AND SUPPLIES

Calcium carbide pressure moisture tester (see "Speedy Moisture Tester" in Figure 12–1)

Tared scale

Two 1¼-in. steel balls

Cleaning brush and cloth

Scoop for measuring calcium carbide reagent

Calcium carbide reagent

Sieve No. 4 (4.75 mm)

SAFETY HAZARDS [1]

(1) When combined with water, the calcium carbide reagent produces a highly flammable or explosive acetylene gas. Testing should not be carried out in confined spaces or in the vicinity of an open flame, embers, or other source of heat that can cause combustion. Care should be exercised when releasing the gas from the apparatus to direct it away from the body. Lighted cigarettes, hot objects, or open flames are extremely dangerous in the area of testing.

FIGURE 12–1   Speedy Moisture Tester (Courtesy of Soiltest, Inc.)

(2) As an added precaution, the operator should use a dust mask, clothing with long sleeves, gloves, and goggles to keep the reagent from irritating the eyes, respiratory system, or hands and arms.

(3) Attempts to test excessively wet soils or improper use of the equipment, such as adding water to the testing chamber, could cause pressures to exceed the safe level for the apparatus. This may cause damage to the equipment and an unsafe condition for the operator.

(4) Care should be taken not to dispose or place a significant amount of the calcium carbide reagent where it may contact water because it will produce an explosive gas.

CALIBRATION [1]

(1) The manufacturer-supplied equipment set, including the testing chamber with attached gage and the balance scales, are calibrated as a unit and paired together for the testing procedure.

(2) Calibration curves must be developed for each equipment set using the general soil types to be tested and the expected water content range of the soil. As new materials are introduced, further calibration is needed to extend the curve data for the specific instrument. If tests are made over a long period of time on the same soil,

a new calibration curve should be made periodically, not exceeding 12 months. Before a new batch of reagent is used for testing, two checkpoints shall be compared to the existing curve. If variation is exceeded by more than 1.0% of moisture, a new calibration curve shall be established.

(3) Calibration curves are produced by selecting several samples representing the range of soil materials to be tested and having a relatively wide range of water content. Each sample is carefully divided into two specimens by quartering procedures or use of a sample splitter. Taking care to not lose any moisture, one specimen is tested in accordance with the procedure of this test method [see (1) through (6) in the "Procedure" section] without using a calibration curve, and the other specimen is tested in accordance with ASTM Test Method D 2216 (Chapter 3).

(4) The results of the oven-dry water content determined by Test Method D 2216 from all the selected samples are plotted versus the gage reading from the calcium carbide tester for the corresponding test specimen pair. A best fit curve is plotted through the points to form a calibration curve for each soil type. Comparisons should be

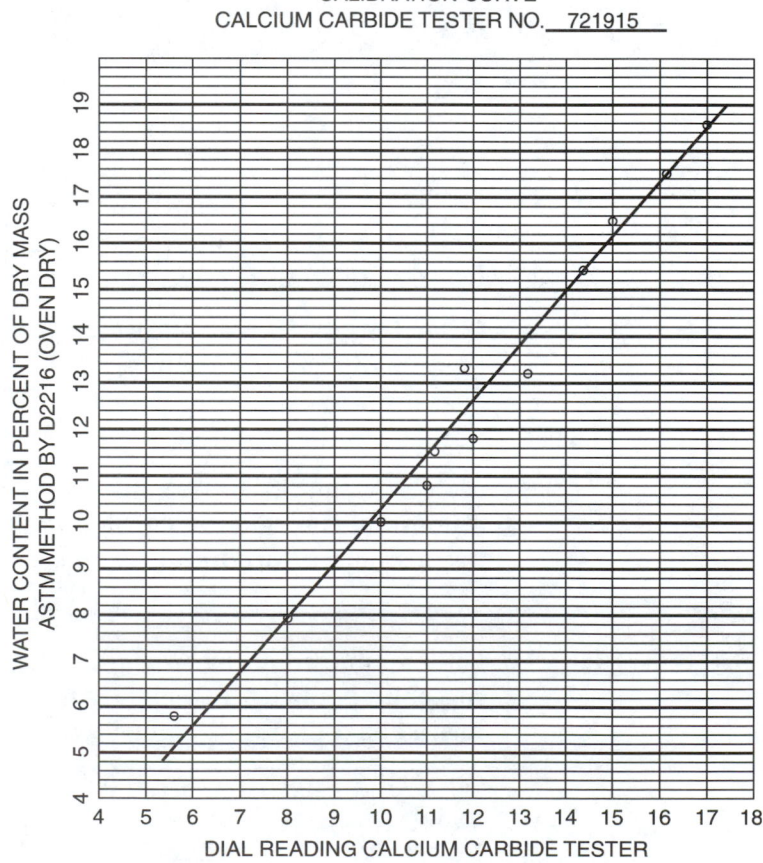

FIGURE 12–2 Typical Calibration Curve [1]

relatively consistent. A wide scatter in data indicates that either this test method or Test Method D 2216 is not applicable to the soil or conditions. Figure 12–2 shows a typical calibration curve.

(5) A comparison of this test method with Test Method D 2216 for a given soil can be made by using the calibration curve. Points that plot off the curve indicate deviations. Standard and maximum deviations can be determined if desired.

PROCEDURE [1]

The general procedure for determining moisture content by means of a calcium carbide gas moisture tester is to place a measured volume of calcium carbide in the testing apparatus along with two steel balls and a representative specimen of soil having all particles smaller than the No. 4 sieve size and having a mass equal to that specified by the manufacturer of the instrument or equipment. The apparatus is shaken vigorously in a rotating motion so the calcium carbide reagent can contact all the available water in the soil. Acetylene gas is produced proportionally to the amount of available water present. The apparent water content is read from a pressure gage on the apparatus calibrated to read in percent water content for the mass of soil specified.

The actual step-by-step procedure is as follows (ASTM D 4944-98 [1]):

(1) Remove the cap from the testing chamber of the apparatus and place the recommended amount of calcium carbide reagent along with the two steel balls into the testing chamber. Most equipment built to test 20-g samples requires approximately 22 g of reagent (measured using the supplied scoop, which is filled two times).

(2) Use the balance to obtain a specimen of soil that has a mass recommended for the equipment and contains particles smaller than the No. 4 sieve size. One-half specimen size should be used when the water content is expected to exceed the limits of the gage on the gas pressure chamber or when it actually reaches or exceeds the gage limit in any test [see (6)].

(3) Place the soil specimen in the testing chamber cap; then, with the apparatus in the horizontal position, insert the cap in the testing chamber and tighten the clamp to seal the cap to the unit. Take care that no calcium carbide comes in contact with the soil until a complete seal is achieved.

> *Note*—The soil specimen may be placed in the chamber with the calcium carbide in the cap if desired.

(4) Raise the apparatus to the vertical (upright) position so that the contents of the cap fall into the testing chamber. Strike the side of the apparatus with an open hand to assure that all the material falls out of the cap.

(5) Shake the apparatus vigorously with a rotating motion so that the steel balls roll around the inside circumference and impact a grinding effect on the soil and reagent. This motion also prevents the steel balls from striking the orifice that leads to the pressure gage. Shake the apparatus for at least 1 min for sands, increasing the time for silts, and up to 3 min for clays. Some highly plastic clay soils may take more than 3 min. Periodically check the progress of the needle on the pressure gage dial. Allow time for the needle to stabilize as the heat from the chemical reaction is dissipated.

(6) When the pressure gage dial needle stops moving, read the dial while holding the apparatus in the horizontal position. If the dial goes to the limit of the gage, (1) through (6) should be repeated using a new specimen having a mass half as large as the recommended specimen. When a half size specimen is used, the final dial reading is multiplied by two for use with the calibration curve.

(7) Record the final pressure gage dial reading and use the appropriate calibration curve to determine the corrected water content in percent of dry mass of soil and record.

(8) With the cap of the testing chamber pointed away from the operator, slowly release the gas pressure (see "Safety Hazards" section). Empty the chamber and examine the specimen for lumps. If the material is not completely pulverized, the test should be repeated using a new specimen.

(9) Clean the testing chamber and cap with a brush or cloth and allow the apparatus to cool before performing another test. Repeated tests can cause the apparatus to heat up, which will affect the results of the test. The apparatus should be at about the same temperature as it was during calibration (determined by touch). This may require warming the instrument up to calibration temperature before use when the temperature is cold.

(10) Discard the specimen where it will not contact water and produce an explosive gas. It is recommended that the specimen soil not be used for further testing as it is contaminated with the reagent.

DATA

The only datum recorded in this test is the dial reading from the moisture tester.

CALCULATIONS

The percent of moisture by dry weight of soil may be determined from the observed dial reading of the calcium carbide moisture tester by using a calibration curve such as the one given in Figure 12–2.

CONCLUSIONS

The calcium carbide gas moisture tester provides a quick, simple means of finding the moisture content of soil. It is useful when rapid results are needed, when testing is done in field locations, and when an oven (or microwave) is not available. It is particularly useful for field determinations of moisture contents in conjunction with field compaction testing. This test method is not, however, intended as a replacement for the "oven method" (ASTM D 2216; see Chapter 3); in fact, Method D 2216 is to be used as the test method to compare for accuracy of the calcium carbide tester.

Interestingly, this tester can be used to determine the moisture content of many materials other than soil (e.g., ores, solid fuels, ash, oil, textiles, paper, ceramics, cosmetics, processed food) in many other fields (e.g., agriculture, brewing, chemicals, construction, pharmaceuticals, waste disposal).

REFERENCE

[1]   ASTM, *2001 Annual Book of ASTM Standards,* West Conshohocken, PA, 2001. Copyright, American Society for Testing and Materials, 100 Barr Harbor Drive, West Conshohocken, PA 19428-2959. Reprinted with permission.

CHAPTER THIRTEEN

Determining the Density and Unit Weight of Soil in Place by the Sand-Cone Method

(Referenced Document: ASTM D 1556)

INTRODUCTION   In certain cases, it is necessary to determine the density and/or unit weight of a soil either as it exists naturally in the ground or as it may be compacted in a fill. In the former case, the density of soil in place may be used to evaluate the comparative strength of the soil; in the latter, the unit weight may be used to document results of field compaction to meet contract specifications. The principal use of the in-place soil unit weight test is in documentation of field compaction.

In Chapter 11 the laboratory compaction test was described. In practice, representative samples of the soil to be used for fill material are subjected to the laboratory compaction test to determine the optimum moisture content and maximum dry unit weight. The maximum dry unit weight is used by designers in specifying design shear strength, resistance to future settlement, and permeability characteristics. The fill soil is then compacted mechanically in the field by field compaction methods to achieve the laboratory maximum dry unit weight (or a percentage of it). In order to determine whether the laboratory maximum dry unit weight (or an acceptable percentage thereof) has been achieved, in-place soil unit weight tests must be performed in the field on compacted soil.

There are several methods for finding the density/unit weight of soil in place. This chapter discusses the "sand-cone" method, and Chapter 14 considers the "rubber-balloon" method. In-place soil density and unit weight can also be determined with nuclear equipment, utilizing radioactive materials. This method, which is called "density of soil and soil-aggregate in place by nuclear methods," is covered in Chapter 15.

In documenting field compaction, it is necessary to determine the moisture content of each sample, in addition to in-place unit weight of the soil. The moisture content is needed to compute the dry unit weight of the compacted soil. A relatively quick method for determining moisture content was described in Chapter 12. Thus, Chapters 11, 12, and either 13, 14, or 15 are generally all used in an overall compaction project.

APPARATUS AND SUPPLIES

Density/unit weight apparatus: composed of a jar and detachable appliance consisting of a cylindrical valve with an orifice and having a small funnel continuing to a standard G mason jar top on one end and a large funnel on the other end. Figure 13–1 gives specific requirements for a standard density/unit weight apparatus suitable

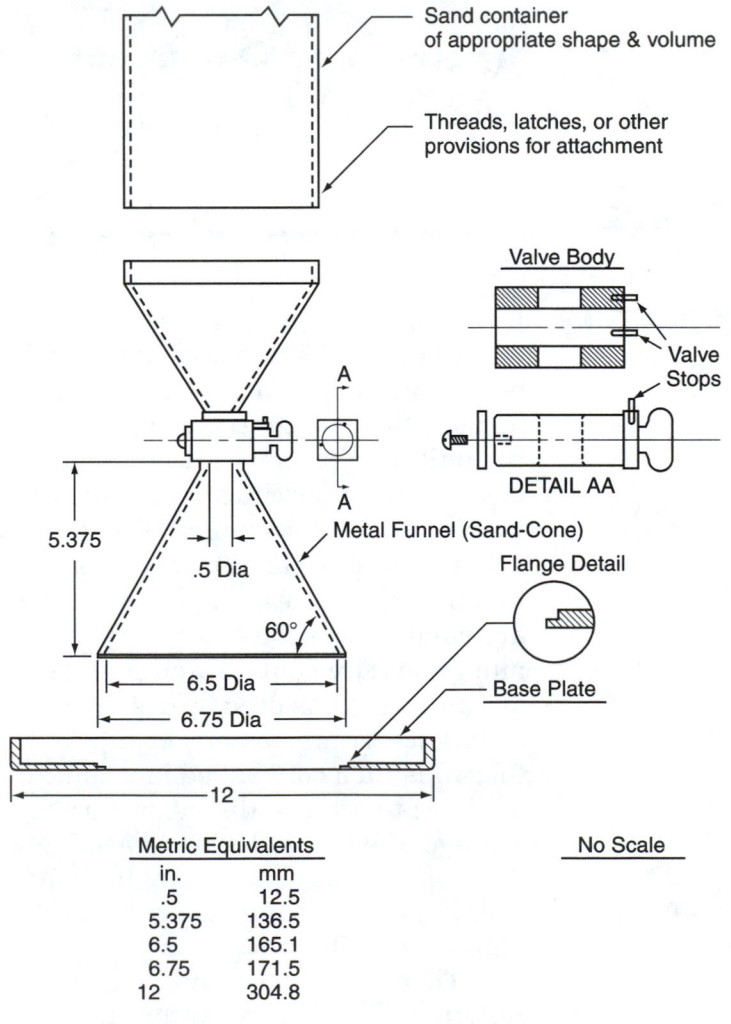

Metric Equivalents	
in.	mm
.5	12.5
5.375	136.5
6.5	165.1
6.75	171.5
12	304.8

FIGURE 13–1 Density/Unit Weight Apparatus [1]

FIGURE 13–2 Density/Unit Weight Apparatus (Courtesy of Soiltest, Inc.)

for testing soils having maximum particle sizes of approximately 1½ in. (37.5 mm) and test hole volumes of approximately 0.1 ft^3 (2830 cm^3); Figure 13–2 shows a photograph of the apparatus.

Base plate: a square or rectangular metal plate with a flanged center hole cast or machined to receive the large funnel (cone) of the sand-cone apparatus (Figure 13–1)

Sand: a clean, dry, free-flowing, uncemented sand having a maximum particle size smaller than 2.0 mm (No. 10) sieve and less than 3% by weight passing 250 μm (No. 60) sieve; the uniformity coefficient ($C_u = D_{60}/D_{10}$) must be less than 2.0

Balances: one with a 10-kg capacity and accuracy to 1.0 g, and one with a 500-g capacity and accuracy to 0.1 g

Drying equipment: calcium carbide gas moisture tester or oven or other suitable equipment for drying moisture content samples

Digging tools: chisels, hammers, picks, and spoons

Miscellaneous equipment: suitable containers for retaining density samples, moisture samples, and salvaged density sand

CALIBRATION OF MECHANICAL DEVICE

Two calibration procedures are required for this test—one to determine the density and unit weight of the sand used in the test and the other to determine the mass of sand required to fill the sand cone (the funnel). These procedures are as follows (see ASTM D 1556-90 (Reapproved 1996) and ASTM D 4253-00 [1]):

[A] Determination of Bulk Density and Unit Weight of Sand

The bulk density of sand to be used in the field test is determined by first determining the volume of the container and then by using the container to measure a volume and mass of sand as follows:

(1) Select a container of known volume. The $\frac{1}{30}$ ft^3 (944 cm^3) and 1/13.33 ft^3 (2,124 cm^3) molds specified in ASTM Test Method D 698 (Chapter 11), or the 0.1 ft^3 (2,830 cm^3) mold specified in Test Method D 4253 are recommended.

(2) Determine the volume of the container by the water-filling method, as described in ASTM D 4253. Completely fill the mold with water. Slide a glass plate carefully over the top surface (rim) of the container to ensure that the mold is completely filled with water. A thin film of grease or silicone lubricant on the rim of the container will make a watertight joint between the glass plate and the rim of the container. Determine the mass and temperature of the water required to fill the container.

(3) Fill the assembled sand-cone apparatus with sand. Invert and support the apparatus over the calibrated container so that the sand falls approximately the same distance and location as in a field test, and fully open the valve.

(4) Fill the container until it just overflows and close the valve. Using a minimum number of strokes and taking care not to jar or densify the sand, carefully strike off excess sand to a smooth, level surface.

(5) Clean any sand from outside the calibrated container. Determine the mass of the container and sand. Determine the net mass of the sand by subtracting the mass of the empty container.

(6) Perform at least three bulk-density determinations and calculate the average. The maximum variation between any one determination and the average will not exceed 1%.

[B] Determination of Mass of Sand Required to Fill Sand Cone and Base Plate

(1) Fill the assembled apparatus with sand and determine the mass of apparatus and sand.

(2) Place the base plate on a clean, level, plane surface. Invert the apparatus and seat the large funnel into the flanged center hole in the base plate, and mark and identify the funnel and plate so that the same funnel and plate can always be matched and reseated in the same position during testing.

(3) Open the valve fully and keep open until the sand stops running, making sure the apparatus, base plate, or plane surface are not jarred or vibrated before the valve is closed.

(4) Close the valve sharply, remove the apparatus, determine the mass of the apparatus with remaining sand, and calculate the loss of sand. This loss represents the mass of sand required to fill the funnel and base plate.

(5) Repeat the procedures in (1) to (4) at least three times. The mass of sand used in the calculations shall be the average of three determinations. The maximum variation between any one determination and the average shall not exceed 1%.

PROCEDURE

The general procedure for determining in-place soil density/unit weight is to find the mass/weight and volume of an in-place soil sample, from which the density/unit weight can be computed. In the sand-cone method, a quantity of soil is removed from the ground or compacted fill. The mass of the soil removed is determined directly, whereas the volume is obtained indirectly by finding how much sand is required to fill the hole. The mass/weight of soil removed from the hole divided by the volume gives the density/unit weight.

The actual step-by-step procedure is as follows (ASTM D 1556-00 [1]):

(1) Select a location/elevation that is representative of the area to be tested, and determine the density of the soil in-place as follows:

(1.1) Inspect the cone apparatus for damage, free rotation of the valve, and properly matched baseplate. Fill the cone container with conditioned sand for which the bulk-density has been determined and determine the total mass.

(1.2) Prepare the surface of the location to be tested so that it is a level plane. The base plate may be used as a tool for striking off the surface to a smooth level plane.

(1.3) Seat the base plate on the plane surface, making sure there is contact with the ground surface around the edge of the flanged center hole. Mark the outline of the base plate to check for movement during the test, and if needed, secure the plate against movement using nails pushed into the soil adjacent to the edge of the plate, or by other means, without disturbing the soil to be tested.

(1.4) In soils where leveling is not successful, or surface voids remain, the volume horizontally bounded by the funnel, plate, and ground surface must be determined by a preliminary test. Fill the space with sand from the apparatus, determine the mass of sand used to fill the space, refill the apparatus, and determine a new initial mass of apparatus and sand before proceeding with the test. After this measurement is completed, carefully brush the sand from the prepared surface (see Note 1).

> *Note 1*—A second calibrated apparatus may be taken to the field when this condition is anticipated (instead of refilling and making a second determination). The procedure in (1.4)

may be used for each test when the best possible accuracy is desired; however, it is usually not needed for most production testing where a relatively smooth surface is obtainable.

(1.5) The test hole volume will depend on the anticipated maximum particle size in the soil to be tested. Test hole volumes are to be as large as practical to minimize the errors and shall not be less than the volumes indicated in Table 13–1. A hole depth should be selected that will provide a representative sample of the soil. For construction control, the depth of the hole should approximate the thickness of one, or more, compacted lift(s). The procedure for calibrating the sand must reflect this hole depth.

(1.6) Dig the test hole through the center hole in the base plate, being careful to avoid disturbing or deforming the soil that will bound the hole. The sides of the hole should slope slightly inward, and the bottom should be reasonably flat or concave. The hole should be kept as free as possible of pockets, overhangs, and sharp obtrusions since these affect the accuracy of the test. Soils that are essentially granular require extreme care and may require digging a conical-shaped test hole. Place all excavated soil, and any soil loosened during digging, in a moisture tight container that is marked to identify the test number. Take care to avoid losing any materials. Protect this material from any loss of moisture until the mass has been determined and a specimen has been obtained for a water content determination.

Table 13–1   Minimum Test Hole Volumes Based on Maximum Size of Included Particle [1]

Maximum Particle Size		Minimum Test Hole Volumes	
in.	*(mm)*	cm^3	ft^3
½	(12.7)	1,415	0.05
1	(25.4)	2,125	0.075
1½	(38)	2,830	0.1

(1.7) Clean the flange of the base plate hole, invert the sand-cone apparatus, and seat the sand-cone funnel into the flanged hole at the same position as marked during calibration. Eliminate or minimize vibrations in the test area due to personnel or equipment. Open the valve and allow the sand to fill the hole, funnel, and base plate. Take care to avoid jarring or vibrating the apparatus while the sand is running. When the sand stops flowing, close the valve.

(1.8) Determine the mass of the apparatus with the remaining sand, record, and calculate the mass of sand used.

(1.9) Determine and record the mass of the moist material that was removed from the test hole. When oversize material corrections are required, determine the mass of the oversize material on the appropriate sieve and record, taking care to avoid moisture losses. When required, make appropriate corrections for the oversize material using ASTM Practice D 4718.

(1.10) Mix the material thoroughly, and either obtain a representative specimen for water content determination or use the entire sample.

(1.11) Determine the water content in accordance with ASTM Test Method D 2216 (Chapter 3), D 4643 (Chapter 4), D 4944 (Chapter 12), or D 4959. Correlations to Method D 2216 will be performed when required by other test methods.

(2) Water content specimens must be large enough and selected in such a way that they represent all the material obtained from the test hole. The minimum mass of the water content specimens is that required to provide water content values accurate to 1%.

DATA   Data collected in this test should include the following:

Calibration Data

[A] Determination of Bulk Density and Unit Weight of Sand

Volume of container

   Mass of container

   Mass of container plus water required to fill it

   Temperature of water

Bulk density of sand

   Mass of container plus sand required to fill it

[B] Determination of Mass of Sand Required to Fill Funnel and Base Plate

Mass of sand-cone apparatus plus sand required to fill it (before filling funnel and base plate)

Mass of sand-cone apparatus plus remaining sand (after filling funnel and base plate)

Determination of Density of Soil in Place

[A] Moisture Content of Material from Test Hole

Mass of container

Mass of container plus moist mass of moisture sample

Mass of container plus dry mass of moisture sample

Note—These data are not collected if moisture content is determined using a calcium carbide gas moisture tester (Chapter 12).

[B] In-Place Density and Unit Weight

Mass of apparatus plus mass of remaining sand (after filling hole)

Mass of pan

Mass of pan plus wet mass of soil from hole

CALCULATIONS

Calibration Data

[A] Determination of Bulk Density and Unit Weight of Sand

The volume of the container can be calculated using the equation [1]

$$V_1 = GT \qquad \text{(13–1)}$$

where:
V_1 = volume of container, mL
G = mass of water required to fill container, g
T = water temperature-volume correction shown in column 3 of Table 13–2

Table 13–2 Volume of Water per Gram Based on Temperature [1]

Temperature		Volume of Water
°C	°F	(mL/g)
12	53.6	1.00048
14	57.2	1.00073
16	60.8	1.00103
18	64.4	1.00138
20	68.0	1.00177
22	71.6	1.00221
24	75.2	1.00268
26	78.8	1.00320
28	82.4	1.00375
30	86.0	1.00435
32	89.6	1.00497

The bulk density of the sand (ρ_1 in g/cm^3 or g/mL) can be calculated by dividing the mass of sand required to fill the calibration container (M_1 in g) by the volume of the calibration container (V_1 in cm^3 or mL). Hence,

$$\rho_1 = \frac{M_1}{V_1} \qquad \text{(13–2)}$$

The unit weight of the sand can be determined by:

$$\gamma_1 = 62.427\ \rho_1 \tag{13-3}$$

where:
   γ_1 = unit weight in lb/ft^3
   ρ_1 = density in g/mL

[B] Determination of Mass of Sand Required to Fill Funnel and Base Plate

The mass of sand required to fill the funnel and base plate can be found by subtracting the mass of the remaining sand (after filling the funnel and base plate) from the mass of sand required to fill the apparatus.

Determination of Density of Soil in Place

[A] Moisture Content of Material from Test Hole

Note—If moisture content is determined by a calcium carbide gas tester, the calculations that follow are not required.
   Moisture content can be calculated using equation [1]

$$w = \frac{M_2 - M_3}{M_3} \times 100 \tag{13-4}$$

where:
   w = moisture content of material from test hole, %
   M_2 = moist mass of moisture sample, g
   M_3 = dry mass of moisture sample, g

Note—Equation (13–4) is an alternative form of Eq. (3–1).

[B] In-Place Density and Unit Weight

The dry mass of material from the test hole can be calculated using the equation

$$M_5 = \frac{M_4}{(0.01)(w + 100)} \tag{13-5}$$

where:
   M_5 = dry mass of material from test hole, g
   M_4 = moist mass of material from test hole, g
   w = moisture content of material from test hole, %

The volume of the test hole can be calculated using the equation

$$V = \frac{M_6 - M_7}{\rho_1} \qquad (13\text{–}6)$$

where:
  V = volume of test hole, mL
  M_6 = mass of sand used, g
  M_7 = mass of sand in funnel and base plate, g
  ρ_1 = density of sand, g/mL

Dry density (in place) of the tested material can be calculated using the equation

$$\rho_2 = \frac{M_5}{V} \qquad (13\text{–}7)$$

where:
  ρ_2 = dry density (in place) of tested material, g/mL
  M_5 = dry mass of material from test hole, g
  V = volume of test hole, mL

The dry unit weight (in place) of the tested material can be found by

$$\gamma_2 = 62.43\,\rho_2 \qquad (13\text{–}8)$$

where:
  γ_2 = unit weight in lb/ft^3
  ρ_2 = density in g/mL

NUMERICAL EXAMPLE   A field test was conducted according to the preceding procedure. The following data were obtained:

Calibration Data

[A] Determination of Bulk Density of Sand

Volume of container (i.e., volume of 6-in. mold specified in ASTM D 698, Chapter 11)

Mass of container

Trial no. 1	**2,783 g**
Trial no. 2	**2,780 g**
Trial no. 3	**2,783 g**
Average	**2,782 g**

Mass of container plus water required to fill it:

Trial no. 1	**4,922 g**
Trial no. 2	**4,919 g**
Trial no. 3	**4,922 g**
Average	**4,921 g**

Temperature of water = **24°C**

Bulk density of sand

Mass of container plus sand required to fill it = **6,139 g**

[B] Determination of Mass of Sand Required to Fill Funnel and Base Plate

Mass of sand-cone apparatus plus sand required to fill it = **8,045 g**

Mass of sand-cone apparatus plus remaining sand (after filling funnel and base plate) = **6,378 g**

Determination of Density and Unit Weight of Soil in Place

[A] Moisture Content of Material from Test Hole

Mass of container = **42.6 g**

Mass of container plus moist mass of moisture sample = **295.6 g**

Mass of container plus dry mass of moisture sample = **250.7 g**

[B] In-Place Density and Unit Weight

Mass of apparatus plus mass of remaining sand = **4,867 g**

Mass of pan = **815 g**

Mass of pan plus wet mass of soil from hole = **2,669 g**

With these data known, calculations proceed as follows:

Calibration Data

[A] Determination of Bulk Density of Sand

Volume of container:

$$V_1 = GT \tag{13-1}$$

$$G = \mathbf{4,921} - \mathbf{2,782} = 2,139 \text{ g}$$

$$T = 1.00268 \text{ (from Table 13-2 for a temperature of } \mathbf{24°C})$$

$$V_1 = (2,139)(1.00268) = 2,145 \text{ mL}$$

Bulk density of sand:

$$\rho_1 = \frac{M_1}{V_1} \tag{13–2}$$

$$M_1 = \mathbf{6,139} - \mathbf{2,782} = 3,357 \text{ g}$$

$$V_1 = 2,145 \text{ mL}$$

$$\rho_1 = \frac{3,357}{2,145} = 1.565 \text{ g/mL}$$

$$\gamma_1 = 62.43 \, \rho_1 \tag{13–3}$$

$$\gamma_1 = (62.43)(1.565) = 97.7 \text{ lb/ft}^3$$

[B] Determination of Mass of Sand Required to Fill Funnel and Base Plate (M_7)

$$M_7 = \mathbf{8,045} - \mathbf{6,378} = 1,667 \text{ g}$$

Determination of Density and Unit Weight of Soil in Place

[A] Moisture Content of Material from Test Hole

$$w = \frac{M_2 - M_3}{M_3} \times 100 \tag{13–4}$$

$$M_2 = \mathbf{295.6} - \mathbf{42.6} = 253.0 \text{ g}$$

$$M_3 = \mathbf{250.7} - \mathbf{42.6} = 208.1 \text{ g}$$

$$w = \frac{253.0 - 208.1}{208.1} \times 100 = 21.6\%$$

[B] In-Place Density and Unit Weight

$$M_5 = \frac{M_4}{(0.01)(w + 100)} \tag{13–5}$$

$$M_4 = \mathbf{2,669} - \mathbf{815} = 1,854 \text{ g}$$

$$w = 21.6\%$$

$$M_5 = \frac{1,854}{(0.01)(21.6 + 100)} = 1,525 \text{ g}$$

$$V = \frac{M_6 - M_7}{\rho_1} \tag{13–6}$$

$$M_6 = \mathbf{8,045 - 4,867} = 3,178 \text{ g}$$

$$M_7 = 1,667 \text{ g}$$

$$\rho_1 = 1.565 \text{ g/mL}$$

$$V = \frac{3,178 - 1,667}{1.565} = 965.5 \text{ mL}$$

$$\rho_2 = \frac{M_5}{V} \qquad\qquad (13\text{--}7)$$

$$\rho_2 = \frac{1,525}{965.5} = 1.579 \text{ g/mL}$$

$$\gamma_2 = 62.43 \, \rho_2 \qquad\qquad (13\text{--}8)$$

$$\gamma_2 = (62.43)(1.579) = 98.6 \text{ lb/ft}^3$$

All data—both given and calculated—are presented on the form shown on pages 196–197. (At the end of the chapter, two blank copies of this form are included for the reader's use.) Careful study of this form should facilitate understanding of the calculations required to determine in-place dry density and unit weight.

CONCLUSIONS

Because test holes are relatively small, it is important that no soil be lost while excavating and that volume determinations be done very carefully to ensure accurate evaluations of in-place density. To get an accurate moisture content of the soil, excavation should be done as quickly as possible. Also, vibration of the ground and the jar should be avoided to prevent too much sand from entering the hole.

As explained in the introduction to this chapter, unit weight of soil in place is used primarily in documenting field compaction. The sand-cone method provides a convenient and accurate means of determining in-place density and unit weight, although the volume of the sample is determined indirectly. An alternative procedure, the rubber-balloon method, makes a more direct measurement of the volume of the sample. This procedure is described in Chapter 14.

REFERENCE

[1] ASTM, *2001 Annual Book of ASTM Standards*, West Conshohocken, PA, 2001. Copyright, American Society for Testing and Materials, 100 Barr Harbor Drive, West Conshohocken, PA 19428-2959. Reprinted with permission.

Soils Testing Laboratory
In-Place Density and Unit Weight Determination:
Sand-Cone Method

Project _____ SR 2828 _____   *Location of Test* _____ Newell, N.C. _____

Tested by _____ John Doe _____   *Date* _____ 5/7/02 _____

Calibration Data

[A] Determination of Bulk Density and Unit Weight of Sand

Volume of container, V_1
- (1) Mass of water required to fill container + mass of container
    - (a) Trial no. 1 __4,922__ g
    - (b) Trial no. 2 __4,919__ g   Average mass __4,921__ g
    - (c) Trial no. 3 __4,922__ g
- (2) Mass of container
    - (a) Trial no. 1 __2,783__ g
    - (b) Trial no. 2 __2,780__ g   Average mass __2,782__ g
    - (c) Trial no. 3 __2,783__ g
- (3) Mass of water required to fill container, G __2,139__ g
- (4) Temperature of water __24°C__ , T __1.00268__ (see Table 13–2)
- (5) Volume of container, $V_1 = GT$ __2,145__ mL

Bulk density and unit weight of sand, ρ_1
- (6) Mass of sand required to fill apparatus + mass of container __6,139__ g
- (7) Mass of container __2,782__ g
- (8) Mass of sand required to fill container, M_1 __3,357__ g
- (9) Bulk density of sand, $\rho_1 = \dfrac{M_1}{V_1}$ __1.565__ g/mL
- (10) Unit weight of sand, $\gamma_1 = 62.43\, \rho_1$ __97.7__ lb/ft³

[B] Determination of Mass of Sand Required to Fill Funnel and Base Plate, M_7

- (1) Mass of sand required to fill apparatus + mass of apparatus __8,045__ g
- (2) Mass of remaining sand (after filling funnel and base plate) + mass of apparatus __6,378__ g
- (3) Mass of sand required to fill funnel and base plate, M_7 __1,667__ g

Determination of Density and Unit Weight of Soil in Place

[A] Moisture Content of Material from Test Hole

Note—If moisture content is determined by a calcium carbide gas tester, enter moisture content directly in appropriate blank below and skip intervening steps.
- (1) Container no. __10A__
- (2) Moist mass of moisture sample + mass of container __295.6__ g
- (3) Dry mass of moisture sample + mass of container __250.7__ g
- (4) Mass of container __42.6__ g
- (5) Moist mass of moisture sample, M_2 __253.0__ g

(6) Dry mass of moisture sample, M_3 ___208.1___ g

(7) Moisture content of material from test hole, $w = \dfrac{M_2 - M_3}{M_3} \times 100$ ___21.6___ %

[B] In-Place Density and Unit Weight

(1) Mass of apparatus + mass of sand (before use) ___8,045___ g

(2) Mass of apparatus + mass of remaining sand (after use) ___4,867___ g

(3) Mass of sand used in test, M_6 ___3,178___ g

(4) Mass of sand to fill funnel and base plate, M_7 ___1,667___ g
(see "Calibration Data")

(5) Mass of sand used in test hole, $M_6 - M_7$ ___1,511___ g

(6) Density of sand, ρ_1 ___1.565___ g/mL (see "Calibration Data")

(7) Volume of test hole, $V = \dfrac{M_6 - M_7}{\rho_1}$ ___965.5___ mL

(8) Wet mass of soil from hole + mass of pan ___2,669___ g

(9) Mass of pan ___815___ g

(10) Wet mass of soil from hole, M_4 ___1,854___ g

(11) Moisture content in soil from test hole, w ___21.6___ %

(12) Dry mass of soil from test hole, $M_5 = \dfrac{M_4}{(0.01)(w + 100)}$ ___1,525___ g

(13) Dry density of soil in place, $\rho_2 = \dfrac{M_5}{V}$ ___1.579___ g/mL

(14) Dry unit weight of soil in place, $\gamma_2 = 62.43\,\rho_2$ ___98.6___ lb/ft³

Soils Testing Laboratory
In-Place Density and Unit Weight Determination:
Sand-Cone Method

Project _____   *Location of Test* _____

Tested by _____   *Date* _____

Calibration Data

[A] Determination of Bulk Density and Unit Weight of Sand

Volume of container, V_1

    (1) Mass of water required to fill container + mass of container
        (a) Trial no. 1 _____ g
        (b) Trial no. 2 _____ g   Average mass _____ g
        (c) Trial no. 3 _____ g
    (2) Mass of container
        (a) Trial no. 1 _____ g
        (b) Trial no. 2 _____ g   Average mass _____ g
        (c) Trial no. 3 _____ g
    (3) Mass of water required to fill container, G _____ g
    (4) Temperature of water _____ , T _____ (see Table 13–2)
    (5) Volume of container, $V_1 = GT$ _____ mL

Bulk density and unit weight of sand, ρ_1

    (6) Mass of sand required to fill apparatus + mass of container _____ g
    (7) Mass of container _____ g
    (8) Mass of sand required to fill container, M_1 _____ g
    (9) Bulk density of sand, $\rho_1 = \dfrac{M_1}{V_1}$ _____ g/mL
    (10) Unit weight of sand, $\gamma_1 = 62.43\, \rho_1$ _____ lb/ft^3

[B] Determination of Mass of Sand Required to Fill Funnel and Base Plate, M_7

    (1) Mass of sand required to fill apparatus + mass of apparatus _____ g
    (2) Mass of remaining sand (after filling funnel and base plate) + mass of apparatus _____ g
    (3) Mass of sand required to fill funnel and base plate, M_7 _____ g

Determination of Density and Unit Weight of Soil in Place

[A] Moisture Content of Material from Test Hole

Note—If moisture content is determined by a calcium carbide gas tester, enter moisture content directly in appropriate blank below and skip intervening steps.

    (1) Container no. _____
    (2) Moist mass of moisture sample + mass of container _____ g
    (3) Dry mass of moisture sample + mass of container _____ g
    (4) Mass of container _____ g
    (5) Moist mass of moisture sample, M_2 _____ g

(6) Dry mass of moisture sample, M_3 _____ g

(7) Moisture content of material from test hole, $w = \dfrac{M_2 - M_3}{M_3} \times 100$ _____ %

[B] In-Place Density and Unit Weight

(1) Mass of apparatus + mass of sand (before use) _____ g

(2) Mass of apparatus + mass of remaining sand (after use) _____ g

(3) Mass of sand used in test, M_6 _____ g

(4) Mass of sand to fill funnel and base plate, M_7 _____ g
(see "Calibration Data")

(5) Mass of sand used in test hole, $M_6 - M_7$ _____ g

(6) Density of sand, ρ_1 _____ g/mL (see "Calibration Data")

(7) Volume of test hole, $V = \dfrac{M_6 - M_7}{\rho_1}$ _____ mL

(8) Wet mass of soil from hole + mass of pan _____ g

(9) Mass of pan _____ g

(10) Wet mass of soil from hole, M_4 _____ g

(11) Moisture content in soil from test hole, w _____ %

(12) Dry mass of soil from test hole, $M_5 = \dfrac{M_4}{(0.01)(w + 100)}$ _____ g

(13) Dry density of soil in place, $\rho_2 = \dfrac{M_5}{V}$ _____ g/mL

(14) Dry unit weight of soil in place, $\gamma_2 = 62.43\,\rho_2$ _____ lb/ft^3

Soils Testing Laboratory
In-Place Density and Unit Weight Determination: Sand-Cone Method

Project _____   Location of Test _____

Tested by _____   Date _____

Calibration Data

[A] Determination of Bulk Density and Unit Weight of Sand

Volume of container, V_1
- (1)   Mass of water required to fill container + mass of container
  - (a)   Trial no. 1 _____ g
  - (b)   Trial no. 2 _____ g   Average mass _____ g
  - (c)   Trial no. 3 _____ g
- (2)   Mass of container
  - (a)   Trial no. 1 _____ g
  - (b)   Trial no. 2 _____ g   Average mass _____ g
  - (c)   Trial no. 3 _____ g
- (3)   Mass of water required to fill container, G _____ g
- (4)   Temperature of water _____ , T _____ (see Table 13–2)
- (5)   Volume of container, $V_1 = GT$ _____ mL

Bulk density and unit weight of sand, ρ_1
- (6)   Mass of sand required to fill apparatus + mass of container _____ g
- (7)   Mass of container _____ g
- (8)   Mass of sand required to fill container, M_1 _____ g
- (9)   Bulk density of sand, $\rho_1 = \dfrac{M_1}{V_1}$ _____ g/mL
- (10)   Unit weight of sand, $\gamma_1 = 62.43\,\rho_1$ _____ lb/ft^3

[B] Determination of Mass of Sand Required to Fill Funnel and Base Plate, M_7

- (1)   Mass of sand required to fill apparatus + mass of apparatus _____ g
- (2)   Mass of remaining sand (after filling funnel and base plate) + mass of apparatus _____ g
- (3)   Mass of sand required to fill funnel and base plate, M_7 _____ g

Determination of Density and Unit Weight of Soil in Place

[A] Moisture Content of Material from Test Hole

Note—If moisture content is determined by a calcium carbide gas tester, enter moisture content directly in appropriate blank below and skip intervening steps.
- (1)   Container no. _____
- (2)   Moist mass of moisture sample + mass of container _____ g
- (3)   Dry mass of moisture sample + mass of container _____ g
- (4)   Mass of container _____ g
- (5)   Moist mass of moisture sample, M_2 _____ g
- (6)   Dry mass of moisture sample, M_3 _____ g
- (7)   Moisture content of material from test hole, $w = \dfrac{M_2 - M_3}{M_3} \times 100$ _____ %

[B] In-Place Density and Unit Weight

(1) Mass of apparatus + mass of sand (before use) _____ g

(2) Mass of apparatus + mass of remaining sand (after use) _____ g

(3) Mass of sand used in test, M_6 _____ g

(4) Mass of sand to fill funnel and base plate, M_7 _____ g
(see "Calibration Data")

(5) Mass of sand used in test hole, $M_6 - M_7$ _____ g

(6) Density of sand, ρ_1 _____ g/mL (see "Calibration Data")

(7) Volume of test hole, $V = \dfrac{M_6 - M_7}{\rho_1}$ _____ mL

(8) Wet mass of soil from hole + mass of pan _____ g

(9) Mass of pan _____ g

(10) Wet mass of soil from hole, M_4 _____ g

(11) Moisture content in soil from test hole, w _____ %

(12) Dry mass of soil from test hole, $M_5 = \dfrac{M_4}{(0.01)(w + 100)}$ _____ g

(13) Dry density of soil in place, $\rho_2 = \dfrac{M_5}{V}$ _____ g/mL

(14) Dry unit weight of soil in place, $\gamma_2 = 62.43\,\rho_2$ _____ lb/ft^3

CHAPTER FOURTEEN

Determining the Density and Unit Weight of Soil in Place by the Rubber-Balloon Method

(Referenced Document: ASTM D 2167)

INTRODUCTION

The necessity for finding the density and unit weight of soil in place was mentioned in the introduction to Chapter 13, which presented the sand-cone method for determining in-place density and unit weight. The rubber-balloon method differs from the sand-cone method primarily in the manner in which the volume of soil removed from the ground or compacted fill is determined.

APPARATUS AND SUPPLIES

Balloon apparatus: a calibrated vessel designed to contain a liquid within a relatively thin, flexible, elastic membrane (rubber balloon) for measuring the volume of the test hole (see Figures 14–1 and 14–2)

Base plate (see Figures 14–1 and 14–2)

Balances: one with a 20-kg capacity and accuracy to 1.0 g, and one with a 500-g capacity and accuracy to 0.1 g

Drying equipment: calcium carbide gas moisture tester or oven, or other suitable equipment for drying moisture content samples

Digging tools: chisels, hammers, picks, and spoons

Miscellaneous equipment: suitable containers for retaining density/unit weight samples, moisture samples, and salvaged density sand

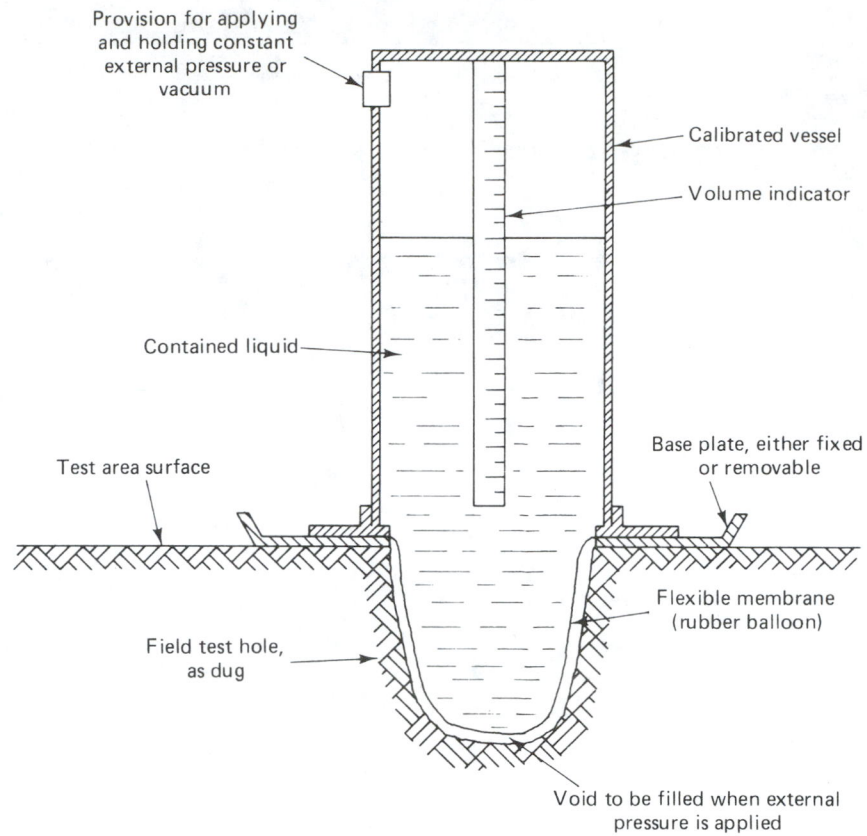

Provision for applying
and holding constant
external pressure or
vacuum

Calibrated vessel

Volume indicator

Contained liquid

Test area surface

Base plate, either fixed
or removable

Field test hole,
as dug

Flexible membrane
(rubber balloon)

Void to be filled when external
pressure is applied

FIGURE 14–1 Schematic Drawing of Calibrated Vessel Indicating Principle (not to scale) [1]

FIGURE 14–2 Rubber-Balloon Apparatus for Determining Density/Unit Weight of Soil in Place (Courtesy of Soiltest, Inc.)

Calibration equipment: thermometer, accurate to 1°F (0.5°C); glass plate, ¼ in. (6 mm) or thicker; grease

CALIBRATION OF APPARATUS [1]

(1) Verify the procedure to be used and the accuracy of the volume indicator by using the apparatus to measure containers or molds of known volume that dimensionally simulate test holes that will be used in the field (Note 1). The apparatus and procedures shall be such that these containers will be measured to within 1% of the actual volumes (Note 1). Containers of different volumes shall be used so that the calibration of the volume indicator covers the range of anticipated test volumes.

> *Note 1*—The 4- and 6-in. (102- and 152-mm) molds described in ASTM Test Methods D 698 and Test Methods D 1557 or any other molds prepared to simulate actual test hole diameters and volumes may be used. When several sets of balloon apparatus are used, or long-term use is anticipated, it may be desirable to cast duplicates of actual test holes. This can be accomplished by forming plaster of paris negatives in actual test holes over a range of volumes, and using these as forms for portland cement concrete castings. They should be cast against a flat plane surface and, after the removal of the negative, sealed water-tight.

(2) *Volume Determination*—Determine the mass of water, in grams, required to fill the containers or hole molds. Using a glass plate and a thin film of grease, if needed for sealing, determine the mass of the container or mold and glass plate to the nearest gram. Fill the container or mold with water, carefully sliding the glass plate over the opening in such a manner as to ensure that no air bubbles are entrapped and that the mold is filled completely with water. Remove excess water and determine the mass of the glass plate, water, and mold or container to the nearest gram. Determine the temperature of the water. Calculate the volume of the mold or container. Repeat this procedure for each container or mold until three consecutive volumes having a maximum variation of 0.0001 ft³ (2.8×10^{-6} m³) are obtained. Record the average of the three trials as the mold or container volume, V_t. Repeat the procedure for each of the containers or molds to be used.

(3) *Calibration Check Tests*—Place the rubber balloon apparatus and base plate on a smooth horizontal surface. Applying an operating pressure, take an initial reading on the volume indicator (Note 2). Transfer the apparatus to one of the previously calibrated molds or containers with a horizontally leveled bearing surface. Apply the operating pressure as necessary until there is no change indicated on the volume indicator. Depending on the type of apparatus, the operating pressure may be as high as 5 psi (34.5 kPa), and it may be necessary to apply a downward load (surcharge) to the

apparatus to keep it from rising (Note 3). Record the readings, pressures, and surcharge loads used. The difference between the initial and final readings is the indicated volume. Determine the volumes of the other molds or containers. A satisfactory calibration check of an apparatus has been achieved when the difference between the indicated and calibrated volume of the container or mold is 1%, or less, for all volumes measured. Select the optimum operating pressure and record it for use with the apparatus during field testing operations.

> *Note 2*—Before any measurements are taken, it may be necessary to distend the rubber balloon and, by kneading, remove the air bubbles adhering to the inside of the membrane. If the calibration castings or molds are airtight, it may be necessary to provide an air escape to prevent erroneous results caused by the trapping of air by the membrane. One means of providing air escape is to place small diameter strings over the edge of and down the inside, slightly beyond bottom center of the mold or casting. This will allow trapped air to escape during the measurement of the calibrated mold or container.

> *Note 3*—It is recommended that the operating pressure of the apparatus be kept as low as possible while maintaining the 1% volume accuracy. The use of higher pressures than necessary may require the use of an additional load or surcharge weight to prevent uplift of the apparatus. The combined pressure and surcharge loads may result in stressing the unsupported soil surrounding the test hole, causing it to deform.

(4) Calculate the volume of the calibration containers or molds as follows:

$$V = (M_2 - M_1) \times V_w \qquad \qquad \textbf{(14–1)}$$

where:

V = volume of the container or mold, mL,
M_2 = mass of mold or container, glass, and water, g,
M_1 = mass of mold or container and glass, g, and
V_w = volume of water based on temperature taken from Table 14–1, mL/g.

> *Note 4*—Multiply milliliters by 3.5315×10^{-5} for ft^3 if needed for equipment usage.

Table 14–1 Volume of Water per Gram Based on Temperature[A] [1]

| Temperature | | Volume of Water, |
°C	°F	mL/g
12	53.6	1.00048
14	57.2	1.00073
16	60.8	1.00103
18	64.4	1.00138
20	68.0	1.00177
22	71.6	1.00221
24	75.2	1.00268
26	78.8	1.00320
28	82.4	1.00375
30	86.0	1.00435
32	89.6	1.00497

[A]Values other than shown may be obtained by referring to *Handbook of Chemistry and Physics,* Chemical Rubber Publishing Co., Cleveland, OH.

PROCEDURE

As explained in Chapter 13, the general procedure for determining in-place soil density/unit weight is to obtain the mass/weight and volume of an in-place soil sample, from which the density/unit weight can be computed. In the rubber-balloon method, a quantity of soil is removed from the ground or compacted fill. The mass/weight of the removed soil is determined directly, whereas its volume is found by measuring the volume of water required to fill the hole.

The actual step-by-step procedure is as follows (ASTM D 2167-94 [1]):

(1) Prepare the surface at the test location so that it is reasonably plane and level. Dependent on the water (moisture) content and texture of the soil, the surface may be leveled using a bulldozer or other heavy equipment blades, provided the test area is not deformed, compressed, torn, or otherwise disturbed.

(2) Assemble the base plate and rubber balloon apparatus on the test location. Using the same pressure and surcharge determined during the calibration of the apparatus, take an initial reading on the volume indicator and record. The base plate shall remain in place through completion of the test.

(3) Remove the apparatus from the test hole location. Using spoons, trowels, and other tools necessary, dig a hole within the base plate. Exercise care in digging the test hole so that soil around the top edge of the hole is not disturbed. The test hole shall be of the minimum volume shown in Table 14–2 based on the maximum

particle size in the soil being tested. When material being tested contains a small amount of oversize, and isolated large particles are encountered, either move the test to a new location or change to another test method, such as ASTM Test Method D 4914 or D 5030. When particles larger than 1½ in. (37.5 mm) are prevalent, larger test apparatus and test volumes are required. Larger test-hole volumes will provide improved accuracy and shall be used where practical. The optimum dimensions of the test hole are related to the design of the apparatus and the pressure used. In general, the dimensions shall approximate those used in the calibration check procedure. The test hole shall be kept as free of pockets and sharp obtrusions as possible, since they may affect accuracy or may puncture the rubber membrane. Place all soil removed from the test hole in a moisture-tight container for later mass and water (moisture) content determination.

Table 14–2   Minimum Test Hole Volumes Based on Maximum Size of Included Particles [1]

| Maximum Particle Size | | Minimum Test Hole Volumes | |
in.	*(mm)*	*cm³*	*ft³*
½	(12.5)	1,420	0.05
1	(25.0)	2,120	0.075
1½	(37.5)	2,840	0.1

(4) After the test hole has been dug, place the apparatus over the base plate in the same position as used for the initial reading. Applying the same pressure and surcharge load as used in the calibration check, take and record the reading on the volume indicator. The difference between the initial and final readings is the volume of the test hole, V_h.

(5) Determine the mass of all the moist soil removed from the test hole to the nearest 5 g. Mix all the soil thoroughly and select a representative water (moisture) content sample and determine the water (moisture) content in accordance with ASTM Method D 2216 (Chapter 3), D 4643 (Chapter 4), D 4959, or D 4944 (Chapter 12). If oversize particles are present, perform field corrections in accordance with ASTM Test Method D 4718.

DATA   Data collected in this test should include the following:

[A] Calibration of Balloon Apparatus

Mass of calibration mold or container and glass

Mass of calibration mold or container, glass, and water

Temperature of water

Operating pressure

Surcharge load

Initial reading of volume indicator

Final reading of volume indicator

[B] Field Data

Initial reading of volume indicator on rubber-balloon apparatus

Final reading of volume indicator on rubber-balloon apparatus

Mass of pan

Mass of pan plus all moist soil from test hole

Data for moisture content determination

Mass of container

Mass of container plus moist weight of moisture sample

Mass of container plus dry weight of moisture sample

Note—These data for moisture content determination would not be collected if the moisture content is determined using a calcium carbide gas moisture tester (Chapter 12).

CALCULATIONS [A] Calibration of Balloon Apparatus

Calculate the volume of the calibration containers or molds from Eq. (14–1).

[B] Field Data

The volume of the test hole, V_h, can be determined by subtracting the initial volume indicator reading on the rubber-balloon apparatus from the final reading. The mass of all moist soil from the test hole, M_{wet}, can be computed by subtracting the mass of the pan from the mass of the pan plus moist soil. The (in-place) wet density, ρ_{wet}, can then be found using the equation

$$\rho_{\text{wet}} = \frac{M_{\text{wet}}}{V_h} \tag{14–2}$$

If M_{wet} is in grams and V_h in cubic centimeters, ρ_{wet} will be in grams per cubic centimeter. To get (in-place) wet unit weight, γ_{wet}, in pounds per cubic foot, merely multiply by 62.427.

The moisture content, w, can be determined in the usual manner [see Eq. (13–4)] and (in-place) dry unit weight, γ_d, can be computed using the equation

$$\gamma_d = \frac{\gamma_{\text{wet}}}{w + 10} \times 100 \tag{14–3}$$

A field test was conducted according to the previous procedure. The following field data were obtained:

Initial reading of volume indicator on rubber-balloon apparatus = **51 mL**

Final reading of volume indicator on rubber-balloon apparatus = **1,317 mL**

Mass of pan = **815 g**

Mass of pan plus all moist soil from test hole = **3,172 g**

Data for moisture content determination:

Mass of container = **45.2 g**

Mass of container plus moist weight of moisture sample = **293.8 g**

Mass of container plus dry weight of moisture sample = **258.8 g**

With these data known, calculations proceed as follows:

$$\rho_{\text{wet}} = \frac{M_{\text{wet}}}{V_h} \qquad\qquad (14\text{--}2)$$

$$M_{\text{wet}} = \mathbf{3{,}172} - \mathbf{815} = 2{,}357 \text{ g}$$

$$V_h = \mathbf{1{,}317} - \mathbf{51} = 1{,}266 \text{ mL}$$

$$\rho_{\text{wet}} = \frac{2{,}357}{1{,}266} = 1.862 \text{ g/cm}^3$$

$$\gamma_{\text{wet}} = (62.427)(1.862) = 116.2 \text{ lb/ft}^3$$

$$w = \frac{\mathbf{293.8} - \mathbf{258.8}}{\mathbf{258.8} - \mathbf{45.2}} \times 100 = 16.4\%$$

$$\gamma_d = \frac{\gamma_{\text{wet}}}{w + 100} \times 100 \qquad\qquad (14\text{--}3)$$

$$\gamma_d = \frac{116.2}{16.4 + 100} \times 100 = 99.8 \text{ lb/ft}^3$$

All data—both given and calculated—are presented on the form shown on page 212. (At the end of the chapter, two blank copies of this form are included for the reader's use.) Careful study of this form should facilitate understanding of the calculations required to determine in-place dry density/unit weight by the rubber-balloon method.

CONCLUSIONS   Again, as stated in Chapter 13, test holes are relatively small, and it is important that no soil be lost during excavation and that volume determinations be done very carefully to ensure accurate evaluations of in-place density and unit weight. To get an accurate moisture content of the soil, excavation should be done as rapidly as possible, and the soil container should be capped and sealed quickly to prevent loss of natural soil moisture.

The methods for determining the in-place dry unit weight of soil in Chapters 13 and 14 are destructive testing methods, in that a sizable hole must be dug in the ground or compacted soil. Chapter 15 presents a nondestructive method for making such determinations.

REFERENCE   [1] ASTM, *2001 Annual Book of ASTM Standards,* West Conshohocken, PA, 2001. Copyright, American Society for Testing and Materials, 100 Barr Harbor Drive, West Conshohocken, PA 19428-2959. Reprinted with permission.

Soils Testing Laboratory
In-Place Density and Unit Weight Determination:
Rubber-Balloon Method

Sample No. _____18_____   Project No. _____SR 2828_____

Location _____Newell, N.C._____   Date _____5/22/02_____

Description of Soil _____Brown silty clay_____

Tested by _____John Doe_____

(1) Initial reading of volume indicator on rubber-balloon apparatus on job site
___51___ mL

(2) Final reading of volume indicator on rubber-balloon apparatus on test hole
___1,317___ mL

(3) Volume of test hole [(2) − (1)] ___1,266___ mL

(4) Mass of all moist soil from test hole + of pan ___3,172___ g

(5) Mass of pan ___815___ g

(6) Mass of all moist soil from test hole [(4) − (5)] ___2,357___ g

(7) Wet density $\left[\dfrac{(6)}{(3)}\right]$ ___1.862___ g/cm^3

(8) Wet unit weight [62.427 × (7)] ___116.2___ lb/ft^3

(9) Field moisture content determination

   Note—If the moisture content is determined by a calcium carbide gas tester, enter the moisture content directly in the appropriate blank below, and skip the intervening steps.

   (a) Can no. ___2-A___

   (b) Mass of moist soil + can ___293.8___ g

   (c) Mass of dry soil + can ___258.8___ g

   (d) Mass of can ___45.2___ g

   (e) Mass of water ___35.0___ g

   (f) Mass of dry soil ___213.6___ g

   (g) Moisture content $\left[\dfrac{(e)}{(f)} \times 100\right]$ ___16.4___ %

(10) In-place dry unit weight of soil $\left[\dfrac{(8)}{(g) + 100} \times 100\right]$ ___99.8___ lb/ft^3

Soils Testing Laboratory
In-Place Density and Unit Weight Determination:
Rubber-Balloon Method

Sample No. _____   Project No. _____

Location _____   Date _____

Description of Soil _____

Tested by _____

    (1) Initial reading of volume indicator on rubber-balloon apparatus on job site
        _____ mL

    (2) Final reading of volume indicator on rubber-balloon apparatus on test hole
        _____ mL

    (3) Volume of test hole [(2) − (1)] _____ mL

    (4) Mass of all moist soil from test hole + of pan _____ g

    (5) Mass of pan _____ g

    (6) Mass of all moist soil from test hole [(4) − (5)] _____ g

    (7) Wet density $\left[\dfrac{(6)}{(3)}\right]$ _____ g/cm^3

    (8) Wet unit weight [62.427 × (7)] _____ lb/ft^3

    (9) Field moisture content determination

       Note—If the moisture content is determined by a calcium carbide gas tester, enter the moisture content directly in the appropriate blank below, and skip the intervening steps.

        (a) Can no. _____

        (b) Mass of moist soil + can _____ g

        (c) Mass of dry soil + can _____ g

        (d) Mass of can _____ g

        (e) Mass of water _____ g

        (f) Mass of dry soil _____ g

        (g) Moisture content $\left[\dfrac{(e)}{(f)} \times 100\right]$ _____ %

    (10) In-place dry unit weight of soil $\left[\dfrac{(8)}{(g) + 100} \times 100\right]$ _____ lb/ft^3

Soils Testing Laboratory
In-Place Density and Unit Weight Determination:
Rubber-Balloon Method

Sample No. _____   Project No. _____

Location _____   Date _____

Description of Soil _____

Tested by _____

(1) Initial reading of volume indicator on rubber-balloon apparatus on job site
_____ mL

(2) Final reading of volume indicator on rubber-balloon apparatus on test hole
_____ mL

(3) Volume of test hole [(2) − (1)] _____ mL

(4) Mass of all moist soil from test hole + of pan _____ g

(5) Mass of pan _____ g

(6) Mass of all moist soil from test hole [(4) − (5)] _____ g

(7) Wet density $\left[\dfrac{(6)}{(3)}\right]$ _____ g/cm^3

(8) Wet unit weight [62.427 × (7)] _____ lb/ft^3

(9) Field moisture content determination

   Note—If the moisture content is determined by a calcium carbide gas tester, enter the moisture content directly in the appropriate blank below, and skip the intervening steps.

   (a) Can no. _____

   (b) Mass of moist soil + can _____ g

   (c) Mass of dry soil + can _____ g

   (d) Mass of can _____ g

   (e) Mass of water _____ g

   (f) Mass of dry soil _____ g

   (g) Moisture content $\left[\dfrac{(e)}{(f)} \times 100\right]$ _____ %

(10) In-place dry unit weight of soil $\left[\dfrac{(8)}{(g) + 100} \times 100\right]$ _____ lb/ft^3

CHAPTER FIFTEEN

Determining the Density and Unit Weight of Soil in Place by Nuclear Methods

(Referenced Document: ASTM D 2922)

INTRODUCTION

As previously discussed in Chapter 13, after a fill layer of soil has been placed by the contractor, the compacted soil's in-place dry unit weight must be determined to ascertain whether the maximum laboratory dry unit weight has been attained. If the maximum dry unit weight (or an acceptable percentage thereof) has not been attained, additional compaction is required.

The sand-cone and rubber-balloon methods for determining density and unit weight of soil in place were presented in Chapters 13 and 14, respectively. Although widely used, these are destructive testing methods, in that a sizable hole must be dug in the ground or compacted fill. They are also fairly time consuming, a significant factor when numerous tests must be performed as quickly as possible at a construction site.

A *nondestructive* and relatively quick method for determining density and unit weight of soil in place utilizes a nuclear apparatus (see Figure 15–1). This apparatus contains a radioactive source and a radiation detector.

The detailed test procedure for this method is described later in this chapter, but the basic premise of the test is that the nuclear apparatus, when placed on the ground or compacted fill, emits gamma rays through the soil. Some of the rays are absorbed; others reach the detector. The amount of radiation reaching the detector varies inversely with soil unit weight; thus, through proper calibration, nuclear count rates received at the detector can be translated into values of soil (wet) density/unit weight.

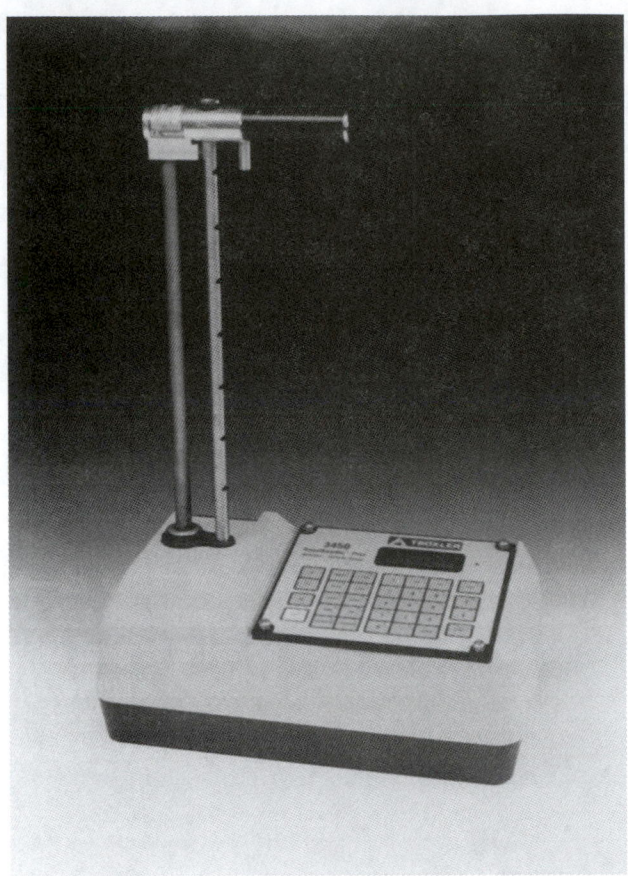

FIGURE 15–1   Nuclear Moisture-Density Apparatus (Courtesy of Troxler Electronic Laboratories, Inc., North Carolina).

This method for determining *in situ* unit weight is known as "density of soil and soil-aggregate in-place by nuclear methods (shallow depth)" and is designated as ASTM D 2922.

There are several modes for measuring soil (wet) density by nuclear methods. One is the *direct transmission method,* where the gamma radiation source is placed at a known depth up to 12 in. (300 mm) while the detector or detectors remain on the surface (see Figure 15–2). Another mode is the *backscatter method,* in which both source and detector(s) remain on the surface (see Figure 15–2).

APPARATUS AND SUPPLIES

Nuclear gage containing a sealed source of high-energy gamma radiation and a gamma detector (see Figure 15–1)

Reference standard

Site-preparation device (a plate, straightedge, or other leveling tool)

Drive pin and drive pin extractor

Slide hammer

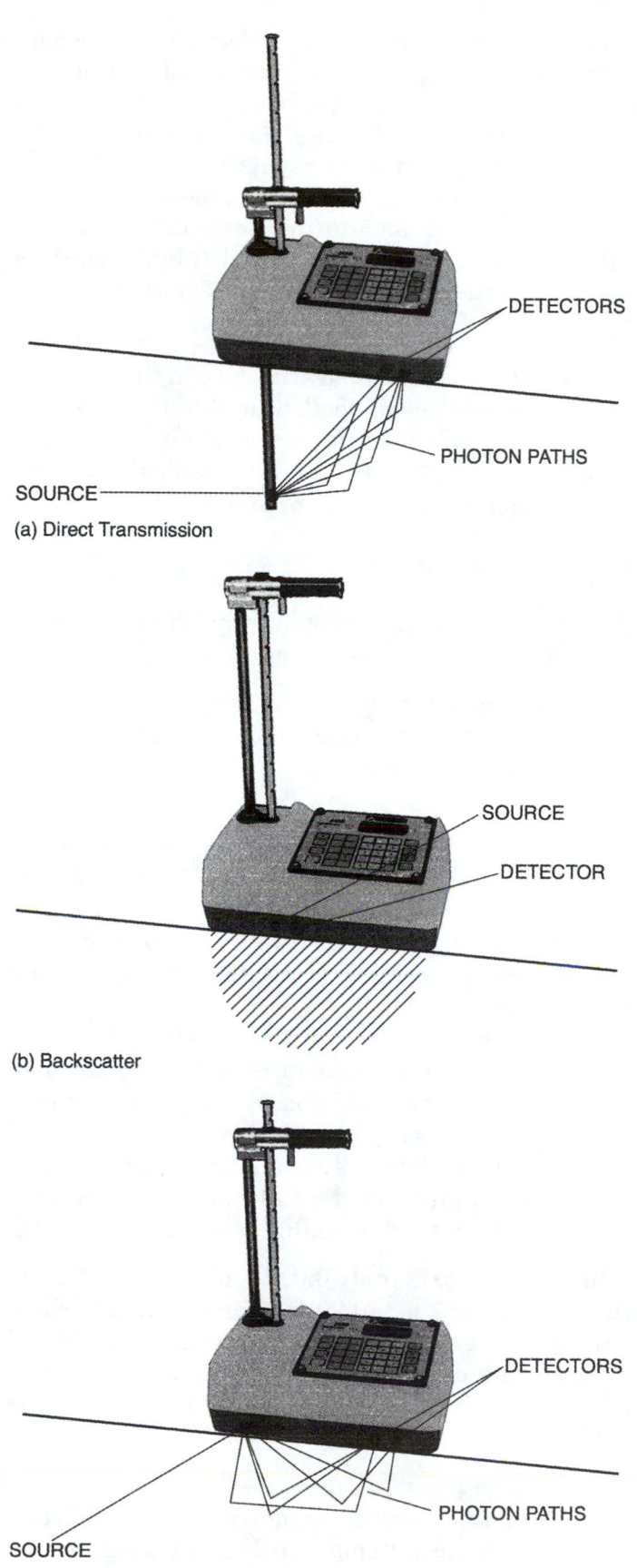

(a) Direct Transmission

DETECTORS

PHOTON PATHS

SOURCE

(b) Backscatter

SOURCE

DETECTOR

(c) Backscatter Moisture Unit

DETECTORS

PHOTON PATHS

SOURCE

FIGURE 15–2 Different Modes for Measuring Soil Density and Moisture Content by Nuclear Methods: (a) Direct Transmission Density Measurement; (b) Backscatter Density Measurement; (c) Backscatter Moisture Measurement (Courtesy of Troxler Electronic Laboratories, Inc., North Carolina).

HAZARDS

This test method presents serious potential hazards. As noted previously, the equipment used in the test contains radioactive material that may be hazardous to the user's health. No one should approach this equipment or attempt to use it without proper training. Users should be thoroughly familiar with safety procedures and government regulations. Additionally, safety procedures, such as proper storage of the equipment, testing for leaks, and recording, evaluating, and monitoring personal radioactive badge data, should be routinely and rigorously followed. (Many states require certification of operators and periodic inspection of equipment.)

CALIBRATION

Ordinarily, manufacturers of the nuclear apparatus used in this test will provide applicable calibration curves or tables. These should be verified every year or so and after any significant equipment repairs. Results of rubber-balloon tests or sand-cone tests may be used as a basis for comparison. More specific calibration instructions are presented in ASTM D 2922 in its Annex.

INTERFERENCES [1]

(1) The chemical composition of the sample may affect the measurement, and adjustments may be necessary.

(2) The test methods exhibit spatial bias in that the instrument is more sensitive to the density of the material in close proximity to the surface (backscatter method only).

> *Note 1*—The nuclear gage density measurements are somewhat biased to the surface layers of the soil being tested. This bias has largely been corrected out of the direct transmission method, and any remaining bias is insignificant. The backscatter method is still more sensitive to the material within the first several inches from the surface.

(3) Oversize rocks or large voids in the source-detector path may cause higher or lower density determination. Where lack of uniformity in the soil due to layering, rock, or voids is suspected, the test volume site should be dug up and visually examined to determine if the test material is representative of the full material in general and if rock correction [see (6) in the "Procedure" section] is required.

(4) The sample volume is approximately 0.0028 m^3 (0.10 ft^3) for the backscatter method and 0.0057 m^3 (0.20 ft^3) for the direct transmission method when the test depth is 15 cm (6 in.). The actual sample volume is indeterminate and varies with the apparatus and the density of the material. In general, the higher the density the smaller the volume.

STANDARDIZATION AND REFERENCE CHECK [1]

(1) Nuclear gages are subject to long-term aging of the radioactive source, detectors, and electronic systems, which may change the relationship between count rate and material density. To offset this

aging, the gage may be calibrated as the ratio of the measured count rate to a count rate made on a reference standard or to an air-gap count [for the backscatter air-gap technique, see (5.1.3) in the "Procedure" section]. The reference count rate should be of the same order of magnitude as the measured count rate over the useful density range of the instrument.

(2) Standardization of the gage shall be performed at the start of each day's work, and a permanent record of these data shall be retained. Perform the standardization with the gage located at least 8 m (25 ft) away from other sources of radioactive material, and clear of large masses or other items which may affect the reference count rate.

(2.1) If recommended by the instrument manufacturer to provide more stable and consistent results: (*1*) turn on the gage prior to use to allow it to stabilize, (*2*) leave the power on during the use of the gage for that day.

(2.2) Using the reference standard, take at least four repetitive readings at the normal measurement period and determine the mean. If available on the gage, one measurement period of four or more times the normal period is acceptable. This constitutes one standardization check.

(2.3) If the value obtained above is within the limits stated below, the gage is considered to be in satisfactory condition, and the value may be used to determine the count ratios for the day of use. If the value is outside these limits, allow additional time for the gage to stabilize, make sure the area is clear of sources of interference, and then conduct another standardization check. If the second standardization check is within the limits, the gage may be used, but if it also fails the test, the gage shall be adjusted or repaired as recommended by the manufacturer. The limits are as follows:

$$|N_s - N_o| \leq 2.0 \sqrt{N_o/F} \qquad (15\text{--}1)$$

where:

N_s = value of current standardization count,

N_o = average of the past four values of N_s taken for prior usage, and

F = value of prescale. [The prescale value (F) is a divisor that reduces the actual value for the purpose of display. The manufacturer will supply this value if other than 1.0.] Some instruments may have provisions to compute and display these values.

(2.3.1) If the instrument standardization has not been checked within the previous three months, perform at least four new standardization checks, and use the mean as the value for N_o.

(3) Use the value of N_s to determine the count ratios for the current day's use of the instrument. If for any reason the measured density

becomes suspect during the day's use, perform another standardization check.

PROCEDURE   As indicated previously, the general procedure for determining (wet)
density/unit weight of soil in place by nuclear methods is carried out by
placing a nuclear apparatus on the ground or compacted fill and causing it to emit gamma rays through the soil. Some of the rays will be absorbed; others will reach a detector. Through proper calibration, nuclear
count rates received at the detector can be translated into values of
(wet) density/unit weight.

The actual step-by-step procedure is as follows (ASTM D 2922-96 [1]):

(1) Standardize the gage. (See the "Standardization and Reference
Check" section.)

(2) Select a test location. If the gage will be closer than 250 mm
(10 in.) to any vertical mass that might influence the result, such
as in a trench or alongside a pipe, follow the manufacturer's correction procedure.

(3) Remove all loose and disturbed material. Remove additional
material as necessary to reach the material that represents a valid
sample of the zone or stratum to be tested. Surface drying and spatial bias should be considered in determining the depth of material
to be removed.

(4) Plane or scrape a smooth horizontal surface so as to obtain maximum contact between the gage and the material being tested. The
placement of the gage on the surface of the material to be tested is
always important, but is especially critical to the successful determination of density when using the backscatter method. The optimum condition in all cases is total contact between the bottom surface of the gage and the surface of the material being tested. To
correct for surface irregularities, use of native fines or fine sand as
a filler may be necessary. The depth of the filler should not exceed
approximately 3 mm (⅛ in.), and the total area filled should not exceed 10% of the bottom area of the instrument. The maximum
depth of any void beneath the gage that can be tolerated without
filling shall not exceed approximately 3 mm (⅛ in.). Several trial
seatings may be required to achieve these conditions.

(5) Proceed with the test in the following manner:

(5.1) *Backscatter Procedure:*

(5.1.1) Seat the gage firmly on the prepared test site.

(5.1.2) Keep all other radioactive sources away from the gage to
avoid affecting the measurement so as not to affect the readings.

(5.1.3) Secure and record one or more readings for the normal measurement period in the backscatter position.

> *Note 2*—When using the backscatter air-gap procedure, follow the instrument manufacturer's instructions regarding apparatus set up. Take the same number of readings for the normal measurement period in the air-gap position as in the standard backscatter position. Determine the air-gap ratio by dividing counts per minute obtained in the air-gap position by counts per minute obtained in standard backscatter position.

(5.1.4) Determine the ratio of the reading to the standard count or to the air-gap count. From this count ratio and the appropriate calibration and adjustment data, determine the in-place wet density.

(5.2) *Direct Transmission Procedure:*

(5.2.1) Make a hole perpendicular to the prepared surface using the guide and the hole-forming device, or by drilling if necessary. The hole shall be of such depth and alignment that insertion of the probe will not cause the gage to tilt from the plane of the prepared area. The depth of the hole must be deeper than the depth to which the probe will be placed. The guide shall be the same size as the base of the gage, with the hole in the same location on the guide as the probe on the gage. The corners of the guide are marked by scoring the surface of the soil. The guide plate is then removed, and any necessary repairs are made to the prepared surface.

(5.2.2) Proceed with testing in the following manner:

(5.2.3) Set the gage on the soil surface, carefully aligning it with the marks on the soil so that the probe will be directly over the pre-formed hole.

(5.2.4) Insert the probe in the hole.

(5.2.5) Seat the gage firmly by rotating it about the probe with a back-and-forth motion.

(5.2.6) Pull gently on the gage in the direction that will bring the side of the probe against the side of the hole that is closest to the detector (or source) location in the gage housing.

(5.2.7) Keep all other radioactive sources away from the gage to avoid affecting the measurement.

(5.2.8) Secure and record one or more readings for the normal measurement period.

(5.2.9) Determine the ratio of the reading to the standard count. From this count ratio and the appropriate calibration and adjustment data, determine the in-place wet density.

> *Note 3*—Some instruments have built-in provisions to compute the ratio and wet density and to enter an adjustment bias. Additionally, some instruments may have provisions to measure and compute moisture content and dry density.

(6) If the volume tested as defined in (4) in the "Interferences" section has excess oversize material with respect to the limitations in the appropriate Test Methods D 698, D 1557, or D 4253, then a correction for wet density (unit weight) and water content must be applied. This correction will be done in accordance with Practice D 4718. This test method requires sampling from the actual test volume.

(6.1) If samples of the measured material are to be taken for purposes of correlation with other test methods or rock correction, the volume measured can be approximated by a 200-mm (8-in.) diameter cylinder located directly under the center line of the radioactive source and detector(s). The height of the cylinder to be excavated will be the depth setting of the source rod when using the direct transmission method or approximately 75 mm (3 in.) when using the backscatter method.

(6.2) An alternative to the correction for oversize particles that can be used with mass density methods or minimal oversize situations involves multiple tests. Tests may be taken at adjacent locations and the results averaged to get a representative value. Comparisons need to be made to evaluate whether the presence of a single large rock or void in the soil is producing unrepresentative values of density. Whenever values obtained are questionable, the test volume site should be dug up and visually examined.

CALCULATIONS

The density/unit weight determined by nuclear methods is the in-place, wet density/unit weight. If dry unit weight is needed, as is usually the case, the soil's moisture content can be determined by the conventional oven method (Chapter 3), the microwave oven method (Chapter 4), or a calcium carbide gas moisture tester (Chapter 12); the dry unit weight can then be found using Eq. (15–2):

$$\gamma_d = \frac{\gamma_{\text{wet}}}{w + 100} \times 100 \qquad (15\text{–}2)$$

where:

γ_d = dry unit weight

γ_{wet} = wet unit weight

w = water content, in percent.

Alternatively, a nuclear instrument that determines moisture content by neutron thermalization may be used to find the soil's moisture content. If this method (ASTM Test Method D 3017) is used, dry unit weight is computed simply by subtracting the lb/ft^3 of moisture found from the lb/ft^3 of wet unit weight.

Some models of the nuclear density device give readouts not only of wet density/unit weight but also of moisture content and dry density/unit weight, in which case a separate moisture content determination is not needed.

NUMERICAL EXAMPLE

A field test was conducted to determine soil densities by the nuclear method. The following direct readings from a nuclear density device were obtained:

Test No.	Test Depth (in.)	Wet Unit Weight (lb/ft³)	Moisture Content (%)	Dry Unit Weight (lb/ft³)
1	8	124.9	9.6	114.0
2	8	126.9	8.4	117.1
3	8	122.3	10.1	111.1
4	8	124.2	9.8	113.1
5	8	126.3	8.4	116.5
6	8	128.2	8.2	118.5

Additionally, the following information was known from a previous laboratory compaction test:

Test No.	Max. Lab. Dry Unit Weight (lb/ft³)	Optimum Moisture Content (%)
1	119.6	8.9
2	119.6	8.9
3	120.5	8.8
4	120.5	8.8
5	121.5	8.3
6	121.5	8.3

The specified minimum compaction is **95.0** percent for each test. The percent compaction for each test is determined by dividing the dry unit weight, as determined by the nuclear density device, by the maximum laboratory dry unit weight. For example, for test no. 1, the percent compaction is **114.0/119.6** = 0.953, or 95.3 percent. This value is compared to the specified minimum compaction to see if the compaction effort is satisfactory. In this example, a specified minimum compaction of **95.0** percent was indicated; hence, this particular compaction effort (i.e., test no. 1) is acceptable. The remaining tests (nos. 2 through 6) are treated in the same manner.

All data are presented on the form shown on page 227. (At the end of the chapter, two blank copies of this form are included for the reader's use.)

CONCLUSIONS

The nuclear method is considerably faster to perform than the sand-cone and rubber-balloon methods. It has the disadvantage, however, of potential hazards to individuals handling radioactive materials. The nuclear apparatus is also considerably more costly than the apparatuses used in the other two methods. Additionally, certain soil types (e.g., micaceous ones) may not yield accurate results by the nuclear method.

Recall that Chapters 11 through 15 are all used in an overall compaction operation. Chapter 11 describes the laboratory compaction test, which determines optimum moisture content and maximum dry unit weight of the soil that is to be used for fill material. Chapter 12 discusses the method for rapid evaluation of moisture content. Chapters 13, 14, and 15 cover three methods of finding the in-place dry unit weight of soil, which in turn is used to determine whether and when the laboratory maximum dry unit weight (or an acceptable percentage thereof) has been achieved in the field.

REFERENCE   [1] ASTM, *2001 Annual Book of ASTM Standards,* West Conshohocken, PA, 2001. Copyright, American Society for Testing and Materials, 100 Barr Harbor Drive, West Conshohocken, PA 19428-2959. Reprinted with permission.

Soils Testing Laboratory
Nuclear Density Test

Project <u>SR 1180</u>

Location of Test <u>Charlotte, NC</u>

Test Instrument <u>Troxler 3440</u>

Tested By <u>John Doe</u>

Date <u>March 22, 2002</u>

Test No.	Test Depth (in.)	Elevation	Wet Unit Weight (lb/ft³) (1)	Moisture Content (%) (2)	Dry Unit Weight (lb/ft³) (3)	Max. Lab. Dry Unit Weight (lb/ft³) (4)	Optimum Moisture Content (%) (5)	Percent Compaction (%) (6)	Specified Compaction (min.) (%) (7)	Remarks
1	8	Subgrade	124.9	9.6	114.0	119.6	8.9	95.3	95.0	O.K.
2	8	Subgrade	126.9	8.4	117.1	119.6	8.9	97.9	95.0	O.K.
3	8	Subgrade	122.3	10.1	111.1	120.5	8.8	92.2	95.0	N.G.
4	8	Subgrade	124.2	9.8	113.1	120.5	8.8	93.9	95.0	N.G.
5	8	Subgrade	126.3	8.4	116.5	121.5	8.3	95.9	95.0	O.K.
6	8	Subgrade	128.2	8.2	118.5	121.5	8.3	97.5	95.0	O.K.
7										
8										
9										
10										

Note: (1), (2), and (3) are direct readings from a nuclear density device.
(4) and (5) are obtained from a laboratory compaction test. (See Chapter 11.)
(6) = [(3)/(4)] × 100%.*
(7) = contract specifications.

*With some models, the maximum laboratory dry unit weight [column (4)] can be entered into the nuclear density device prior to the test and the percent compaction [column (6)] will be output, eliminating the need for a manual calculation.

Soils Testing Laboratory
Nuclear Density Test

Project _____

Location of Test _____

Test Instrument _____

Tested By _____

Date _____

Test No.	Test Depth (in.)	Elevation	Wet Unit Weight (lb/ft³) (1)	Moisture Content (%) (2)	Dry Unit Weight (lb/ft³) (3)	Max. Lab. Dry Unit Weight (lb/ft³) (4)	Optimum Moisture Content (%) (5)	Percent Compaction (%) (6)	Specified Compaction (min.) (%) (7)	Remarks
1										
2										
3										
4										
5										
6										
7										
8										
9										
10										

Note: (1), (2), and (3) are direct readings from a nuclear density device.

(4) and (5) are obtained from a laboratory compaction test. (See Chapter 11.)

(6) = [(3)/(4)] × 100%.*

(7) = contract specifications.

*With some models, the maximum laboratory dry density [column (4)] can be entered into the nuclear density device prior to the test and the percent compaction [column (6)] will be output, eliminating the need for a manual calculation.

Soils Testing Laboratory
Nuclear Density Test

Project _____

Location of Test _____

Test Instrument _____

Tested By _____

Date _____

Test No.	Test Depth (in.)	Elevation	Wet Unit Weight (lb/ft³) (1)	Moisture Content (%) (2)	Dry Unit Weight (lb/ft³) (3)	Max. Lab. Dry Unit Weight (lb/ft³) (4)	Optimum Moisture Content (%) (5)	Percent Compaction (%) (6)	Specified Compaction (min.) (%) (7)	Remarks
1										
2										
3										
4										
5										
6										
7										
8										
9										
10										

Note: (1), (2), and (3) are direct readings from a nuclear density device.
 (4) and (5) are obtained from a laboratory compaction test. (See Chapter 11.)
 (6) = [(3)/(4)] × 100%.*
 (7) = contract specifications.

*With some models, the maximum laboratory dry density [column (4)] can be entered into the nuclear density device prior to the test and the percent compaction [column (6)] will be output, eliminating the need for a manual calculation.

[B] Permeability Test

    (1)   Cross-sectional area of standpipe, a _____ cm^2

    (2)   Length of soil specimen in permeameter, L _____ cm

    (3)   Cross-sectional area of soil specimen, A _____ cm^2

Trial no.	Head, h_1 (cm)	Head, h_2 (cm)	Time, t (s)	Temperature, T (°C)	Permeability at T °C, k_T (cm/s)	Ratio of $\dfrac{\text{Viscosity at T °C}}{\text{Viscosity at 20°C}}$	Permeability at 20°C, k_{20} (cm/s)
(1)	(2)	(3)	(4)	(5)	$(6) = \dfrac{2.3aL}{At} \log \dfrac{h_1}{h_2}$	(7)	(8)
						(see Table 16–2)	

16

CHAPTER SIXTEEN

Percolation Test

INTRODUCTION   *Percolation* refers to movement of water through soil. Percolation tests are performed to determine the rate at which percolation occurs (i.e., the rate at which water moves through soil). Percolation rates are needed to determine whether or not given job sites will be suitable for certain engineering projects, such as reservoirs, sewage lagoons, sanitary landfills, and septic tank drain fields. In the first three of these, bottom soil should not have a high percolation rate. If it did, considerable water could be lost through the bottom of a reservoir, lagoon, or landfill. Such water loss would, of course, be undesirable for reservoirs; and in the case of lagoons and landfills, water loss could carry waste material with it and thereby contaminate groundwater in the area. On the other hand, relatively high percolation rates are needed for soils to be used for septic tank drain fields, so that the liquid part of sewage will percolate readily through the soil and thereby be removed from the septic tank and drainage pipes.

Percolation tests are performed in the field, in place, at locations where seepage is a matter of concern. They are performed, essentially, by digging a hole and measuring how long it takes a given depth of water to drain (seep) out of the bottom of the hole. For septic tank drain fields, the depth of the test hole should generally be about the same as the depth at which the drain pipe is to be placed. To ensure adequate percolation, some local ordinances require that septic tank drains be located at least 4 ft above underlying bedrock.

APPARATUS AND SUPPLIES

Hand auger

Timing device

Percolation tester: a particular tester known as a "Martin Perk-Tester" consists of a base stand, a 36-in. metal rule, and an adjustable metal rod. Electronic circuitry, powered by a dry-cell battery, provides meter readings on an instrument panel to indicate the water level in a percolation test hole (see Figure 16–1).

PREPARATION OF TEST HOLE

Test holes should be bored with an auger to the required depth. If a Martin Perk-Tester is used, the diameter of the hole may be as small as 4 in. After the hole is bored and cleaned of loose material, the bottom should be covered with 1 to 2 in. of gravel so that it will be less likely to become clogged with soil from the sides of the hole.

In order that "worst-case" conditions may be simulated more accurately, holes should be filled with water and allowed to stand for a long period of time (preferably overnight) before beginning a percolation test. This allows time for the soil to become completely saturated and for soils containing clay to swell [1].

PROCEDURE

After a test hole has been prepared in the manner just described, the percolation test is conducted by placing water in the hole and, in effect,

FIGURE 16–1   Percolation Tester (Martin Perk-Tester. Courtesy of Soiltest, Inc.)

determining how long it takes the water surface to fall 6 in. The percolation rate is normally reported in either inches per hour or minutes per inch, or both.

One procedure, given by Soiltest, Inc., manufacturer of the Martin Perk-Tester, is as follows [1]:

(1) Following the presoaking period, refill the hole with water and allow it to drop to a point below the topsoil.

(2) Place the test stand in position with the opening directly over the hole and the stand legs pressed firmly into the soil.

(3) Set the Perk-Tester on the test stand with the binding post situated over the curved cutout in the stand (see Figure 16–1). Using the level mounted on the tester, adjust the stand as needed.

(4) Turn the Perk-Tester on and press the test button. The meter should read full scale. This checks the battery and amplifier circuits.

(5) Install the grounding rod down through the binding post and well into the water. This rod must remain in contact with the water at all times during the entire process of testing.

(6) Insert the aluminum rule through the slot in the center of the Perk-Tester and lower it down into the hole until it penetrates the surface of the water. As the water is penetrated, the meter should read full scale. The rule should be adjusted so that the top edge of the rule slot on the Perk-Tester is in alignment with a suitable inch mark on the rule when the lower tip of the rule is only a short distance below the surface of the water. Note that reading.

(7) Prepare a stopwatch for timing the fall of the water and watch the meter pointer. When the water meniscus breaks away from the tip of the rule, the meter pointer will fall to zero. Start the stopwatch at this time.

(8) Slide the rule down into the water until it is 6 in. below the starting point and lock it. Again watch the meter pointer and stop the timer when it falls to zero. Record the elapsed time.

The procedure just described is for use specifically with a Martin Perk-Tester. It is possible, however, to perform a percolation test without using such a tester. All one needs are a timing device (e.g., a stopwatch) and a means of measuring how far the water surface has dropped. The latter can be determined by measuring down to the water surface from a fixed point, using a rule or yardstick. Figure 16–2 shows schematically how such a measurement can be made.

A procedure for conducting percolation tests, as given by the Public Health Service and intended mainly for septic tank drain field design, is as follows [2]:

(1) Six or more tests shall be made in separate test holes spaced uniformly over the proposed absorption-field site.

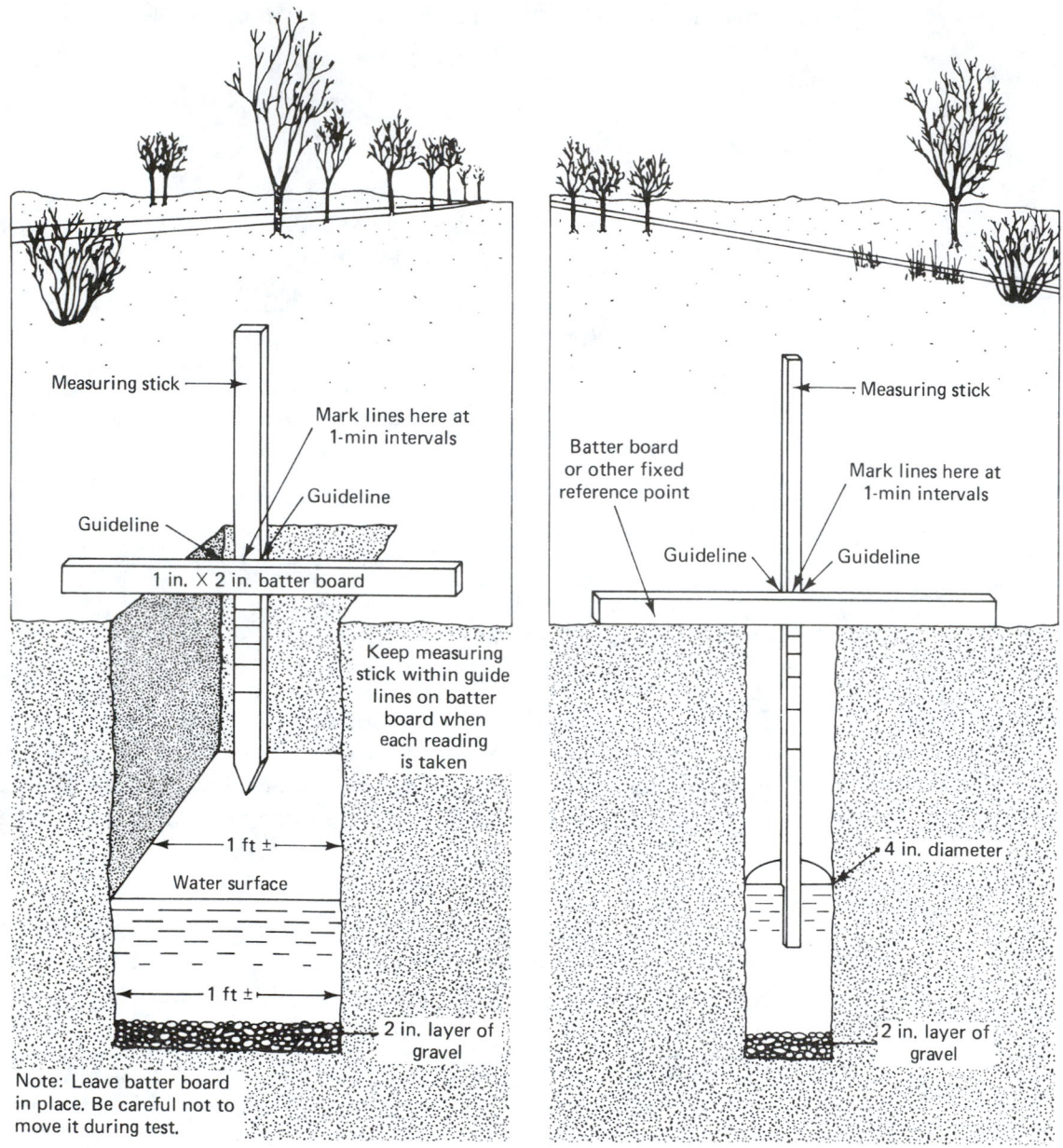

FIGURE 16–2   Schemes for Making Percolation Tests [2]

(2) Dig or bore a hole, with horizontal dimensions of from 4 to 12 in. and vertical sides to the depth of the proposed absorption trench. In order to save time, labor, and volume of water required per test, the holes can be bored with a 4-in. auger (see Figure 16–2).

(3) Carefully scratch the bottom and sides of the hole with a knife blade or sharp-pointed instrument, in order to remove any smeared soil surfaces and to provide a natural soil interface into which water may percolate. Remove all loose material from the hole. Add

2 in. of coarse sand or fine gravel to protect the bottom from scouring and sediment.

(4) It is important to distinguish between saturation and swelling. Saturation means that the void spaces between soil particles are full of water. This can be accomplished in a short period of time. Swelling is caused by intrusion of water into the individual soil particle. This is a slow process, especially in clay-type soil, and is the reason for requiring a prolonged soaking period.

In the conduct of the test, carefully fill the hole with clear water to a minimum depth of 12 in. over the gravel. In most soils, it is necessary to refill the hole by supplying a surplus reservoir of water, possibly by means of an automatic syphon, to keep water in the hole for at least 4 h and preferably overnight. Determine the percolation rate 24 h after water is first added to the hole. This procedure is to ensure that the soil is given ample opportunity to swell and to approach the condition it will be in during the wettest season of the year. Thus, the test will give comparable results in the same soil, whether made in a dry or in a wet season. In sandy soils containing little or no clay, the swelling procedure is not essential, and the test may be made as described under item (5)(C), after the water from one filling of the hole has completely seeped away.

(5) With the exception of sandy soils, percolation-rate measurements shall be made on the day following the procedure described under item (4) above.

A. If water remains in the test hole after the overnight swelling period, adjust the depth to approximately 6 in. over the gravel. From a fixed reference point, measure the drop in water level over a 30-min period. This drop is used to calculate the percolation rate.

B. If no water remains in the hole after the overnight swelling period, add clear water to bring the depth of water in the hole to approximately 6 in. over the gravel. From a fixed reference point, measure the drop in water level at approximately 30-min intervals for 4 h, refilling 6 in. over the gravel as necessary. The drop that occurs during the final 30-min period is used to calculate the percolation rate. The drops during prior periods provide information for possible modification of the procedure to suit local circumstances.

C. In sandy soils (or other soils in which the first 6 in. of water seeps away in less than 30 min, after the overnight swelling period), the time interval between measurements shall be taken as 10 min and the test run for 1 h. The drop that occurs during the final 10 min is used to calculate the percolation rate.

It will be noted that the foregoing method provides a slightly different procedure for computing percolation rate. In any event, percolation rate is reported in either inches per hour or minutes per inch, or both.

DATA   When the Martin Perk-Tester is used, data collected during this test consist of the following:

> Initial reading on the rule, r_1 (in.)
>
> Final reading on the rule, r_2 (in.)
>
> Elapsed time, T (min or h)

(Normally, the difference between r_1 and r_2 is 6 in.; hence, T represents the time taken for water to fall 6 in.)

If the Public Health Service procedure is followed, data collected are the drop in water level (normally in inches) and the associated time interval.

CALCULATIONS   As mentioned, the percolation rate P is normally reported in either inches per hour or minutes per inch. When the Martin Perk-Tester is used, the rate in inches per hour can be calculated by dividing the difference between r_1 and r_2 (normally 6 in.) by the elapsed time T expressed in hours. In equation form,

$$P = \frac{r_2 - r_1}{T} \tag{16-1}$$

The rate in minutes per inch can be calculated by dividing T by $(r_2 - r_1)$, with T expressed in minutes.

If the Public Health Service procedure is followed, the computation is the same except that the term $(r_2 - r_1)$ is replaced by the drop in water level.

NUMERICAL EXAMPLE   A percolation test was conducted according to the procedure described for use with a Martin Perk-Tester. The following data were obtained:

> Initial reading on rule, r_1 = **18.00 in.**
>
> Final reading on rule, r_2 = **24.00 in.**
>
> Elapsed time, T = **1 h 24 min**

Using Eq. (16–1), we obtain

$$P = \frac{r_2 - r_1}{T} \tag{16-1}$$

$$P = \frac{24.00 - 18.00}{1.4} = 4.29 \text{ in./h}$$

or

$$P = \frac{84}{24.00 - 18.00} = 14.0 \text{ min/in.}$$

CONCLUSIONS

As related in the introduction to this chapter, percolation rates are needed to determine whether or not a given job site will be suitable for certain engineering projects, including reservoirs, sewage lagoons, sanitary landfills, and septic tank drain fields. In most cases, several percolation tests are performed at various locations throughout a project site in order to obtain a representative value of the percolation rate. Table 16–1 gives some typical values and limitations of percolation rates, whereas Table 16–2 shows how percolation rates can be used in septic tank drain field design to determine absorption areas.

Table 16–1 Typical Values and Limitations of Percolation Rates [3]

Project	Slight	Moderate	Severe
Reservoir	Less than 0.2 in./h	Between 0.2 and 2.0 in./h	More than 2.0 in./h
Lagoon	Less than 0.6 in./h	Between 0.6 and 2.0 in./h	More than 2.0 in./h
Landfill	Less than 2.0 in./h	Less than 2.0 in./h	More than 2.0 in./h
Septic tank drain field	Faster than 45 min/in.	Between 45 and 60 min/in.	Slower than 60 min/in.

Table 16–2 Absorption-Area Requirements for Private Residences[a] [2]

Percolation Rate (Time Required for Water to Fall 1 in.) (min)	Required Absorption Area (ft^2) per Bedroom,[b] Standard Trench,[c] and Seepage Pits[d]	Percolation Rate (Time Required for Water to Fall 1 in.) (min)	Required Absorption Area (ft^2) per Bedroom,[b] Standard Trench,[c] and Seepage Pits[d]
1 or less	70	10	165
2	85	15	190
3	100	30[e]	250
4	115	45[e]	300
5	125	60[e,f]	330

[a]Provides for garbage grinders and automatic-sequence washing machines.
[b]In every case, sufficient area should be provided for at least two bedrooms.
[c]Absorption area for standard trenches is figured as trench-bottom area.
[d]Absorption area for seepage pits is figured as effective sidewall area beneath the inlet.
[e]Unsuitable for seepage pits if over 30.
[f]Unsuitable for leaching systems if over 60.

REFERENCES

[1] Soiltest, Inc., *Operating Instructions for Martin Perk-Tester,* Evanston, Ill., 1970.

[2] U.S. Department of Health, Education, and Welfare, Public Health Service, *Manual of Septic Tank Practice,* Washington, D.C., 1963.

[3] U.S. Department of Agriculture, Soil Conservation Service, *Guide for Interpreting Engineering Uses of Soils,* Washington, D.C., 1971.

CHAPTER SEVENTEEN

Permeability Test for Granular Soils (Constant-Head Method)

(Referenced Document: ASTM D 2434)

INTRODUCTION *Permeability* refers to the propensity of a material to allow fluid to move through its pores or interstices. In the context of soil, permeability generally relates to the propensity of a soil to allow water to move through its void spaces. According to Darcy's law, the flow rate of water q through a soil of cross-sectional area A is directly proportional to the imposed gradient (slope) $i,$ or

$$\frac{q}{A} \sim i \qquad\qquad (17\text{--}1)$$

If a constant of proportionality k is introduced, we obtain the equation

$$q = kiA \qquad\qquad (17\text{--}2)$$

The constant k is known as the coefficient of permeability, or just permeability. Obviously, it indicates the ease with which water will flow through a given soil. The greater the value of permeability, the greater the flow will be for a given area and gradient.

Permeability is an important soil parameter for any project where flow of water through soil is a matter of concern—for example, seepage through or under a dam and drainage from subgrades or backfills.

There are several factors that influence the permeability of a soil: the viscosity of its water (which is a function of temperature), size and shape of the soil particles, degree of saturation, and void ratio. (The void ratio, customarily denoted by *e*, is the ratio of volume of voids to volume of solids.) The void ratio has a significant influence on permeability. For a given soil, permeability is inversely proportional to soil density. This is intuitively obvious if one considers that the denser a soil, the more tightly its particles are packed, the smaller will be the void space and void ratio, and the lower will be the tendency for the soil to allow water to move through it (i.e., its permeability). Hence, permeability is directly proportional to void ratio.

In view of the foregoing, each value of permeability for a soil should be associated with a particular void ratio. Normally, when a permeability value is needed, it is for the permeability of the soil *in situ*. Permeability is often, however, determined by laboratory tests; and to be representative of the soil's *in situ* permeability, tests should be performed on "undisturbed" samples. In the case of granular soils, however, it is extremely difficult to obtain undisturbed samples. Thus the recommended procedure is to perform permeability tests on three soil specimens of the same sample, with each specimen having a different void ratio. A relationship between void ratio and permeability can be established for a given soil by plotting a graph of permeability on a logarithmic scale versus void ratio on an arithmetic scale. Then, whenever the value of the permeability of that soil *in situ* is needed, a sample can be taken at the project site, its void ratio determined, and associated permeability obtained from the graphical relationship.

This chapter describes how to determine the coefficient of permeability by a constant-head method for laminar flow of water through granular soils only. To limit consolidation influences during testing, this procedure is restricted to disturbed granular soils containing not more than 10% soil passing a No. 200 sieve. Chapter 18 describes the falling-head method for determining the coefficient of permeability; this method may be used to test both fine-grained soils (such as silts and clays) and coarse-grained soils.

APPARATUS AND SUPPLIES

Permeameter (including constant-head filter tank and manometer tubes): a specialized device for determining soil permeability (see Figure 17–1)

Large funnels (with spouts 25 mm in diameter for 9.5-mm maximum particle size and 13 mm in diameter for 2.00-mm maximum particle size)

Timing device

Thermometer

Compaction equipment

Vacuum pump or water-faucet aspirator

Balance (with accuracy to 1.0 g)

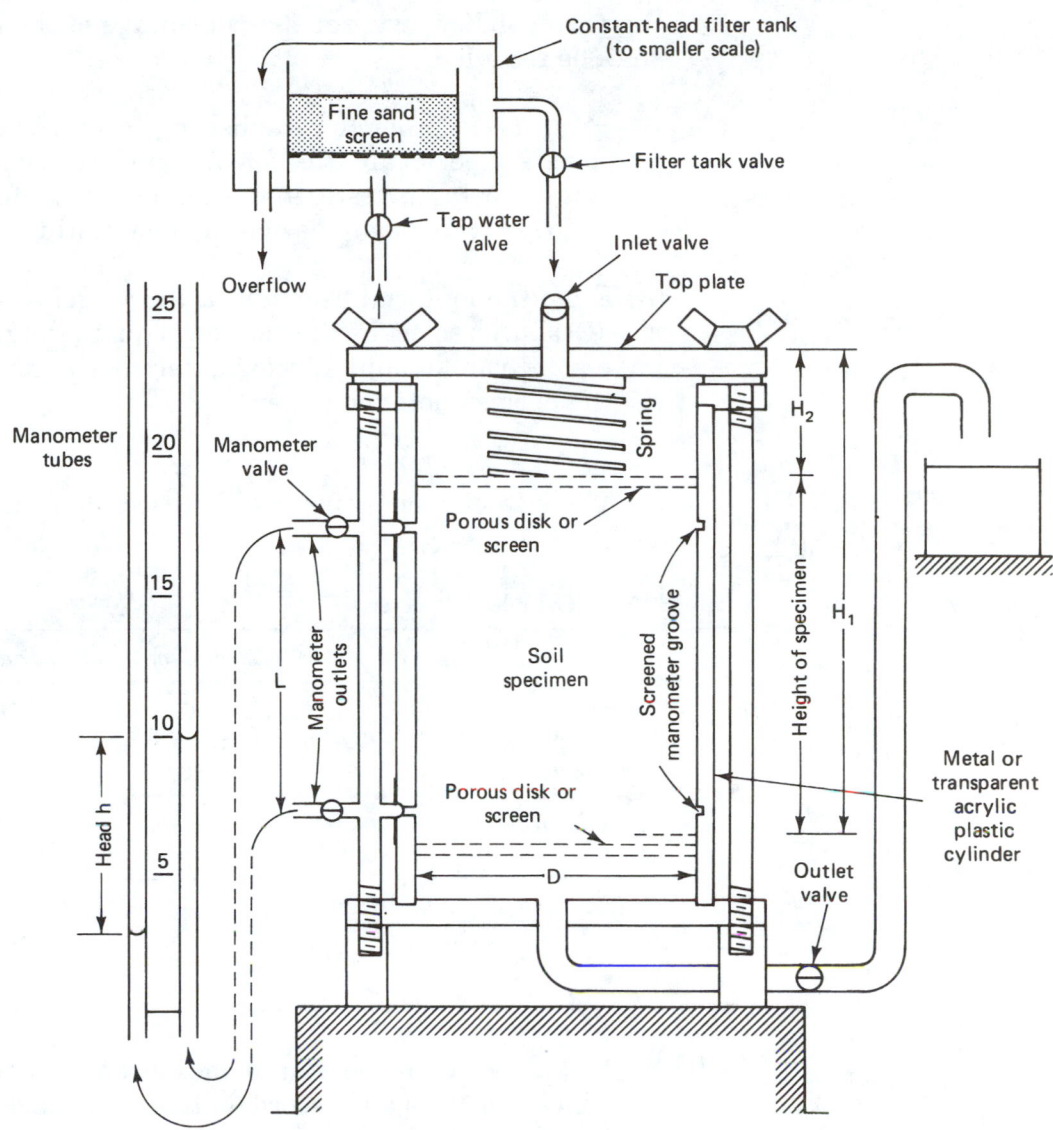

FIGURE 17-1 Constant-Head Permeameter [1]

Miscellaneous apparatus: scoop, 250-ml graduate, jars, mixing pans, etc.

SAMPLE [1]

(1) A representative sample of air-dried granular soil, containing less than 10% of the material passing the 0.075-mm (No. 200) sieve and equal to an amount sufficient to satisfy the requirements prescribed in (2) and (3) below shall be selected by the method of quartering.

(2) A sieve analysis (ASTM Method D 422, for Particle-Size Analysis of Soils; see Chapter 9) shall be made on a representative sample for the complete soil prior to the permeability test. Any particles larger than 19 mm (¾ in.) shall be separated out by sieving (ASTM Method D 422). This oversize material shall not be used for

the permeability test, but the percentage of the oversize material shall be recorded.

> *Note 1*—In order to establish representative values of coefficients of permeabilities for the range that may exist in the situation being investigated, samples of the finer, average, and coarser soils should be obtained for testing.

(3) From the material from which the oversize has been removed [see (2) above], select by the method of quartering a sample for testing equal to an amount approximately twice that required for filling the permeameter chamber.

PREPARATION OF SPECIMENS [1]

(1) The size of permeameter to be used shall be as prescribed in Table 17–1.

Table 17–1   Cylinder Diameter [1]

| Maximum Particle Size Lies between Sieve Openings | Minimum Cylinder Diameter | | | |
| | Less than 35% of Total Soil Retained on Sieve Opening | | More than 35% of Total Soil Retained on Sieve Opening | |
	2.00 mm (No. 10)	9.5 mm (⅜ in.)	2.00 mm (No. 10)	9.5 mm (⅜ in.)
2.00 mm (No. 10) and 9.5 mm (⅜ in.)	76 mm (3 in.)	—	114 mm (4.5 in.)	—
9.5 mm (⅜ in.) and 19.0 mm (¾ in.)	—	152 mm (6 in.)	—	229 mm (9 in.)

(2) Make the following initial measurements in centimeters or square centimeters and record on the data sheet: the inside diameter, D, of the permeameter; the length, L, between manometer outlets; the depth, H_1, measured at four symmetrically spaced points from the upper surface of the top plate of the permeability cylinder to the top of the upper porous stone or screen temporarily placed on the lower porous plate or screen. This automatically deducts the thickness of the upper porous plate or screen from the height measurements used to determine the volume of soil placed in the permeability cylinder. Use a duplicate top plate containing four large symmetrically spaced openings through which the necessary measurements can be made to determine the average value for H_1. Calculate the cross-sectional area, A, of the specimen.

(3) Take a small portion of the sample selected as prescribed in (3) in the section "Sample" for water content determinations. Record the weight of the remaining air-dried sample, W_1, for unit weight determinations.

(4) Place the prepared soil by one of the following procedures in uniform thin layers approximately equal in thickness after com-

paction to the maximum size of particle, but not less than approximately 15 mm (0.60 in.).

(4.1) For soils having a maximum size of 9.5 mm (⅜ in.) or less, place the appropriate size of funnel, as prescribed in the section "Apparatus and Supplies," in the permeability device with the spout in contact with the lower porous plate or screen, or previously formed layer, and fill the funnel with sufficient soil to form a layer, taking soil from different areas of the sample in the pan. Lift the funnel by 15 mm (0.60 in.), or approximately the unconsolidated layer thickness to be formed, and spread the soil with a slow spiral motion, working from the perimeter of the device toward the center, so that a uniform layer is formed. Remix the soil in the pan for each successive layer to reduce segregation caused by taking soil from the pan.

(4.2) For soils with a maximum size greater than 9.5 mm (⅜ in.), spread the soil from a scoop. Uniform spreading can be obtained by sliding a scoopful of soil in a nearly horizontal position down along the inside surface of the device to the bottom or to the formed layer, then tilting the scoop and drawing it toward the center with a single slow motion; this allows the soil to run smoothly from the scoop in a windrow without segregation. Turn the permeability cylinder sufficiently for the next scoopful, thus progressing around the inside perimeter to form a uniform compacted layer of a thickness equal to the maximum particle size.

(5) Compact successive layers of soil to the desired relative density by appropriate procedures, as follows, to a height of about 2 cm (0.8 in.) above the upper manometer outlet.

(5.1) *Minimum Density (0% Relative Density)*—Continue placing layers of soil in succession by one of the procedures described in (4.1) or (4.2) until the device is filled to the proper level.

(5.2) *Maximum Density (100% Relative Density):*

(5.2.1) *Compaction by Vibrating Tamper*—Compact each layer of soil thoroughly with the vibrating tamper, distributing the light tamping action uniformly over the surface of the layer in a regular pattern. The pressure of contact and the length of time of the vibrating action at each spot should not cause soil to escape from beneath the edges of the tamping foot, thus tending to loosen the layer. Make a sufficient number of coverages to produce maximum density, as evidenced by practically no visible motion of surface particles adjacent to the edges of the tamping foot.

(5.2.2) *Compaction by Sliding Weight Tamper*—Compact each layer of soil thoroughly by tamping blows uniformly distributed over the surface of the layer. Adjust the height of drop and give sufficient coverages to produce maximum density, depending on the coarseness and gravel content of the soil.

(5.2.3) *Compaction by Other Methods*—Compaction may be accomplished by other approved methods, such as by vibratory packer equipment, where care is taken to obtain a uniform specimen without segregation of particle sizes.

(5.3) *Relative Density Intermediate between 0 and 100%*—By trial in a separate container of the same diameter as the permeability cylinder, adjust the compaction to obtain reproducible values of relative density. Compact the soil in the permeability cylinder by these procedures in thin layers to a height about 2.0 cm (0.80 in.) above the upper manometer outlet.

> *Note 2*—In order to bracket, systematically and representatively, the relative density conditions that may operate in natural deposits or in compacted embankments, a series of permeability tests should be made to bracket the range of field relative densities.

(6) *Preparation of Specimen for Permeability Test:*

(6.1) Level the upper surface of the soil by placing the upper porous plate or screen in position and by rotating it gently back and forth.

(6.2) Measure and record: the final height of specimen, $H_1 - H_2$, by measuring the depth, H_2, from the upper surface of the perforated top plate employed to measure H_1 to the top of the upper porous plate or screen at four symmetrically spaced points after compressing the spring lightly to seat the porous plate or screen during the measurements; the final mass of air-dried soil used in the test $(M_1 - M_2)$ by weighing the remainder of soil, M_2, left in the pan. Compute and record the unit weights, void ratio, and relative density of the test specimen.

(6.3) With its gasket in place, press down the top plate against the spring and attach it securely to the top of the permeameter cylinder, making an air-tight seal.

(6.4) Using a vacuum pump or suitable aspirator, evacuate the specimen under 50 cm (20 in.) Hg minimum for 15 min to remove air adhering to soil particles and from the voids. Follow the evacuation by a slow saturation of the specimen from the bottom upward (Figure 17–2) under full vacuum in order to free any remaining air in the specimen. Continued saturation of the specimen can be maintained more adequately by the use of (1) de-aired water, or (2) water maintained at an in-flow temperature sufficiently high to cause a decreasing temperature gradient in the specimen during the test. Native water or water of low mineral content (Note 3) should be used for the test, but in any case the fluid should be described on the report form.

> *Note 3*—Native water is the water occurring in the rock or soil *in situ*. It should be used if possible, but it (as well as de-aired water) may be a refinement not ordinarily feasible for large-scale production testing.

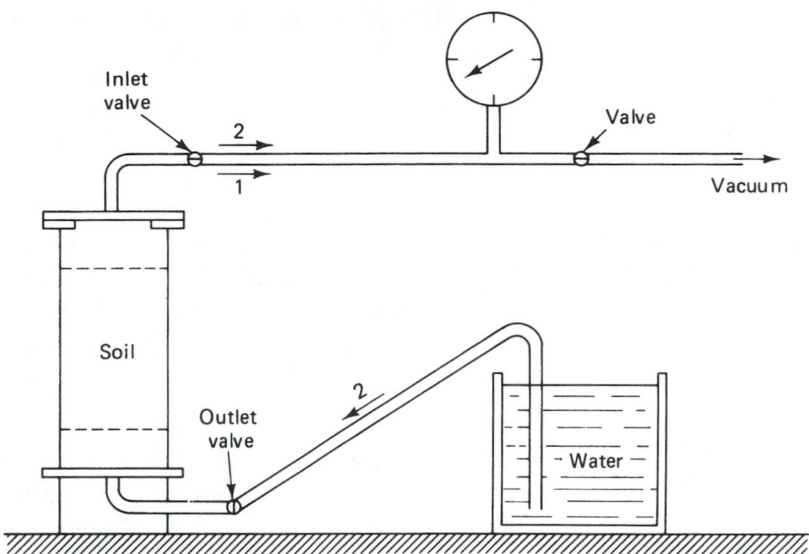

FIGURE 17–2   Device for Evacuating and Saturating Specimen [1]

(6.5) After the specimen has been saturated and the permeameter is full of water, close the bottom valve on the outlet tube (Figure 17–2) and disconnect the vacuum. Care should be taken to ensure that the permeability flow system and the manometer system are free of air and are working satisfactorily. Fill the inlet tube with water from the constant-head tank by slightly opening the filter tank valve. Then connect the inlet tube to the top of the permeameter, open the inlet valve slightly, and open the manometer outlet cocks slightly to allow water to flow, thus freeing them of air. Connect the water manometer tubes to the manometer outlets and fill with water to remove the air. Close the inlet valve and open the outlet valve to allow the water in the manometer tubes to reach their stable water level under zero head.

PROCEDURE

After the preparation of specimens has been completed as described, the general test procedure is to allow water to move through the soil specimen under a stable head condition while determining and recording the time required for a certain quantity of water to pass through the specimen. Using these data together with others obtained in the section "Preparation of Specimens," one can determine the coefficient of permeability. It would be good practice to make several successive determinations of the time required for a certain quantity of water to pass through the soil specimen, so that an average value of permeability can be determined.

The actual step-by-step procedure is as follows (ASTM D 2434-68 Reapproved 2000 [1]):

(1) Open the inlet valve from the filter tank slightly for the first run; delay measurements of quantity of flow and head until a stable head condition without appreciable drift in water manometer

levels is attained. Measure and record the time, t, head, h (the difference in level in the manometers), quantity of flow, Q, and water temperature, T.

(2) Repeat test runs at heads increasing by 0.5 cm in order to establish accurately the region of laminar flow with velocity, v (where $v = Q/At$), directly proportional to hydraulic gradient, i (where $i = h/L$). When departures from the linear relation become apparent, indicating the initiation of turbulent flow conditions, 1-cm intervals of head may be used to carry the test run sufficiently along in the region of turbulent flow to define this region if it is significant for field conditions.

> *Note 4*—Much lower values of hydraulic gradient, h/L, are required than generally recognized, in order to ensure laminar flow conditions. The following values are suggested: loose compactness ratings, h/L from 0.2 to 0.3, and dense compactness ratings, h/L from 0.3 to 0.5, the lower values of h/L applying to coarser soils and the higher values to finer soils.

(3) At the completion of the permeability test, drain the specimen and inspect it to establish whether it was essentially homogeneous and isotropic in character. Any light and dark alternating horizontal streaks or layers are evidence of segregation of fines.

DATA   Data collected during this test consist of the following:

Diameter of specimen (inside diameter of permeameter), D (cm)

Length between manometer outlets, L (cm)

Height, H_1 (see "Preparation of Specimens" and Figure 17–1) (cm)

Height, H_2 (see "Preparation of Specimens" and Figure 17–1) (cm)

For water content determination:

Mass of air-dried soil plus can (g)

Mass of oven-dried soil plus can (g)

Mass of can (g)

Mass of air-dried soil before compaction, M_1 (g)

Mass of unused remaining portion of air-dried soil left in pan after compaction, M_2 (g)

Manometer readings, h_1 and h_2 (cm)

Quantity of water discharged through soil specimen, Q (cm^3)

Total time of discharge, t (s)

Water temperature, T (°C)

CALCULATIONS   [A] Unit Weight Determination

By using the preceding data, values of the unit weight of the soil specimen (air dried), water content of the air-dried soil, dry unit weight of the soil specimen, and void ratio e can be computed. The void ratio is the ratio of volume of voids to volume of solids; the other (computed) parameters listed have been covered in detail in previous chapters.

[B] Permeability Test

The coefficient of permeability can be evaluated by modifying Eq. (17–2),

$$q = kiA \qquad (17\text{–}2)$$

In the experimental procedure, the value of i, the gradient, can be replaced by h/L, where h is the head (difference in manometer levels) and L is the length between manometer outlets. Additionally, the value of q, the rate of water flow, can be replaced by Q/t, where Q is the quantity (volume) of water discharged and t the time required for that quantity to be discharged. If these substitutions are made in Eq. (17–2) and the resulting equation solved for k, the result is [1]

$$k = \frac{QL}{Ath} \qquad (17\text{–}3)$$

where:
  k = coefficient of permeability, cm/s
  Q = quantity (volume) of water discharged during test, cm^3
  L = length between manometer outlets, cm
  A = cross-sectional area of specimen, cm^2
  t = time required for quantity Q to be discharged during test, s
  h = head (difference in manometer levels) during test, cm

The permeability computed using Eq. (17–3) is the value for the particular water temperature at which the test was conducted. It is necessary to correct this permeability to that for 20°C by multiplying the computed value by the ratio of viscosity of water at the test temperature to viscosity of water at 20°C (see Table 17–2).

NUMERICAL EXAMPLE

A laboratory test was conducted according to the procedure described previously. The following data were obtained:

Diameter of specimen, D = **10.16 cm**

Length between manometer outlets, L = **11.43 cm**

Height, H_1 = **20.0 cm**

Height, H_2 = **4.5 cm**

Table 17-2   Viscosity Corrections for n_r/n_{20} [2]

°C	0	0.1	0.2	0.3	0.4	0.5	0.6	0.7	0.8	0.9
10	1.3012	1.2976	1.2940	1.2903	1.2867	1.2831	1.2795	1.2759	1.2722	1.2686
11	1.2650	1.2615	1.2580	1.2545	1.2510	1.2476	1.2441	1.2406	1.2371	1.2336
12	1.2301	1.2268	1.2234	1.2201	1.2168	1.2135	1.2101	1.2068	1.2035	1.2001
13	1.1968	1.1936	1.1905	1.1873	1.1841	1.1810	1.1777	1.1746	1.1714	1.1683
14	1.1651	1.1621	1.1590	1.1560	1.1529	1.1499	1.1469	1.1438	1.1408	1.1377
15	1.1347	1.1318	1.1289	1.1260	1.1231	1.1202	1.1172	1.1143	1.1114	1.1085
16	1.1056	1.1028	1.0999	1.0971	1.0943	1.0915	1.0887	1.0859	1.0803	1.0802
17	1.0774	1.0747	1.0720	1.0693	1.0667	1.0640	1.0613	1.0586	1.0560	1.0533
18	1.0507	1.0480	1.0454	1.0429	1.0403	1.0377	1.0351	1.0325	1.0300	1.0274
19	1.0248	1.0223	1.0198	1.0174	1.0149	1.0124	1.0099	1.0074	1.0050	1.0025
20	1.0000	0.9976	0.9952	0.9928	0.9904	0.9881	0.9857	0.9833	0.9809	0.9785
21	0.9761	0.9738	0.9715	0.9692	0.9669	0.9646	0.9623	0.9600	0.9577	0.9554
22	0.9531	0.9509	0.9487	0.9465	0.9443	0.9421	0.9399	0.9377	0.9355	0.9333
23	0.9311	0.9290	0.9268	0.9247	0.9225	0.9204	0.9183	0.9161	0.9140	0.9118
24	0.9097	0.9077	0.9056	0.9036	0.9015	0.8995	0.8975	0.8954	0.8934	0.8913
25	0.8893	0.8873	0.8853	0.8833	0.8813	0.8794	0.8774	0.8754	0.8734	0.8714
26	0.8694	0.8675	0.8656	0.8636	0.8617	0.8598	0.8579	0.8560	0.8540	0.8521
27	0.8502	0.8484	0.8465	0.8447	0.8428	0.8410	0.8392	0.8373	0.8355	0.8336
28	0.8318	0.8300	0.8282	0.8264	0.8246	0.8229	0.8211	0.8193	0.8175	0.8157
29	0.8139	0.8122	0.8105	0.8087	0.8070	0.8053	0.8036	0.8019	0.8001	0.7984
30	0.7967	0.7950	0.7934	0.7917	0.7901	0.7884	0.7867	0.7851	0.7834	0.7818
31	0.7801	0.7785	0.7769	0.7753	0.7737	0.7721	0.7705	0.7689	0.7673	0.7657
32	0.7641	0.7626	0.7610	0.7595	0.7579	0.7564	0.7548	0.7533	0.7517	0.7502
33	0.7486	0.7471	0.7456	0.7440	0.7425	0.7410	0.7395	0.7380	0.7364	0.7349
34	0.7334	0.7320	0.7305	0.7291	0.7276	0.7262	0.7247	0.7233	0.7218	0.7204
35	0.7189	0.7175	0.7161	0.7147	0.7133	0.7120	0.7106	0.7092	0.7078	0.7064

For water content determination:

Mass of air-dried soil plus can = **299.41 g**

Mass of oven-dried soil plus can = **295.82 g**

Mass of can = **59.39 g**

Mass of air-dried soil before compaction, M_1 = **3,200.0 g**

Mass of unused remaining portion of air-dried soil after compaction, M_2 = **1,102.5 g**

Manometer readings, h_1 and h_2 = **10.9 cm** and **5.4 cm** (same manometer readings for each of three trials)

Quantity of water discharged, Q = **250 cm^3** (same quantity for each of three trials)

Total time of discharge, t = **65 s, 63 s, 64 s** (respective values for each of three trials)

Water temperature, T = **23°C, 24°C, 24°C** (respective values for each of three trials)

Additionally, specific gravity of the solids is known to be **2.71** from a previous test.

[A] Unit Weight Determination

With the diameter of the soil specimen known, the cross-sectional area A can be determined:

$$A = \frac{\pi D^2}{4} = \frac{\pi (\mathbf{10.16})^2}{4} = 81.07 \text{ cm}^2$$

With H_1 and H_2 known, the height of the specimen can be determined, and then its volume can be found:

$$\text{Height of specimen} = H_1 - H_2 = \mathbf{20.0} - \mathbf{4.5} = 15.5 \text{ cm}$$

$$\text{Volume of specimen} = (15.5)(81.07) = 1{,}256.6 \text{ cm}^3$$

With M_1 and M_2 known, the mass of the soil specimen (air dried) can be determined, and then its unit weight can be found:

$$\text{Mass of soil specimen (air dried)} = \mathbf{3{,}200.0} - \mathbf{1{,}102.5} = 2{,}097.5 \text{ g}$$

$$\text{Unit weight of soil specimen (air dried)} = \frac{2{,}097.5}{1{,}256.6} \times 62.4$$

$$= 104.2 \text{ lb/ft}^3$$

The water content of the air-dried soil can be computed as follows:

$$\text{Mass of water} = \mathbf{299.41} - \mathbf{295.82} = 3.59 \text{ g}$$

$$\text{Mass of oven-dried soil} = \mathbf{295.82} - \mathbf{59.39} = 236.43 \text{ g}$$

$$\text{Water content} = \frac{3.59}{236.43} \times 100 = 1.52\%$$

The dry unit weight of the soil specimen can then be found:

$$\text{Dry unit weight} = \frac{104.2}{1.52 + 100} \times 100 = 102.6 \text{ lb/ft}^3$$

To determine the volume of solids and of voids to be used in computing the void ratio, it is convenient to consider 1 ft^3 of dry soil. This cubic foot of soil consists of only solid material and air. Because the weight of air is negligible, the weight of the solid material will be 102.6 lb. The volume of solid material can be computed by dividing this weight by the unit weight of the solid material. The unit weight of the solid material can be obtained by multiplying specific gravity of the solids by the unit weight of water (62.4 lb/ft^3). Hence,

$$\text{Volume of solids} = \frac{102.6}{(\mathbf{2.71})(62.4)} = 0.6067 \text{ ft}^3$$

Subtracting this volume from 1.0 ft^3 gives the volume of air, which is the same as the volume of voids (because no water is present). Hence,

$$\text{Volume of voids} = 1 - 0.6067 = 0.3933 \text{ ft}^3$$

The void ratio e, which is the ratio of volume of voids to volume of solids, can now be computed:

$$e = \frac{0.3933}{0.6067} = 0.648$$

[B] Permeability Test

For trial no. 1 of the permeability test, the head, h, is $\mathbf{10.9} - \mathbf{5.4}$, or 5.5 cm; the quantity of water discharged, Q, is $\mathbf{250}$ $\mathbf{cm}^3$; and the total time of discharge, t, is $\mathbf{65}$ $\mathbf{s}$. Furthermore, the length between manometer outlets, L, was determined to be $\mathbf{11.43}$ $\mathbf{cm}$, and area, A, was computed to be 81.07 cm^2. With these data known, the coefficient of permeability, k, can be computed using Eq. (17–3):

$$k = \frac{QL}{Ath} \tag{17–3}$$

$$k = \frac{(\mathbf{250})(\mathbf{11.43})}{(81.07)(\mathbf{65})(5.5)} = 0.0986 \text{ cm/s}$$

To correct for permeability at 20°C, the ratio of viscosity at 23°C to that at 20°C is determined from Table 17–2 to be 0.9311. The permeability at 20°C is therefore

$$k_{20°C} = (0.0986)(0.9311) = 0.0918 \text{ cm/s}$$

In like manner, the value of $k_{20°C}$ for trials no. 2 and 3 can be calculated to be 0.0925 and 0.0911 cm/s, respectively. Taking an average gives

$$\text{Average } k_{20°C} = \frac{0.0918 + 0.0925 + 0.0911}{3} = 0.0918 \text{ cm/s}$$

These results, together with the initial data, are summarized in the form on pages 255 and 256. At the end of the chapter, two blank copies of this form are included for the reader's use.

GRAPH As mentioned previously, because permeability varies with void ratio, it is recommended that permeability tests be performed on three different specimens of the same soil sample, with each specimen having a different void ratio. A relationship between void ratio and permeability can be established for a given soil by plotting, on semilogarithmic graph paper, a graph of permeability (logarithmic scale) versus void ratio (arithmetic scale). This graph will usually approximate a straight line. Then, whenever the permeability of that soil *in situ* is needed, a sample can be taken at the project site, its void ratio determined, and the associated permeability obtained from the graphical relationship.

Such a graph of permeability versus void ratio is shown in Figure 17–3. The graph was obtained by plotting the results of three different permeability tests corresponding to three different void ratios on three different specimens of the same soil sample. The plotted point labeled "A" represents the results of the permeability test related in the section "Numerical Example." The other plotted points represent the results (not covered here) of two other permeability tests at different void ratios on the same sample.

In practice (such as in a commercial laboratory), it will, of course, be necessary to perform three permeability tests at different void ratios on the same sample. In a college laboratory, however, a single laboratory period is generally not enough time for students to perform the required three tests. To expedite matters, it is recommended that each student group take part of the sample, compact it to a dry density different from that of any other group (and therefore to a different void ratio), and perform a permeability test on it. When all groups have completed their tests, the results (permeability and void ratio) can be pooled to plot the graph of permeability versus void ratio.

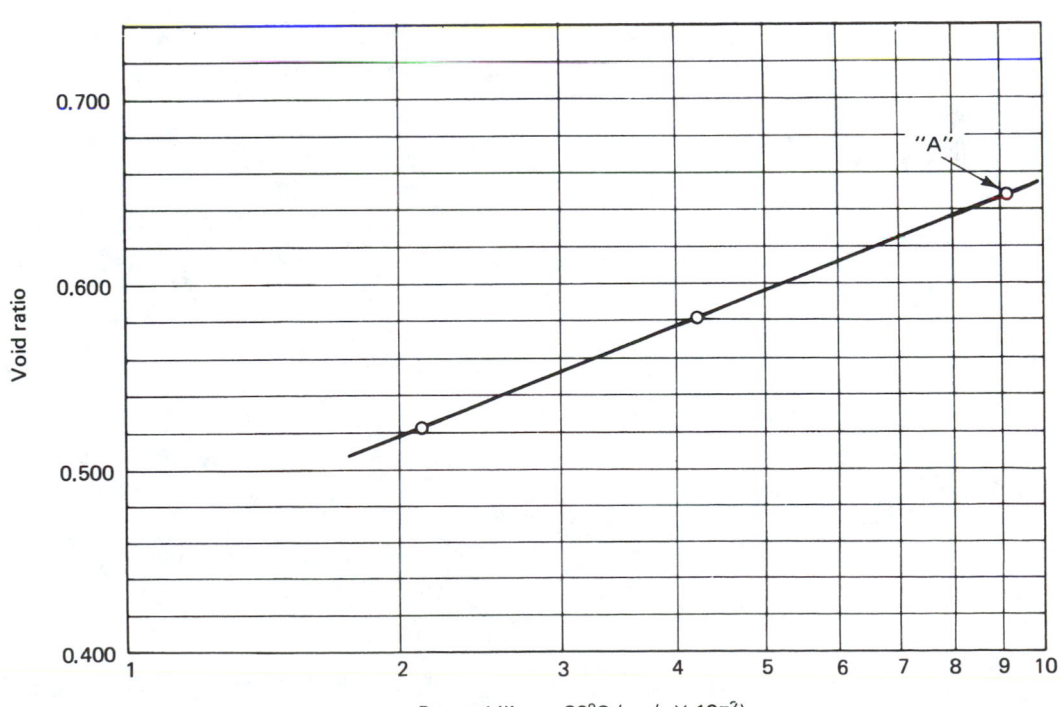

FIGURE 17–3 Permeability versus Void Ratio Curve

CONCLUSIONS Comprehensive reports of permeability tests normally include not only a graph of permeability versus void ratio for the given soil, but also other pertinent information, such as sample identifications, densities and dry unit weights of tested specimens, and so on. Generally, these data are contained in the form on page 255.

It should be noted that permeability determined in a laboratory may not be truly indicative of *in situ* permeability. There are several reasons for this, in addition to the fact that soil in the permeameter does not exactly duplicate the condition of soil *in situ,* at least not for granular soils. For one thing, flow of water in a permeameter is downward, whereas flow in soil *in situ* may be more nearly horizontal or in a direction between horizontal and vertical. Indeed, permeability of natural soils in the horizontal direction can be considerably greater than in the vertical direction. For another thing, naturally occurring strata in *in situ* soils will not be duplicated in a permeameter. Also, the relatively smooth walls of a permeameter afford different boundary conditions from *in situ* soil. Finally, the hydraulic head in a permeameter may differ from the field gradient.

Another concern is any effect on the permeability test from entrapped air in the water. To avoid this, water to be used in the test should be de-aired by boiling distilled water and keeping it covered and nonagitated until used.

REFERENCES [1] ASTM, *2001 Annual Book of ASTM Standards,* West Conshohocken, PA, 2001. Copyright, American Society for Testing and Materials, 100 Barr Harbor Drive, West Conshohocken, PA 19428-2959. Reprinted with permission.

[2] Joseph E. Bowles, *Engineering Properties of Soils and Their Measurement,* 2d ed., McGraw-Hill Book Company, New York, 1978.

Soils Testing Laboratory
Permeability Test:
Constant-Head Method

Sample No. _____2_____   Project No. _____SR1820_____

Location _____Southport_____   Boring No. _____B-5_____

Depth of Sample _____3 ft_____   Date of Test _____5/26/02_____

Tested by _____John Doe_____

Description of Soil _____Brown sand with trace of mica_____

[A] Unit Weight Determination

(1) Length between manometer outlets, L __11.43__ cm
(2) Diameter of soil specimen, D __10.16__ cm

(3) Cross-sectional area of soil specimen $\left[\text{i.e., } \dfrac{\pi D^2}{4}\right]$ __81.07__ cm^2

(4) Height, H_1 __20.0__ cm
(5) Height, H_2 __4.5__ cm
(6) Height of specimen (i.e., $H_1 - H_2$) __15.5__ cm
(7) Volume of specimen [(3) × (6)] __1,256.6__ cm^3
(8) Mass of air dried soil before compaction, M_1 __3,200__ g
(9) Mass of unused remaining portion of soil after compaction, M_2 __1,102.5__ g
(10) Mass of soil specimen (air dried) [(8) − (9)] __2,097.5__ g

(11) Unit weight of soil specimen (air-dried) $\left[\dfrac{(10)}{(7)} \times 62.4\right]$ __104.2__ lb/ft^3

(12) Water content of air dried soil __1.52__ %
    (a) Can no. __3-B__
    (b) Mass of air-dried soil + can __299.41__ g
    (c) Mass of oven-dried soil + can __295.82__ g
    (d) Mass of can __59.39__ g
    (e) Mass of water __3.59__ g
    (f) Mass of oven-dried soil __236.43__ g

    (g) Water content $\left[\dfrac{(e)}{(f)} \times 100\right]$ __1.52__ %

(13) Dry unit weight of soil specimen $\left[\dfrac{(11)}{(12) + 100} \times 100\right]$ __102.6__ lb/ft^3

(14) Specific gravity of soil __2.71__

(15) Volume of solids $\left[\dfrac{(13)}{(14) \times 62.4}\right]$ __0.6067__ ft^3

(16) Volume of voids [1 − (15)] __0.3933__ ft^3

(17) Void ratio, e, of soil specimen $\left[\dfrac{(16)}{(15)}\right]$ __0.648__

[B] Permeability Test

(1) Length between manometer outlets, L __11.43__ cm

(2) Cross-sectional area of soil specimen, A __81.07__ ·cm²

Trial No.	Manometer Readings		Head, h (cm)	Quantity of Water Discharged, Q (cm³)	Time, t (s)	Temper-ature, T (°C)	Permeability at T°C, k_T (cm/s)	Ratio of Viscosity at T°C / Viscosity at 20°C	Permeability at 20°C, k_{20} (cm/s)
	h_1 (cm)	h_2 (cm)							
(1)	(2)	(3)	(4) = (2) − (3)	(5)	(6)	(7)	(8) $= \dfrac{QL}{Ath}$	(9)	(10) = (8) × (9)
1	10.9	5.4	5.5	250	65	23	0.0986	0.9311	0.0918
2	10.9	5.4	5.5	250	63	24	0.1017	0.9097	0.0925
3	10.9	5.4	5.5	250	64	24	0.1001	0.9097	0.0911

Average $k_{20} = \frac{1}{3}(0.0918 + 0.0925 + 0.0911)$

$\qquad\qquad = 0.0918$ cm/s

$\qquad$ or 9.18×10^{-2} cm/s

Soils Testing Laboratory
Permeability Test:
Constant-Head Method

Sample No. _____

Project No. _____

Location _____

Boring No. _____

Depth of Sample _____

Date of Test _____

Tested by _____

Description of Soil _____

[A] Unit Weight Determination

(1) Length between manometer outlets, L _____ cm

(2) Diameter of soil specimen, D _____ cm

(3) Cross-sectional area of soil specimen $\left[\text{i.e., } \dfrac{\pi D^2}{4} \right]$ _____ cm^2

(4) Height, H_1 _____ cm

(5) Height, H_2 _____ cm

(6) Height of specimen (i.e., $H_1 - H_2$) _____ cm

(7) Volume of specimen [(3) × (6)] _____ cm^3

(8) Mass of air dried soil before compaction, M_1 _____ g

(9) Mass of unused remaining portion of soil after compaction, M_2 _____ g

(10) Mass of soil specimen (air dried) [(8) − (9)] _____ g

(11) Unit weight of soil specimen (air dried) $\left[\dfrac{(10)}{(7)} \times 62.4 \right]$ _____ lb/ft^3

(12) Water content of air dried soil _____ %

  (a) Can no. _____

  (b) Mass of air-dried soil + can _____ g

  (c) Mass of oven-dried soil + can _____ g

  (d) Mass of can _____ g

  (e) Mass of water _____ g

  (f) Mass of oven-dried soil _____ g

  (g) Water content $\left[\dfrac{(e)}{(f)} \times 100 \right]$ _____ %

(13) Dry unit weight of soil specimen $\left[\dfrac{(11)}{(12) + 100} \times 100 \right]$ _____ lb/ft^3

(14) Specific gravity of soil _____

(15) Volume of solids $\left[\dfrac{(13)}{(14) \times 62.4} \right]$ _____ ft^3

(16) Volume of voids [1 − (15)] _____ ft^3

(17) Void ratio, e, of soil specimen $\left[\dfrac{(16)}{(15)} \right]$ _____

[B] Permeability Test

(1) Length between manometer outlets, L _____ cm

(2) Cross-sectional area of soil specimen, A _____ cm^2

Trial No.	Manometer Readings		Head, h (cm)	Quantity of Water Discharged, Q (cm³)	Time, t (s)	Temper-ature, T (°C)	Permeability at T°C, k_T (cm/s)	Ratio of Viscosity at T°C	Permeability at 20°C, k_{20} (cm/s)
	h_1 (cm)	h_2 (cm)						Viscosity at 20°C	
(1)	(2)	(3)	(4) = (2) − (3)	(5)	(6)	(7)	$(8) = \dfrac{QL}{Ath}$	(9)	(10) = (8) × (9)

Soils Testing Laboratory
Permeability Test:
Constant-Head Method

Sample No. _____   Project No. _____

Location _____   Boring No. _____

Depth of Sample _____   Date of Test _____

Tested by _____

Description of Soil _____

[A] Unit Weight Determination

(1) Length between manometer outlets, L _____ cm

(2) Diameter of soil specimen, D _____ cm

(3) Cross-sectional area of soil specimen $\left[\text{i.e., } \dfrac{\pi D^2}{4}\right]$ _____ cm²

(4) Height, H_1 _____ cm

(5) Height, H_2 _____ cm

(6) Height of specimen (i.e., $H_1 - H_2$) _____ cm

(7) Volume of specimen [(3) × (6)] _____ cm³

(8) Mass of air-dried soil before compaction, M_1 _____ g

(9) Mass of unused remaining portion of soil after compaction, M_2 _____ g

(10) Mass of soil specimen (air dried) [(8) − (9)] _____ g

(11) Unit weight of soil specimen (air dried) $\left[\dfrac{(10)}{(7)} \times 62.4\right]$ _____ lb/ft³

(12) Water content of air dried soil _____ %

    (a) Can no. _____

    (b) Mass of air-dried soil + can _____ g

    (c) Mass of oven-dried soil + can _____ g

    (d) Mass of can _____ g

    (e) Mass of water _____ g

    (f) Mass of oven-dried soil _____ g

    (g) Water content $\left[\dfrac{(e)}{(f)} \times 100\right]$ _____ %

(13) Dry unit weight of soil specimen $\left[\dfrac{(11)}{(12) + 100} \times 100\right]$ _____ lb/ft³

(14) Specific gravity of soil _____

(15) Volume of solids $\left[\dfrac{(13)}{(14) \times 62.4}\right]$ _____ ft³

(16) Volume of voids [1 − (15)] _____ ft³

(17) Void ratio, e, of soil specimen $\left[\dfrac{(16)}{(15)}\right]$ _____

[B] Permeability Test

(1) Length between manometer outlets, L _____ cm

(2) Cross-sectional area of soil specimen, A _____ cm^2

Trial No.	Manometer Readings h_1 (cm)	Manometer Readings h_2 (cm)	Head, h (cm)	Quantity of Water Discharged, Q (cm^3)	Time, t (s)	Temperature, T (°C)	Permeability at T°C, k_T (cm/s)	Ratio of Viscosity at T°C / Viscosity at 20°C	Permeability at 20°C, k_{20} (cm/s)
(1)	(2)	(3)	(4) = (2) − (3)	(5)	(6)	(7)	(8) $= \dfrac{QL}{Ath}$	(9)	(10) = (8) × (9)

18

CHAPTER EIGHTEEN

Permeability Test for Fine-Grained and Granular Soils (Falling-Head Method)

INTRODUCTION The general definition, theory, influencing factors, and so on for permeability as a soil parameter were given in the introduction to Chapter 17, which described the constant-head method for determining permeability. As indicated in that chapter, the constant-head method is applicable only to granular soils. The falling-head method, covered in this chapter, may be used to determine the permeability of both fine-grained soils (such as silts and clays) and coarse-grained, or granular, soils.

APPARATUS AND SUPPLIES

Permeability device

Ring stand with test tube clamp

Burette (100 mL)

Timing device

Thermometer

A falling-head permeability setup, including the first three items, is shown schematically in Figure 18–1.

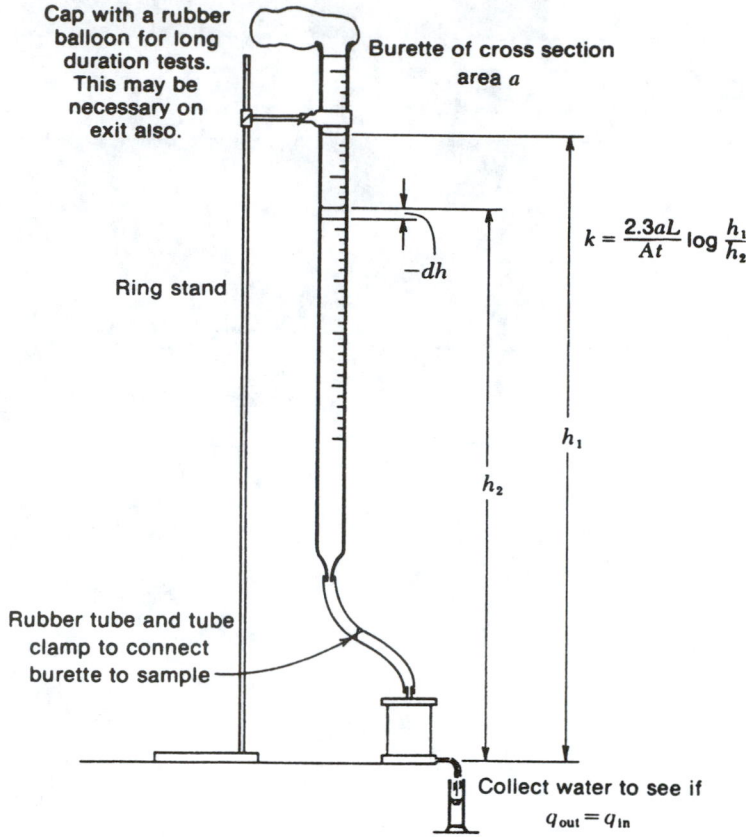

Cap with a rubber balloon for long duration tests. This may be necessary on exit also.

Burette of cross section area a

$$k = \frac{2.3aL}{At} \log \frac{h_1}{h_2}$$

Ring stand

$-dh$

h_1

h_2

Rubber tube and tube clamp to connect burette to sample

Collect water to see if $q_{out} = q_{in}$

FIGURE 18–1   Schematic of the Falling-Head Permeability Setup [1]

PROCEDURE   Fabricate or compact the soil sample to form three specimens with different dry densities. Calculate the dry density of each specimen by finding the mass of the permeameter and of the permeameter plus compacted soil and taking a sample for water content determination. Also, measure each specimen's diameter and length.

After obtaining these preliminary data, permeability tests can be performed. The general test procedure does not differ a great deal from the constant-head method. The soil specimen is first saturated with water. Water is then allowed to move through the specimen under a falling-head condition (rather than a stable-head condition), while the time required for a certain quantity of water to pass through the specimen is measured and recorded. Using these data together with others described previously, one can determine the coefficient of permeability.

A total of three permeability tests should be performed, one for each test specimen with different dry densities. Each specimen's void ratio can be computed using the specific gravity of the soil and the corresponding dry density. A curve of permeability k versus void ratio e can then be plotted on semilogarithmic paper, with void ratio on the arithmetic scale and permeability on the logarithmic scale.

The actual step-by-step procedure is as follows [1]:

(1) Weigh the permeameter (mold) with base plate and gasket attached. Measure the inside diameter of the permeameter (mold). Note that the area a of the standpipe must be evaluated. Since the burette is graduated in cubic centimeters, measuring the distance between graduations will yield a as a simple, direct computation. Take a small portion of the soil sample for water content determination.

(2) Place the air-dried soil sample into the permeameter (mold) and compact it to a desirable density. Weigh the permeameter (mold) with base plate and gasket attached plus compacted soil and also measure the length of the specimen in centimeters. Determine the dry density and void ratio of the specimen.

(3) Place a piece of porous disk on the top of the specimen and a spring on the porous disk. Carefully clean the permeameter (mold) rim. With its gasket in place, press down the top (cover) plate against the spring and attach it securely to the top of the permeameter cylinder, making an air-tight seal. The spring should be compressed and should apply a pressure to the compacted soil specimen to keep it in place when it is saturated with water.

(4) Place the permeameter in a sink in which the water is about 2 in. above the cover. Be sure the outlet pipe is open so that water can back up through the specimen. This procedure will saturate the sample with a minimum amount of entrapped air. When water in the plastic inlet tube on top of the mold reaches equilibrium with water in the sink (allowing for capillary rise in the tube), the specimen may be assumed to be saturated. (A soaking period of 24 h might provide better results.)

(5) With the water level stabilized in the inlet tube of the permeability mold, take a hose clamp and clamp the exit tube. Remove the permeameter from the sink and attach it to the rubber tube at the base of the burette, which has been fastened to a ring stand (Figure 18–1). Fill the burette with water from a supply, which should be temperature-stabilized (and de-aired if desired).

(6) Now de-air the lines at the top of the specimen by opening the hose clamp from the burette and opening the petcock on top of the cover plate. Allow water to flow (but keep adding water to the burette so it does not become empty) from the petcock. When no more air comes out, close the petcock. Do not close the inlet tube from the burette. Remember that the exit tube is still clamped shut.

(7) Fill the burette to a convenient height, and measure the hydraulic head across the sample to obtain h_1 (see Figure 18–1).

(8) Open the exit tube (and petcock) and simultaneously start timing the test. Allow water to flow through the sample until the burette is almost empty. Simultaneously record the elapsed time and clamp only the exit tube. Measure the hydraulic head across the

sample at this time to obtain h_2 (see Figure 18–1). Take the temperature each time.

(9) Refill the burette and repeat (8) two additional times. Take the temperature each time.

(10) To check on whether the sample is saturated, one may collect the water coming out of the exit tube and compare this volume with that entering the sample. Obviously, if

$$q_{out} < q_{in}$$

the specimen was not saturated.

DATA Data collected during this test consist of the following:

Mass of permeameter (mold) with base plate and gasket attached (g)

Mass of permeameter (mold) with base plate and gasket attached plus soil (g)

Length of specimen, L (cm)

Diameter of specimen, D (cm)

For water content determination:

Mass of air-dried soil plus can (g)

Mass of oven-dried soil plus can (g)

Mass of can (g)

Cross-sectional area of standpipe (burette), a (cm^2)

Hydraulic head at beginning of test [see "Procedure," step (7), and Figure 18–1], h_1 (cm)

Hydraulic head at end of test [see "Procedure," step (8), and Figure 18–1], h_2 (cm)

Total time for water in burette to drop from h_1 to h_2, t (s)

Temperature of water, T (°C)

CALCULATIONS **[A] Unit Weight Determination**

Values of the unit weight of the air-dried soil specimen, water content of the air-dried soil, dry unit weight of the soil specimen, and void ratio can be computed in the same manner as related in Chapter 17.

[B] Permeability Test

The coefficient of permeability can be computed using equation [1]

$$k = \frac{2.3 \, aL}{At} \log \frac{h_1}{h_2} \qquad\qquad (18\text{--}1)$$

where:

$\qquad k$ = coefficient of permeability, cm/s

$\qquad a$ = cross-sectional area of standpipe (burette), cm^2

$\qquad L$ = length of specimen, cm

$\qquad A$ = cross-sectional area of soil specimen, cm^2

$\qquad h_1$ = hydraulic head at beginning of test, cm

$\qquad h_2$ = hydraulic head at end of test, cm

$\qquad t$ = total time for water in burette to drop from h_1 to h_2, s

The permeability computed using Eq. (18–1) is the value for the particular water temperature at which the test was conducted. It is necessary to correct this permeability to that for 20°C by multiplying the computed value by the ratio of viscosity of water at the test temperature to viscosity of water at 20°C (see Table 17–2, page 250).

NUMERICAL EXAMPLE

A laboratory test was conducted according to the procedure described previously. The following data were obtained:

Mass of permeameter (mold) with base plate and gasket attached = **1,098.5 g**

Mass of permeameter (mold) with base plate and gasket attached plus soil = **3,401.1 g**

Length of specimen, L = **15.80 cm**

Diameter of specimen, D = **10.16 cm**

For water content determination:

$\qquad$ Mass of air-dried soil plus can = **308.17 g**

$\qquad$ Mass of oven-dried soil plus can = **305.40 g**

$\qquad$ Mass of can = **53.60 g**

Cross-sectional area of standpipe (burette), a = **1.83 cm^2**

Hydraulic head at beginning of test, h_1 = **150.0 cm** (same reading for each of three trials)

Hydraulic head at end of test, h_2 = **20.0 cm** (same reading for each of three trials)

Total times for water in burette to drop from h_1 to h_2, t = **32.3 s, 32.6 s, 31.7 s** (respective values for each of three trials)

Water temperature, T = **22°C** (same value for each of three trials)

Additionally, the specific gravity of the solids is known to be **2.71** from a previous test.

[A] Unit Weight Determination

With the soil specimen's diameter known, its cross-sectional area A can be calculated:

$$A = \frac{\pi D^2}{4} = \frac{\pi (10.16)^2}{4} = 81.07 \text{ cm}^2$$

With the length of the soil specimen known, its volume can be computed:

$$\text{Volume of specimen} = (81.07)(15.80) = 1{,}280.9 \text{ cm}^3$$

With the mass of the permeameter and of the permeameter plus soil known, the air-dried soil specimen's mass and then its unit weight can be determined:

$$\text{Mass of air-dried soil specimen} = 3{,}401.1 - 1{,}098.5 = 2{,}302.6 \text{ g}$$

$$\text{Unit weight of air} - \text{dried soil specimen} = \frac{2{,}302.6}{1{,}280.9} \times 62.4 = 112.2 \text{ lb/ft}^3$$

The water content of the air-dried soil sample can be determined as follows:

$$\text{Mass of water} = 308.17 - 305.40 = 2.77 \text{ g}$$

$$\text{Mass of oven-dried soil} = 305.40 - 53.60 = 251.80 \text{ g}$$

$$\text{Water content} = \frac{2.77}{251.80} \times 100 = 1.10\%$$

The dry unit weight of the soil specimen can now be calculated:

$$\text{Dry unit weight} = \frac{112.2}{1.10 + 100} \times 100 = 111.0 \text{ lb}ft^3$$

The volume of solids, volume of voids, and void ratio can be obtained as follows:

$$\text{Volume of solids} = \frac{111.0}{(2.71)(62.4)} = 0.6564 \text{ ft}^3$$

$$\text{Volume of voids} = 1 - 0.6564 = 0.3436 \text{ ft}^3$$

$$\text{Void ratio}, e = \frac{0.3436}{0.6564} = 0.523$$

(For a more complete explanation of the determination of void ratio, see the section "Numerical Example" in Chapter 17.)

[B] Permeability Test

For trial no. 1 of the permeability test, the standpipe's cross-sectional area, a, is **1.83 cm^2**; length of specimen, L, is **15.80 cm**; specimen's cross-sectional area, A, is **81.07 cm^2**; total time, t, for the water in the standpipe (burette) to drop from h_1 to h_2 is **32.3 s**; hydraulic head at the beginning of the test, h_1, is **150.0 cm**; and hydraulic head at the end of the test, h_2, is **20.0 cm**. With these data known, coefficient of permeability k can be computed using Eq. (18–1):

$$k = \frac{2.3\, aL}{At} \log \frac{h_1}{h_2} \qquad\qquad (18\text{–}1)$$

$$k = \frac{(2.3)(1.83)(15.80)}{(81.07)(32.3)} \log \frac{150.0}{20.0} = 2.222 \times 10^{-2} \text{ cm/s}$$

To correct for permeability at 20°C, the ratio of the viscosity of water at **22°C** to that at 20°C is determined from Table 17–2 to be 0.9531. The permeability at 20°C is therefore

$$k_{20°C} = (2.222 \times 10^{-2})(0.9531) = 2.118 \times 10^{-2} \text{ cm/s}$$

In like manner, calculation of $k_{20°C}$ for trials no. 2 and 3 yields 2.099×10^{-2} and 2.158×10^{-2} cm/s, respectively. Taking an average gives

$$\text{Average } k_{20°C} = \frac{(2.118 + 2.099 + 2.158) \times 10^{-2}}{3} = 2.125 \times 10^{-2} \text{ cm/s}$$

These results, together with the initial data, are summarized in the form on pages 269 and 270. At the end of the chapter, two blank copies of this form are included for the reader's use.

GRAPH As explained in Chapter 17, a graph, on semilogarithmic paper, of permeability (logarithmic scale) versus void ratio (arithmetic scale) is normally included in the report of the results of a permeability test. See the section "Graph" of Chapter 17 for a complete explanation.

CONCLUSIONS As indicated previously, the falling-head method for determining soil permeability may be used for both fine-grained and coarse-grained soils. Table 18–1 gives some general information with regard to permeability and drainage characteristics of soils. A permeability of 10^{-4} cm/s may be considered borderline between pervious and impervious soils. A soil with permeability less than 10^{-4} cm/s might thus be considered for a dam core or impervious blanket, whereas one with permeability greater than 10^{-4} cm/s might be used for a dam shell or pervious backfill [2].

Table 18–1 Permeability and Drainage Characteristics of Soils[a] [3]

	Coefficient of Permeability k (cm/s) (log scale)											
	10^2 10^1 1.0 10^{-1} 10^{-2} 10^{-3} 10^{-4} 10^{-5} 10^{-6} 10^{-7} 10^{-8} 10^{-9}											
Drainage	Good						Poor		Practically Impervious			
Soil Types	Clean gravel	Clean sands, clean sand and gravel mixtures		Very fine sands, organic and inorganic silts, mixtures of sand, silt, and clay, glacial till, stratified clay deposits, etc.					"Impervious" soils, e.g., homogeneous clays below zone of weathering			
				"Impervious" soils modified by effects of vegetation and weathering								
Direct Determination of k	Direct testing of soil in its original position—pumping tests; reliable if properly conducted; considerable experience required											
	Constant-head permeameter; little experience required											
Indirect Determination of k		Falling-head permeameter; reliable; little experience required		Falling-head permeameter; unreliable; much experience required			Falling-head permeameter; fairly reliable; considerable experience necessary					
	Computation from grain-size distribution; applicable only to clean cohesionless sands and gravels								Computation based on results of consolidation tests; reliable; considerable experience required			

[a]After Casagrande and Fadum (1940).

It should be emphasized again that permeability determined in a laboratory may not be truly indicative of *in situ* permeability. Several reasons for this were cited in the section "Conclusions" at the end of Chapter 17.

REFERENCES

[1] Joseph E. Bowles, *Engineering Properties of Soils and Their Measurement*, McGraw-Hill Book Company, New York, 1970.

[2] T. William Lambe, *Soil Testing for Engineers*, John Wiley & Sons, Inc., New York, 1951.

[3] Karl Terzaghi and Ralph B. Peck, *Soil Mechanics in Engineering Practice*, 2d ed., John Wiley & Sons, Inc., New York, 1967.

Soils Testing Laboratory
Permeability Test:
Falling-Head Method

Sample No. _____3_____   Project No. _____SR 1820_____

Location _____Southport_____   Boring No. _____B-6_____

Depth of Sample _____4 ft_____   Date of Test _____5/27/02_____

Tested by_____John Doe_____

Description of Soil _____Light brown sand with some mica_____

[A] Unit Weight Determination

(1) Mass of permeameter (mold) with base plate and gasket attached + soil
___3,401.1___ g

(2) Mass of empty permeameter with base plate and gasket attached
___1,098.5___ g

(3) Mass of soil specimen [(1) − (2)] ___2,302.6___ g

(4) Diameter of soil specimen, D [i.e., inside diameter of permeameter (mold)]
___10.16___ cm

(5) Cross-sectional area of soil specimen $\left[\text{i.e., } \dfrac{\pi D^2}{4}\right]$ ___81.07___ cm^2

(6) Length of soil specimen in permeameter, L ___15.80___ cm

(7) Volume of soil specimen [(5) × (6)] ___1,280.9___ cm^3

(8) Unit weight of soil specimen (air dried) $\left[\dfrac{(3)}{(7)} \times 62.4\right]$ ___112.2___ lb/ft^3

(9) Water content of soil specimen (air dried) ___1.10___ %

    (a) Can no. ___3-C___

    (b) Mass of air-dried soil + can ___308.17___ g

    (c) Mass of oven-dried soil + can ___305.40___ g

    (d) Mass of can ___53.60___ g

    (e) Mass of water ___2.77___ g

    (f) Mass of oven-dried soil ___251.80___ g

    (g) Water content $\left[\dfrac{(e)}{(f)} \times 100\right]$ ___1.10___ %

(10) Dry unit weight of soil specimen $\left[\dfrac{(8)}{(9) + 100} \times 100\right]$ ___111.0___ lb/ft^3

(11) Specific gravity of soil ___2.71___

(12) Volume of solids $\left[\dfrac{(10)}{62.4 \times (11)}\right]$ ___0.6564___ ft^3

(13) Volume of voids [1 − (12)] ___0.3436___ ft^3

(14) Void ratio, e, of soil specimen $\left[\dfrac{(13)}{(12)}\right]$ ___0.523___

[B] Permeability Test

(1) Cross-sectional area of standpipe, a ___1.83___ cm^2
(2) Length of soil specimen in permeameter, L ___15.80___ cm
(3) Cross-sectional area of soil specimen, A ___81.07___ cm^2

Trial no.	Head, h_1 (cm)	Head, h_2 (cm)	Time, t (s)	Temperature, T (°C)	Permeability at T°C, k_T (cm/s)	Ratio of Viscosity at T°C / Viscosity at 20°C	Permeability at 20°C, k_{20} (cm/s)
(1)	(2)	(3)	(4)	(5)	$(6) = \dfrac{2.3aL}{At} \log \dfrac{h_1}{h_2}$	(7)	(8)
						(see Table 17–2)	
1	150.0	20.0	32.3	22	2.222×10^{-2}	0.9531	2.118×10^{-2}
2	150.0	20.0	32.6	22	2.202×10^{-2}	0.9531	2.099×10^{-2}
3	150.0	20.0	31.7	22	2.264×10^{-2}	0.9531	2.158×10^{-2}

$$\text{Average } k_{20°C} = \tfrac{1}{3}(2.118 + 2.099 + 2.158) \times 10^{-2}$$

$$k_{20°C} = 2.125 \times 10^{-2} \text{ cm/s}$$

Soils Testing Laboratory
Permeability Test:
Falling-Head Method

Sample No. _____   Project No. _____

Location _____   Boring No. _____

Depth of Sample _____   Date of Test _____

Tested by_____

Description of Soil _____

[A] Unit Weight Determination

(1) Mass of permeameter (mold) with base plate and gasket attached + soil
_____ g

(2) Mass of empty permeameter with base plate and gasket attached
_____ g

(3) Mass of soil specimen [(1) − (2)] _____ g

(4) Diameter of soil specimen, D [i.e., inside diameter of permeameter (mold)]
_____ cm

(5) Cross-sectional area of soil specimen $\left[\text{i.e., } \dfrac{\pi D^2}{4}\right]$ _____ cm^2

(6) Length of soil specimen in permeameter, L _____ cm

(7) Volume of soil specimen [(5) × (6)] _____ cm^3

(8) Unit weight of soil specimen (air dried) $\left[\dfrac{(3)}{(7)} \times 62.4\right]$ _____ lb/ft^3

(9) Water content of soil specimen (air dried) _____ %
(a) Can no. _____
(b) Mass of air-dried soil + can _____ g
(c) Mass of oven-dried soil + can _____ g
(d) Mass of can _____ g
(e) Mass of water _____ g
(f) Mass of oven-dried soil _____ g

(g) Water content $\left[\dfrac{(e)}{(f)} \times 100\right]$ _____ %

(10) Dry unit weight of soil specimen $\left[\dfrac{(8)}{(9)+100} \times 100\right]$ _____ lb/ft^3

(11) Specific gravity of soil _____

(12) Volume of solids $\left[\dfrac{(10)}{62.4 \times (11)}\right]$ _____ ft^3

(13) Volume of voids [1 − (12)] _____ ft^3

(14) Void ratio, e, of soil specimen $\left[\dfrac{(13)}{(12)}\right]$ _____

271

[B] Permeability Test

(1) Cross-sectional area of standpipe, a _____ cm^2

(2) Length of soil specimen in permeameter, L _____ cm

(3) Cross-sectional area of soil specimen, A _____ cm^2

Trial no.	Head, h_1 (cm)	Head, h_2 (cm)	Time, t (s)	Temperature, T (°C)	Permeability at T °C, k_T (cm/s)	Ratio of $\dfrac{\text{Viscosity at T °C}}{\text{Viscosity at 20°C}}$	Permeability at 20°C, k_{20} (cm/s)
(1)	(2)	(3)	(4)	(5)	$(6) = \dfrac{2.3aL}{At} \log \dfrac{h_1}{h_2}$	(7)	(8)
						(see Table 17–2)	

Soils Testing Laboratory
Permeability Test:
Falling-Head Method

Sample No. _____   Project No. _____

Location _____   Boring No. _____

Depth of Sample _____   Date of Test _____

Tested by_____

Description of Soil _____

[A] Unit Weight Determination

(1) Mass of permeameter (mold) with base plate and gasket attached + soil
_____ g

(2) Mass of empty permeameter with base plate and gasket attached
_____ g

(3) Mass of soil specimen [(1) − (2)] _____ g

(4) Diameter of soil specimen, D [i.e., inside diameter of permeameter (mold)]
_____ cm

(5) Cross-sectional area of soil specimen $\left[\text{i.e., } \dfrac{\pi D^2}{4} \right]$ _____ cm^2

(6) Length of soil specimen in permeameter, L _____ cm

(7) Volume of soil specimen [(5) × (6)] _____ cm^3

(8) Unit weight of soil specimen (air dried) $\left[\dfrac{(3)}{(7)} \times 62.4 \right]$ _____ lb/ft^3

(9) Water content of soil specimen (air dried) _____ %

    (a) Can no. _____

    (b) Mass of air-dried soil + can _____ g

    (c) Mass of oven-dried soil + can _____ g

    (d) Mass of can _____ g

    (e) Mass of water _____ g

    (f) Mass of oven-dried soil_____ g

    (g) Water content $\left[\dfrac{(e)}{(f)} \times 100 \right]$ _____ %

(10) Dry unit weight of soil specimen $\left[\dfrac{(8)}{(9) + 100} \times 100 \right]$ _____ lb/ft^3

(11) Specific gravity of soil _____

(12) Volume of solids $\left[\dfrac{(10)}{62.4 \times (11)} \right]$ _____ ft^3

(13) Volume of voids [1 − (12)] _____ ft^3

(14) Void ratio, e, of soil specimen $\left[\dfrac{(13)}{(12)} \right]$ _____

[B] Permeability Test

(1) Cross-sectional area of standpipe, a _____ cm^2

(2) Length of soil specimen in permeameter, L _____ cm

(3) Cross-sectional area of soil specimen, A _____ cm^2

Trial no.	Head, h_1 (cm)	Head, h_2 (cm)	Time, t (s)	Temperature, T (°C)	Permeability at T°C, k_T (cm/s)	Ratio of Viscosity at T°C / Viscosity at 20°C	Permeability at 20°C, k_{20} (cm/s)
(1)	(2)	(3)	(4)	(5)	$(6) = \dfrac{2.3aL}{At} \log \dfrac{h_1}{h_2}$	(7)	(8)
						(see Table 17–2)	

CHAPTER NINETEEN

Consolidation Test

(Referenced Document: ASTM D 2435)

INTRODUCTION When structures are built on saturated soil, the load is presumed to be carried initially by incompressible water within the soil. Because of additional load on the soil, water will tend to be extruded from voids in the soil, causing a reduction in void volume and settlement of a structure. In soils of high permeability (coarse-grained soils), this process requires a short time interval for completion, with the result that almost all of the settlement has occurred by the time construction is complete. However, in soils of low permeability (fine-grained soils, particularly clayey soils), the process requires a long time interval for completion, with the result that strain occurs very slowly. Thus, settlement takes place slowly and continues over a long period of time.

The phenomenon of compression due to very slow extrusion of water from the voids in a fine-grained soil as a result of increased loading (such as the weight of a structure on a soil) is known as *consolidation*. Associated settlement is referred to as *consolidation settlement*. It is important to be able to predict both the rate and magnitude of the consolidation settlement of structures. Actual prediction (calculation) of consolidation settlement, however, is beyond the scope of this book; as this is a laboratory manual, its purpose is to cover laboratory procedures necessary to obtain results that are needed to estimate the rate and magnitude of settlement of structures on clayey soils.

The method presented in this chapter covers procedures for determining the magnitude and rate of consolidation of soil when it is restrained laterally and drained axially while subjected to incrementally applied, controlled-stress loading. Two alternative procedures are provided. *Test Method A* is performed with constant load increment duration of 24 h, or multiples thereof. Time-deformation readings are required on a minimum of two load increments. In *Test Method B,* time-deformation readings are required on all load increments. Successive load increments are applied after 100% primary consolidation is reached, or at constant time increments as described in Test Method A. The determination of the rate and magnitude of consolidation of soil when it is subjected to controlled-strain loading is covered by ASTM Test Method D 4186.

The test method is most commonly performed on undisturbed samples of fine-grained soils naturally sedimented in water; however, the basic test procedure is applicable as well to specimens of compacted soils and undisturbed samples of soils formed by other processes, such as weathering or chemical alteration. Evaluation techniques specified in this test method are generally applicable to soils naturally sedimented in water. Tests performed on other soils such as compacted and residual (weathered or chemically altered) soils may require special evaluation techniques.

It is the responsibility of the agency requesting this test to specify the magnitude and sequence of each load increment, including the location of a rebound cycle, if required and, for Test Method A, the load increments for which time-deformation readings are desired.

In this test method a soil specimen is restrained laterally and loaded axially with total stress increments. Each stress increment is maintained until excess pore water pressures are completely dissipated. During the consolidation process, measurements are made of change in the specimen height, and these data are used to determine the relationship between the effective stress and void ratio or strain and the rate at which consolidation can occur by evaluating the coefficient of consolidation. [1]

APPARATUS AND SUPPLIES

Load device: a suitable device for applying vertical loads to a specimen (see Figures 19–1 and 19–2)

Consolidometer: a standardized device to hold the sample in a ring that is either fixed to the base or floating. The consolidometer must also provide a means of submerging the sample, applying a vertical load, and measuring the change in thickness of the sample (Figures 19–1 and 19–2).

Porous stone

Moisture room

Specimen ring with cutting edge attached (minimum specimen diameter: 50 mm (2.00 in.); minimum initial specimen height: 12 mm (0.5 in.), but not less than ten times the maximum particle diameter; minimum specimen diameter-to-height ratio: 2.5; to minimize the effects of friction between the specimen's sides and the ring, the

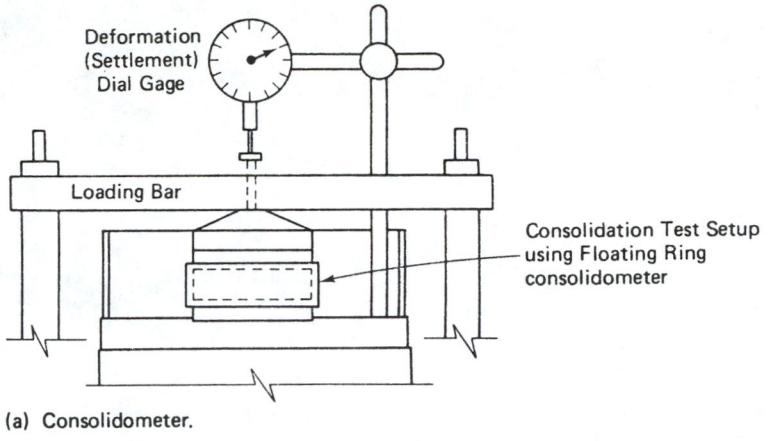

(a) Consolidometer.

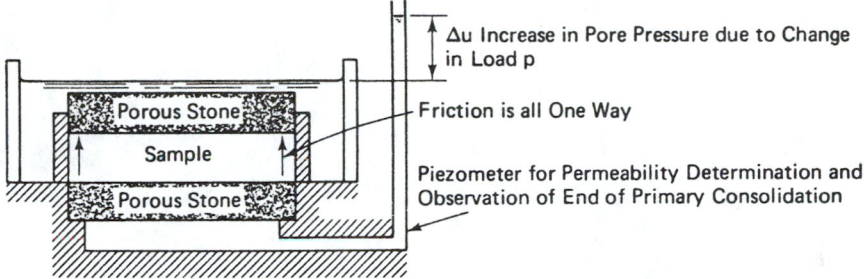

(b) Fixed-Ring Consolidometer. May be used to Obtain Permeability Information during a Consolidation Test if a Piezometer is Installed.

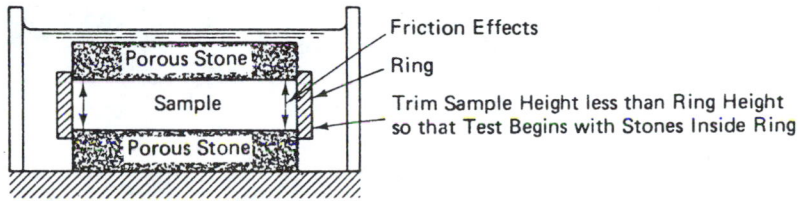

(c) Floating Ring Consolidometer.

FIGURE 19–1   Consolidometers [2]

use of greater diameter-to-height ratios is recommended, with ratios greater than four being preferable.)

Trimming equipment: wire saw, sharp-edged knife, etc.

Balance (with accuracy to 0.1 g)

Drying oven

Extensometer (with accuracy to 0.0001 in. or 0.0025 mm)

Miscellaneous equipment: containers, spatulas, knives, wire saw, etc.

SAMPLING [1]

(1) ASTM Practices D 1587 and D 3550 cover procedures and apparatus that may be used to obtain undisturbed samples generally satisfactory for testing. Specimens may also be trimmed from large

FIGURE 19–2   Consolidation
Device (Courtesy of Soiltest, Inc.)

undisturbed block samples fabricated and sealed in the field.
Finally, remolded specimens may be prepared from bulk samples to
density and moisture conditions stipulated by the agency requesting the test.

(2) Undisturbed samples destined for testing in accordance with
this test method shall be preserved, handled, and transported in accordance with the practices for Group C and D samples in ASTM
Practices D 4220. Bulk samples for remolded specimens should be
handled and transported in accordance with the practice for Group
B samples.

(3) *Storage*—Storage of sealed samples should be such that no
moisture is lost during storage, that is, no evidence of partial drying of the ends of the samples or shrinkage. Time of storage should
be minimized, particularly when the soil or soil moisture is expected to react with the sample tubes.

(4) The quality of consolidation test results diminishes greatly
with sample disturbance. It should be recognized that no sampling
procedure can ensure completely undisturbed samples. Therefore,
careful examination of the sample is essential in selection of specimens for testing.

Note 1—Examination for sample disturbance, stones, or other inclusions, and selection of specimen location is greatly facilitated by x-ray radiography of the samples (see ASTM Methods D 4452).

SPECIMEN PREPARATION [1]

(1) All possible precautions should be taken to minimize disturbance of the soil or changes in moisture and density during specimen preparation. Avoid vibration, distortion, and compression.

(2) Prepare test specimens in an environment where soil moisture change during preparation is minimized.

Note 2—A high humidity environment is usually used for this purpose.

(3) Trim the specimen and insert it into the consolidation ring. When specimens come from undisturbed soil collected using sample tubes, the inside diameter of the tube shall be at least 5 mm (0.25 in.) greater than the inside diameter of the consolidation ring, except as noted in (4) and (5). It is recommended that either a trimming turntable or cylindrical cutting ring be used to cut the soil to the proper diameter. When using a trimming turntable, make a complete perimeter cut, reducing the specimen diameter to the inside diameter of the consolidation ring. Carefully insert the specimen into the consolidation ring, by the width of the cut, with a minimum of force. Repeat until the specimen protrudes from the bottom of the ring. When using a cylindrical cutting ring, trim the soil to a gentle taper in front of the cutting edge. After the taper is formed, advance the cutter a small distance to form the final diameter. Repeat the process until the specimen protrudes from the ring.

(4) Fibrous soils, such as peat, and those soils that are easily damaged by trimming, may be transferred directly from the sampling tube to the ring, provided that the ring has the same diameter as the sample tube.

(5) Specimens obtained using a ring-lined sampler may be used without prior trimming, provided they comply with the requirements of ASTM Practice D 3550 and this test method.

(6) Trim the specimen flush with the plane ends of the ring. The specimen may be recessed slightly below the top of the ring to facilitate centering of the top stone by partial extrusion and trimming of the bottom surface. For soft to medium soils, a wire saw should be used for trimming the top and bottom of the specimen to minimize smearing. A straightedge with a sharp cutting edge may be used for the final trim after the excess soil has first been removed with a wire saw. For stiff soils, a sharpened straightedge alone may be used for trimming the top and bottom. If a small particle is encountered in any surface being trimmed, it should be removed and the resulting void filled with soil from the trimmings.

Note 3—If, at any stage of the test, the specimen swells beyond its initial height, the requirement of lateral restraint of the soil dictates the use of a recessed specimen or the use of a specimen ring equipped with an extension collar of the same inner diameter as the specimen ring. At no time should the specimen extend beyond the specimen ring or extension collar.

(7) Determine the initial wet mass of the specimen, M_{To}, in the consolidation ring by measuring the mass of the ring with specimen and subtracting the tare mass of the ring.

(8) Determine the initial height, H_o, of the specimen to the nearest 0.025 mm (0.001 in.) by taking the average of at least four evenly spaced measurements over the top and bottom surfaces of the specimen using a dial comparator or other suitable measuring device.

(9) Compute the initial volume, V_o, of the specimen to the nearest 0.25 cm^3 (0.015 in.^{3}) from the diameter of the ring and the initial specimen height.

(10) Obtain two or three natural water content determinations of the soil in accordance with ASTM Method D 2216 (Chapter 3) from material trimmed adjacent to the test specimen if sufficient material is available.

(11) When index properties are specified by the requesting agency, store the remaining trimmings taken from around the specimen and determined to be similar material in a sealed container for determination as described in the next section.

SOIL INDEX PROPERTY DETERMINATIONS [1]

(1) The determination of index properties is an important adjunct to but not a requirement of the consolidation test. These determinations when specified by the requesting agency should be made on the most representative material possible. When testing uniform materials, all index tests may be performed on adjacent trimmings collected in (11) of the "Specimen Preparation" section. When samples are heterogeneous or trimmings are in short supply, index tests should be performed on material from the test specimen as obtained in (6) of the "Procedure" section, plus representative trimmings collected in (11) of the "Specimen Preparation" section.

(2) *Specific Gravity*—The specific gravity shall be determined in accordance with ASTM Test Method D 854 (Chapter 5) on material from the sample as specified in (1). The specific gravity from another sample judged to be similar to that of the test specimen may be used for calculation of volume of solid whenever an accurate void ratio is not needed.

(3) *Atterberg Limits*—The liquid limit, plastic limit, and plasticity index shall be determined in accordance with ASTM Test Method D 4318 (Chapters 6 and 7) using material from the sample as spec-

ified in (1). Determinations of the Atterberg limits are necessary for proper material classification but are not a requirement of this test method.

(4) *Particle Size Distribution*—The particle size distribution shall be determined in accordance with ASTM Method D 422 (Chapter 9) (except the minimum sample size requirement shall be waived) on a portion of the test specimen as obtained in (6) of the "Procedure" section. A particle size analysis may be helpful when visual inspection indicates that the specimen contains a substantial fraction of coarse-grained material but is not a requirement of this test method.

PROCEDURE

The first step is to place an undisturbed soil specimen in the consolidometer. One porous stone is placed above the specimen, another below it. The purpose of the porous stones is to allow water to flow into and out of the specimen. This assembly is immersed in water. As load is applied to the upper stone, the specimen is compressed, and deformation is measured by a dial gage.

To begin a particular test, a specific pressure (e.g., 500 lb/ft^2) is applied to the specimen, and deformation dial readings with corresponding time observations are made and recorded until deformation has nearly ceased. Normally, this is done over a 24-h period. From these data, a graph known as the *time curve* is prepared on semilogarithmic graph paper, with time along the abscissa on the logarithmic scale and dial readings along the ordinate on the arithmetic scale.

The foregoing procedure is repeated after doubling the applied pressure, giving another graph of time versus deformation dial readings corresponding to the new pressure. The procedure is then repeated for additional doublings of applied pressure until the final applied pressure is in excess of the total pressure to which the compressible clay formation is expected to be subjected when the proposed structure is built.

Through careful evaluation of each graph of time versus deformation dial readings, it is possible to determine the void ratio e and coefficient of consolidation c_v that correspond to the specific applied pressure or loading p for that graph. With these data, two additional graphs can be prepared: one of void ratio versus logarithm of pressure ($e - \log p$ curve), with pressure along the abscissa on the logarithmic scale and void ratio along the ordinate on the arithmetic scale, and another of coefficient of consolidation versus logarithm of pressure ($c_v - \log p$ curve), with pressure along the abscissa on the logarithmic scale and coefficient of consolidation along the ordinate on the arithmetic scale. The $e - \log p$ curve is used to evaluate the magnitude of settlement, and the $c_v - \log p$ curve is used to estimate the time rate of settlement.

The primary results of a laboratory consolidation test, therefore, are (1) $e - \log p$ curve, (2) $c_v - \log p$ curve, and (3) initial void ratio e_0 of the soil *in situ*.

The actual step-by-step procedure is as follows (ASTM D 2435-96 [1]):

(1) Preparation of the porous disks and other apparatus will depend on the specimen being tested. The consolidometer must be assembled in such a manner as to prevent a change in water content of the specimen. Dry porous disks and filters must be used with dry, highly expansive soils and may be used for all other soils. Damp disks may be used for partially saturated soils. Saturated disks may be used when the specimen is saturated and known to have a low affinity for water. Assemble the ring with specimen, porous disks, filter disks (when needed) and consolidometer. If the specimen will not be inundated shortly after application of the seating load [see (2)], enclose the consolidometer in a loose fitting plastic or rubber membrane to prevent change in specimen volume due to evaporation.

> *Note 4*—In order to meet the stated objectives of this test method, the specimen must not be allowed to swell in excess of its initial height prior to being loaded beyond its preconsolidation pressure. Detailed procedures for the determination of one-dimensional swell or settlement potential of cohesive soils is covered by ASTM Test Method D 4546.

(2) Place the consolidometer in the loading device and apply a seating pressure of 5 kPa (100 lb/ft^2). Immediately after application of the seating load, adjust the deformation indicator and record the initial zero reading, d_0. If necessary, add additional load to keep the specimen from swelling. Conversely, if it is anticipated that a load of 5 kPa (100 lb/ft^2) will cause significant consolidation of the specimen, reduce the seating pressure to 2 or 3 kPa (about 50 lb/ft^2) or less.

(3) If the test is performed on an intact specimen that was either saturated under field conditions or obtained below the water table, inundate shortly after application of the seating load. As inundation and specimen wetting occur, increase the load as required to prevent swelling. Record the load required to prevent swelling and the resulting deformation reading. If specimen inundation is to be delayed to simulate specific conditions, then inundation must occur at a pressure that is sufficiently large to prevent swell. In such cases, apply the required load and inundate the specimen. Take time deformation readings during the inundation period as specified in (5). In such cases, note in the test report the pressure at inundation and the resulting changes in height.

(4) The specimen is to be subjected to increments of constant total stress. The duration of each increment shall conform to guidelines specified in (5). The specific loading schedule will depend on the purpose of the test, but should conform to the following guidelines. If the slope and shape of a virgin compression curve or determination of the preconsolidation pressure is required, the final pressure

shall be equal to or greater than four times the preconsolidation pressure. In the case of overconsolidated clays, a better evaluation of recompression parameters may be obtained by imposing an unload-reload cycle after the preconsolidation pressure has been defined. Details regarding location and extent of an unload-reload cycle are the option of the agency requesting the test; however, unloading shall always span at least two decrements of pressure.

(4.1) The standard loading schedule shall consist of a load increment ratio (LIR) of one which is obtained by doubling the pressure on the soil to obtain values of approximately 12, 25, 50, 100, 200, etc. kPa (250, 500, 1000, 2000, 4000, etc. lb/ft^2).

(4.2) The standard rebound or unloading schedule should be selected by halving the pressure on the soil (that is, use the same increments of (4.1), but in reverse order). However, if desired, each successive load can be only one-fourth as large as the preceding load; that is, skip a decrement.

(4.3) An alternative loading, unloading, or reloading schedule may be employed that reproduces the construction stress changes or obtains better definition of some part of the stress deformation (compression) curve, or aids in interpreting the field behavior of the soil.

> *Note 5*—Small increments may be desirable on highly compressible specimens or when it is desirable to determine the preconsolidation pressure with more precision. It should be cautioned, however, that load increment ratios less than 0.7 and load increments very close to the preconsolidation pressure may preclude evaluation for the coefficient of consolidation, c_v, and the end-of-primary consolidation.

(5) Before each pressure increment is applied, record the height or change in height, d_f, of the specimen. Two alternative procedures are available that specify the time sequence of readings and the required minimum load duration. Longer durations are often required during specific load increments to define the slope of the characteristic straight line secondary compression portion of the deformation versus log of time graph. For such increments, sufficient readings should be taken near the end of the pressure increment to define this straight-line portion. It is not necessary to increase the duration of other pressure increments during the test.

(5.1) *Test Method A*—The standard load increment duration shall be 24 h. For at least two load increments, including at least one load increment after the preconsolidation pressure has been exceeded, record the height or change in height, d, at time intervals of approximately 0.1, 0.25, 0.5, 1, 2, 4, 8, 15 and 30 min, and 1, 2, 4, 8, and 24 h, measured from the time of each incremental pressure application. Take sufficient readings near the end of the pressure increment period to verify that primary consolidation is completed. For some soils, a period of more than 24 h may be required to reach the end-of-primary consolidation. In such cases, load increment durations greater than 24 h are required. The load increment duration for these tests is usually taken at some multiple of 24 h and should be the standard duration for all load increments of the test. The decision to use a time interval greater than 24 h is usually based on experience with particular types of soils. If, however, there is a question as to whether a 24-h period is adequate, a record of height or change in height with time should be made for the initial load increments in order to verify the adequacy of a 24-h period. Load increment durations other than 24 h shall be noted in the report. For pressure increments where time-versus-deformation data are not required, leave the load on the specimen for the same length of time as when time-versus-deformation readings are taken.

(5.2) *Test Method B*—For each increment, record the height or change in height, *d,* at time intervals of approximately 0.1, 0.25, 0.5, 1, 2, 4, 8, 15, 30 min, and 1, 2, 4, 8 and 24 h, measured from the time of each incremental pressure application. The standard load increment duration shall exceed the time required for completion of primary consolidation as determined by [B] of the "Calculations" section or a criterion set by the requesting agency. For each increment where it is impossible to verify the end of primary consolidation (for example, low LIR or rapid consolidation), the load increment duration shall be constant and exceed the time required for primary consolidation of an increment applied after the preconsolidation pressure and along the virgin compression curve. Where secondary compression must be evaluated, apply pressures for longer periods. The report shall contain the load increment duration for each increment.

> *Note 6*—The suggested time intervals for recording height or change in height are for typical soils and load increments. It is often desirable to change the reading frequency to improve interpretation of the data. More rapid consolidation will require more frequent readings. For most soils, primary consolidation during the first load decrements will be complete in less time (typically one-tenth) than would be required for a load increment along the virgin compression curve. However, at very low stresses the rebound time can be longer.

(6) To minimize swell during disassembly, rebound the specimen back to the seating load (5 kPa). Once height changes have ceased (usually overnight), dismantle quickly after releasing the final small load on the specimen. Remove the specimen and the ring from the consolidometer and wipe any free water from the ring and specimen. Determine the mass of the specimen in the ring and subtract the tare mass of the ring to obtain the final wet specimen mass, M_{Tf}. The most accurate determination of the specimen dry

mass and water content is found by drying the entire specimen at the end of the test. If the soil sample is homogeneous and sufficient trimmings are available for the specified index testing, then determine the final water content, w_f, in accordance with ASTM Method D 2216 (Chapter 3) and dry mass of solids, M_d, using the entire specimen. If the soil is heterogeneous or more material is required for the specified index testing, then determine the final water content, w_{fp}, in accordance with Method D 2216 using a small wedge-shaped section of the specimen. The remaining undried material should be used for the specified index testing.

DATA Data collected in the consolidation test should include the following:

[A] Specimen Data

At beginning of test:

Diameter of specimen, D (in.)

Initial height of specimen, H_0 (in.)

Mass of specimen ring plus specimen (g)

Mass of specimen ring (g)

At end of test:

Mass of entire wet specimen plus can (g)

Mass of entire dry specimen plus can (g)

Mass of can (g)

[B] Time-versus-Deformation Data

Successive readings of deformation dial, in inches, as a function of time, in minutes. (As explained in the section "Procedure," separate series of time-versus-deformation data are obtained for a number of different loadings at different applied pressures.)

CALCULATIONS [A] Specimen Parameters

At beginning of test:

1. *Initial wet unit weight:* With the specimen's diameter (D) and initial height (H_0) known, its initial volume (V_0) can be computed. The initial wet mass of the specimen (M_{T0}) can be determined by subtracting the mass of the specimen ring from the mass of the specimen ring plus specimen; and with the initial wet mass and volume known, the initial wet

density, ρ_0, can be calculated by dividing mass by volume. The initial wet unit weight, γ_0, in kN/m^3 can be found by multiplying the initial wet density in g/cm^3 by 9.81; the initial wet unit weight in lb/ft^3 can be determined by multiplying the initial wet density in g/cm^3 by 62.4. In equation form,

$$\rho_0 = \frac{M_{T0}}{V_0} \text{ (in g/cm}^3) \qquad \textbf{(19–1)}$$

$$\gamma_0 \text{ (in kN/m}^3) = \rho_0 \times 9.81 \qquad \textbf{(19–2)}$$

$$\gamma_0 \text{ (in lb/ft}^3) = \rho_0 \times 62.4 \qquad \textbf{(19–3)}$$

2. *Initial moisture content:* The initial moisture content, w_0, can be obtained by dividing mass of water ($M_{T0} - M_d$) by dry mass of total specimen, M_d dry soil. That is

$$w_0 = \frac{M_{T0} - M_d}{M_d} \times 100 \qquad \textbf{(19–4)}$$

3. *Initial degree of saturation:* The initial degree of saturation, S_0, can be computed by dividing volume of water in the soil by volume of void in the specimen, both values obtained at the beginning of the test. That is,

$$S_0 = \frac{M_{T0} - M_d}{A\rho_w(H_0 - H_s)} \times 100 \qquad \textbf{(19–5)}$$

where:
$\quad A$ = specimen's area (cm^2)
$\quad \rho_w$ = density of water (1 g/cm^3)
$\quad H_s$ = height of solid = $\dfrac{V_s}{A}$

where:
$\quad V_s$ = volume of solid = $\dfrac{M_d}{G_s\rho_w}$

where:
$\quad G_s$ = specific gravity of solids

At end of test:

The final moisture content, w_f, can be found by dividing the final mass of water in the specimen ($M_{Tf} - M_d$) by the final dry mass of the specimen (M_d). That is,

$$w_f = \frac{M_{Tf} - M_d}{M_d} \times 100 \qquad \textbf{(19–6)}$$

Initial void ratio:

The initial void ratio, e_0, can be computed by dividing initial volume of void in the specimen, V_v, by volume of solid in the specimen (V_s). That is,

$$e_0 = \frac{V_v}{V_s} = \frac{A(H_0 - H_s)}{AH_s}$$

or

$$e_0 = \frac{H_0 - H_s}{H_s} \qquad\qquad \textbf{(19--7)}$$

[B] Time versus Deformation

1. From those increments of load where time-versus-deformation readings are obtained, plot deformation readings versus logarithm of time (in minutes) for each increment of load or pressure as the test progresses and for any increments of rebound where time-versus-deformation data have been obtained. See Figure 19–3.

2. Find the deformation representing 100% primary consolidation (d_{100}) for each load increment. First, draw a straight line through the points representing final readings that exhibit a straight-line trend and flat slope. Then, draw a second straight line tangent to the steepest part of the curve of deformation versus the logarithm of time. The intersection of these two lines represents the deformation corresponding to 100% primary consolidation. Compression that occurs subsequent to 100% primary consolidation is defined as secondary compression. (See Figure 19–3.)

3. Find the deformation representing 0% primary consolidation (d_0) by selecting deformations at any two times that have a ratio of 1:4. The deformation corresponding to the larger of the two times should be greater than one-fourth but less than one-half of the total change in deformation for the load increment. The deformation corresponding to 0% primary consolidation is equal to the deformation corresponding to the smaller time interval less the difference in the deformations for the two selected times. (See Figure 19–3.)

4. The deformation corresponding to 50% primary consolidation (d_{50}) for each load increment is equal to the average of the deformations corresponding to the 0 and 100% deformations. The time required for 50% consolidation under any load increment may be found graphically from the curve of deformation versus the logarithm of time for that load increment by observing the time that corresponds to 50% of the primary consolidation of the curve [1]. (See Figure 19–3.)

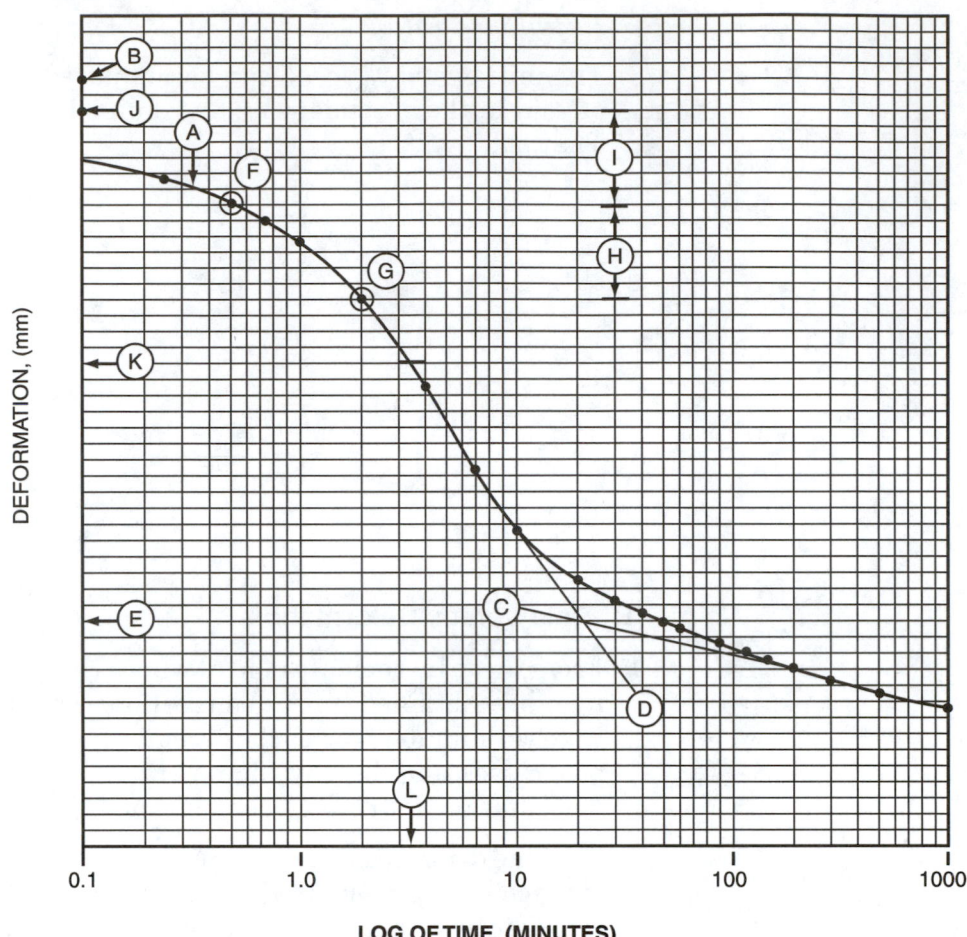

LOG OF TIME, (MINUTES)

A	TIME-DEFORMATION CURVE FROM DATA POINTS
B	DEFORMATION AT TIME = 0 MINUTES
C	EXTENSION OF FINAL LINEAR PORTION OF CURVE
D	EXTENSION OF STEEPEST LINEAR PORTION OF CURVE
E	d_{100} DEFORMATION AT INTERSECTION OF LINES C AND D
F	t_1 SELECTED POINT IN TIME
G	t_2 TIME AT FOUR TIMES t_1 (DEFORMATION AT TIME t_2 SHOULD BE LESS THAN 50% OF THE TOTAL DEFORMATION FOR THE LOAD INCREMENTS)
H	INCREMENT OF DEFORMATION BETWEEN TIMES t_1 AND t_2
I	INCREMENT OF DEFORMATION EQUAL TO H
J	d_0 CALCULATED INITIAL DEFORMATION
K	d_{50} MEAN OF d_0 AND d_{100}
L	t_{50} TIME AT d_{50}

FIGURE 19–3   Time-Deformation Curve From Log of Time Method [1]

[C] Void Ratio

The change in thickness of the specimen for each loading, ΔH, can be computed by subtracting the initial deformation dial reading at the beginning of the very first loading from the deformation dial readings corresponding to 100% primary consolidation for each respective loading. The change in void ratio for each loading, Δe, can then be calculated by dividing each change in thickness of the specimen (ΔH) by the height of solid in the specimen (H_s); hence, $\Delta e = \Delta H / H_s$. The void ratio for each loading, e, can be determined by subtracting the change in void ratio (Δe) from the initial void ratio, e_0; hence, $e = e_0 - \Delta e$.

Values of void ratio for each loading, along with respective applied pressures, can be used to plot the required graph of void ratios versus the logarithm of pressure ($e - \log p$ curve).

[D] Coefficient of Consolidation

For each load increment for which time-versus-deformation readings were obtained, the coefficient of consolidation can be calculated using the equation

$$c_v = \frac{0.196H^2}{t_{50}} \tag{19-8}$$

where:
c_v = coefficient of consolidation, in.^{2}/min
H = half the thickness of the test specimen at 50% consolidation (because the specimen is drained on both top and bottom in this test), in.
t_{50} = time to reach 50% consolidation, min

It should be emphasized that a determination of the coefficient of consolidation must be made for each test loading. The values of coefficient of consolidation for each loading, together with corresponding applied pressures, can be used to plot the required graph of coefficient of consolidation versus the logarithm of pressure ($c_v - \log p$ curve).

NUMERICAL EXAMPLE

A consolidation test was performed in the laboratory, and the following data were obtained:

[A] Specimen Data

At beginning of test:

Diameter of specimen, D = **2.50 in.**

Initial height of specimen, H_0 = **0.780 in.** (i.e., **1.981 cm**)

Mass of specimen ring plus specimen = **208.48 g**

Mass of specimen ring = **100.50 g**

Specific gravity of soil = **2.72**

At end of test:

Mass of entire wet specimen plus can = **234.54 g**

Mass of entire dry specimen plus can = **203.11 g**

Mass of can = **127.17 g**

[B] Time-versus-Deformation Data

Load increment from **0 to 500 lb/ft^2**

Date	Time	Deformation Dial Reading (in.)
6/8/02	9:15 A.M.	0
	9:15.1	0.0067
	9:15.25	0.0069
	9:15.5	0.0071
	9:16	0.0077
	9:17	0.0084
	9:19	0.0095
	9:23	0.0107
	9:30	0.0120
	9:45	0.0132
	10:15	0.0144
	11:15	0.0152
	1:15 P.M.	0.0158
	5:15	0.0160
6/9/02	8:15 A.M.	0.0162
	11:15	0.0162

The time-versus-deformation data given in the preceding table are for a loading increment from **0** to **500 lb/ft^2**. It would, of course, be necessary to have additional sets of time-versus-deformation data for a number of larger loading increments in order to complete a consolidation test. However, in the interest of saving time and space, these additional sets of data are not shown in this example.

All of these (given) data are shown on forms prepared for recording both collected laboratory data and computed results (see pages 298 to 301). The reader is referred to these forms and the explanation of computations that follows, in order to help in understanding the relatively extensive calculations required for a consolidation test. (At the end of the chapter, blank copies of these forms are included for the reader's use.)

[A] Specimen Parameters

At beginning of test:

1. *Initial wet unit weight:*

$$\text{Area } (A) = \frac{\pi D^2}{4} = \frac{\pi \times 2.50^2}{4} = 4.91 \text{ in.}^2 \text{ (i.e., 31.68 cm}^2\text{)}$$

$$\text{Volume } (V_0) = A \times H_0 = (4.91)\,(\mathbf{0.780}) = 3.83 \text{ in.}^3 \text{ (i.e., 62.76 cm}^3\text{)}$$

Initial wet mass of specimen (M_{T0}) =
(mass of specimen ring plus specimen)
− (mass of specimen ring)

$$M_{T0} = \mathbf{208.48} - \mathbf{100.50} = 107.98 \text{ g}$$

Initial wet density:

$$\rho_0 = \frac{M_{T0}}{V_0} \tag{19–1}$$

$$\rho_0 = \frac{M_{T0}}{V_0} = \frac{107.98}{62.76} = 1.721 \text{ g/cm}^3$$

Initial wet unit weight:

$$\gamma_0 \text{ (in lb/ft}^3) = \rho_0 \times 62.4 \tag{19–3}$$

$$\gamma_0 = (1.721)(62.4) = 107.4 \text{ lb/ft}^3$$

2. *Initial moisture content:*

Dry mass of total specimen (M_d)
= (mass of entire dry specimen plus can)
− (mass of can)

$$M_{\text{d}} = \mathbf{203.11} - \mathbf{127.17} = 75.94 \text{ g}$$

$$w_0 = \frac{M_{T0} - M_d}{M_d} \times 100 \tag{19–4}$$

$$w_0 = \frac{107.98 - 75.94}{75.94} \times 100 = 42.2\%$$

3. *Initial degree of saturation:*

$$V_{\text{s}} = \frac{M_{\text{d}}}{G_{\text{s}}\,\rho_w} = \frac{75.94}{(2.72)\,(1)} = 27.92 \text{ cm}^3$$

$$H_{\text{s}} = \frac{V_{\text{s}}}{A} = \frac{27.92}{31.68} = 0.881 \text{ cm}$$

$$S_0 = \frac{M_{T0} - M_d}{A\,\rho_w\,(H_0 - H_s)} \times 100 \qquad (19\text{--}5)$$

$$S_0 = \frac{107.98 - 75.94}{(31.68)(1)(\mathbf{1.981} - 0.881)} \times 100 = 91.9\%$$

At end of test:

$$w_f = \frac{M_{Tf} - M_d}{M_d} \times 100 \qquad (19\text{--}6)$$

$$M_{Tf} = \mathbf{234.54} - \mathbf{127.17} = 107.37 \text{ g}$$

$$M_d = \mathbf{203.11} - \mathbf{127.17} = 75.94 \text{ g}$$

$$w_f = \frac{107.37 - 75.94}{75.94} \times 100 = 41.4\%$$

Initial void ratio:

$$e_0 = \frac{H_0 - H_s}{H_s} \qquad (19\text{--}7)$$

$$e_0 = \frac{\mathbf{1.981} - 0.881}{0.881} = 1.249$$

[B] Time versus Deformation

As indicated previously, a graph of time along the abscissa on a logarithmic scale versus deformation dial readings along the ordinate on an arithmetic scale is essential in order to evaluate the results of a consolidation test. The required graph for the given time-versus-deformation data is presented in Figure 19–4. It must be emphasized again that the curve of the dial readings versus the logarithm of time shown in the figure is for a loading increment from **0** to **500 lb/ft²**. It would be necessary to have additional such curves for a number of larger loading increments; however, these are not shown.

Following the procedure described in the section "Calculations," and referring to Figure 19–4 for a pressure of **500 lb/ft²**, one can observe that the deformations representing 100% and 0% primary consolidations are 0.0158 in. and 0.0058 in., respectively. The deformation corresponding to 50% consolidation is the average of 0.0158 and 0.0058, or 0.0108 in.; and the time required for 50% consolidation (t_{50}) can be determined from the figure to be 8.2 min.

[C] Void Ratio *e*

The change in thickness of the specimen, ΔH, during the **500 lb/ft²** loading can be computed by subtracting the initial dial reading (**0**) from the reading representing 100% primary consolidation of the loading (0.0158). Hence, the change in thickness (ΔH) is 0.0158 in., or 0.0401 cm. The change

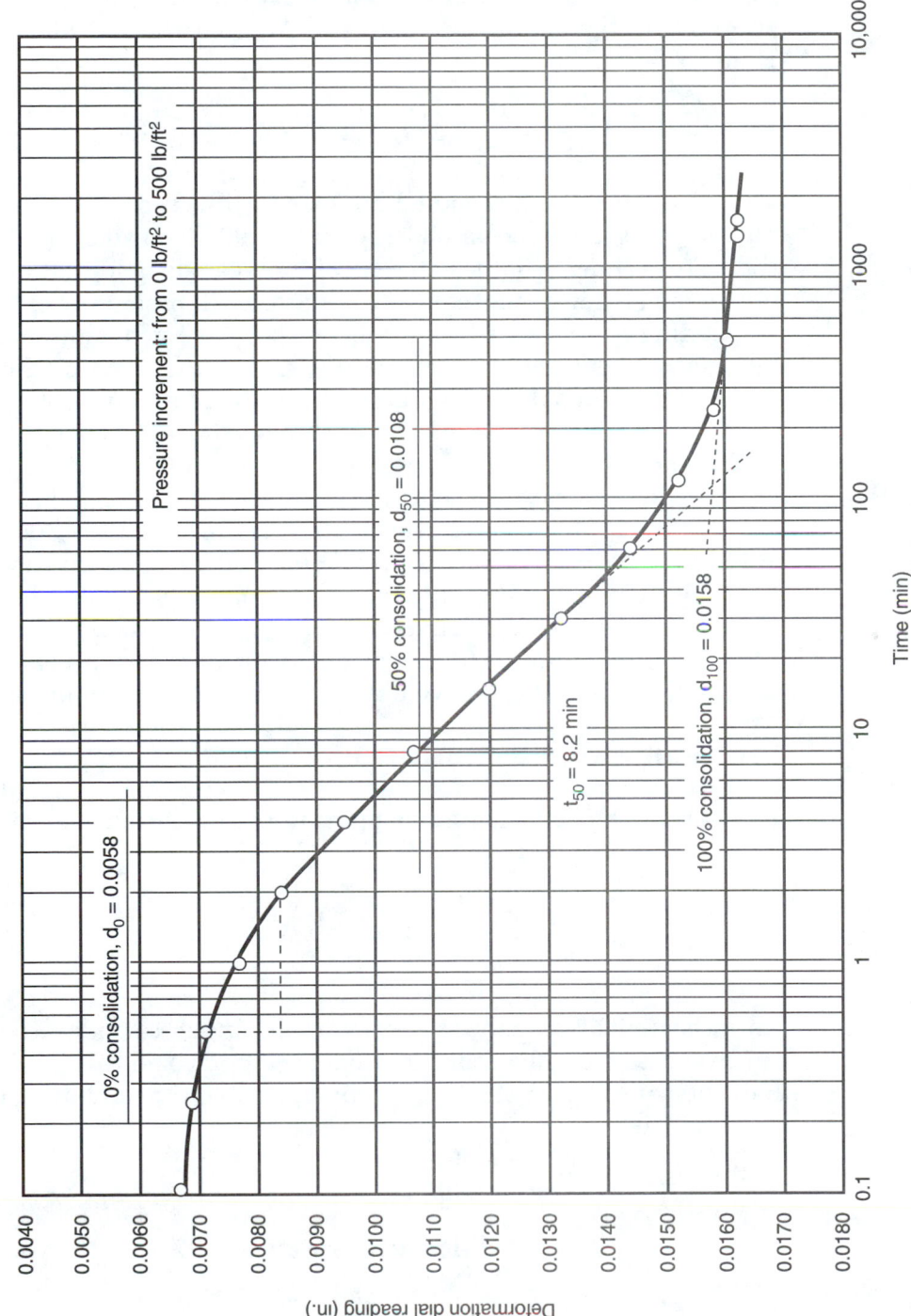

FIGURE 19–4 Plot of Dial Readings versus Logarithm of Time

in void ratio, Δe, can be computed by dividing change in thickness (ΔH) by height of solid in the specimen (H_s). The height of solid in the specimen was determined previously (by dividing volume of solid in the specimen by area of the specimen) to be 0.881 cm. Hence,

$$\Delta e = \frac{\Delta H}{H_s} = \frac{0.0401}{0.881} = 0.046$$

Finally, the void ratio e for the **500 lb/ft²** loading can be determined by subtracting the change in void ratio Δe from the initial void ratio e_0. Hence,

$$e = e_0 - \Delta e = 1.249 - 0.046 = 1.203$$

The values just computed are listed in the second row (corresponding to **500 lb/ft²**) of the data form on page 300. Each additional loading on the test specimen will furnish a set of time-versus-deformation dial readings, which through subsequent evaluation will provide a void ratio e corresponding to a specific loading, or pressure, p. Such additional loadings (and subsequent analyses) would be listed in the remaining rows of the data form. Although no laboratory data are given herein for additional loadings on the test specimen, rows for loadings of **1,000, 2,000, 4,000, 8,000,** and **16,000 lb/ft²** have been included on the form on page 300 in order to demonstrate the complete evaluation of a consolidation test.

Next a graph of pressure (the first column on page 300) along the abscissa on a logarithmic scale versus void ratio (the last column on page 300) along the ordinate on an arithmetic scale can be prepared. This graph, which is known as the $e - \log p$ curve (i.e., void ratio versus the logarithm of pressure), is one of the primary results of a consolidation test. The $e - \log p$ curve for this example is shown in Figure 19–5.

[D] Coefficient of Consolidation, c_v

Recall that the equation for computing the coefficient of consolidation, c_v, is

$$c_v = \frac{0.196H^2}{t_{50}} \tag{19–8}$$

As indicated previously, in the example under consideration, t_{50}, the time required for 50% consolidation, is 8.2 min. The value of H, which is half the thickness of the test specimen at 50% consolidation, can be determined as follows:

$$H = \tfrac{1}{2}[\text{(initial height of specimen at beginning of test)}$$

$$- \text{(deformation dial reading at 50\% consolidation)}]$$

$$H = \tfrac{1}{2}(\textbf{0.780} - 0.0108) = 0.385 \text{ in.}$$

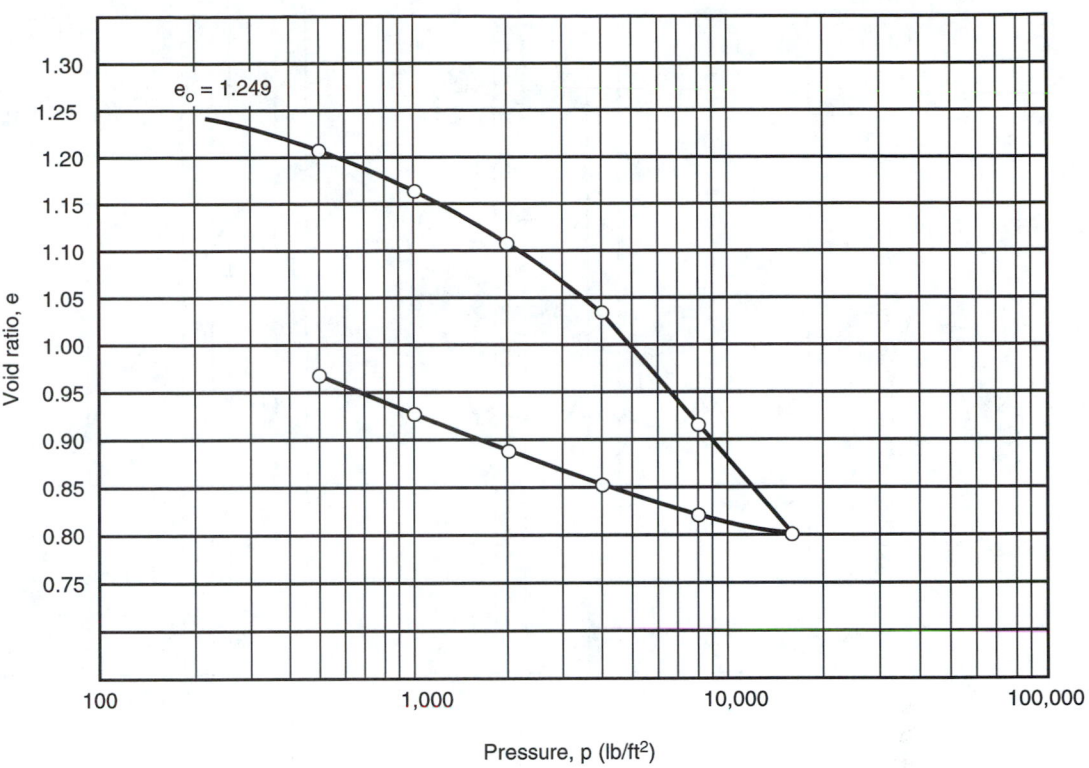

FIGURE 19–5   Void Ratio versus Logarithm of Pressure

The coefficient of consolidation can now be computed by substituting into Eq. (19–8):

$$c_v = \frac{(0.196)(0.385)^2}{8.2} = 3.54 \times 10^{-3} \ \text{in.}^2/\text{min}$$

The values computed are listed in the second row (corresponding to **500 lb/ft²**) of the data form on page 301. Each additional loading on the test specimen will furnish a set of time-versus-deformation dial readings, which through subsequent evaluation will provide a coefficient of consolidation, c_v, corresponding to a specific loading, or pressure, p. Such additional loadings (and subsequent analyses) would be listed in the remaining rows of the data form. Although no laboratory data are given herein for additional loadings on the test specimen, succeeding rows for loadings of **1,000, 2,000, 4,000, 8,000,** and **16,000 lb/ft²** have been included on the form on page 301 in order to demonstrate the complete evaluation of a consolidation test.

A graph of pressure (the first column on page 301) along the abscissa on a logarithmic scale versus coefficient of consolidation (the last column on page 301) along the ordinate on an arithmetic scale can be prepared. This graph, which is known as the $c_v - \log p$ curve (i.e., coefficient of consolidation versus the logarithm of pressure), is one of the

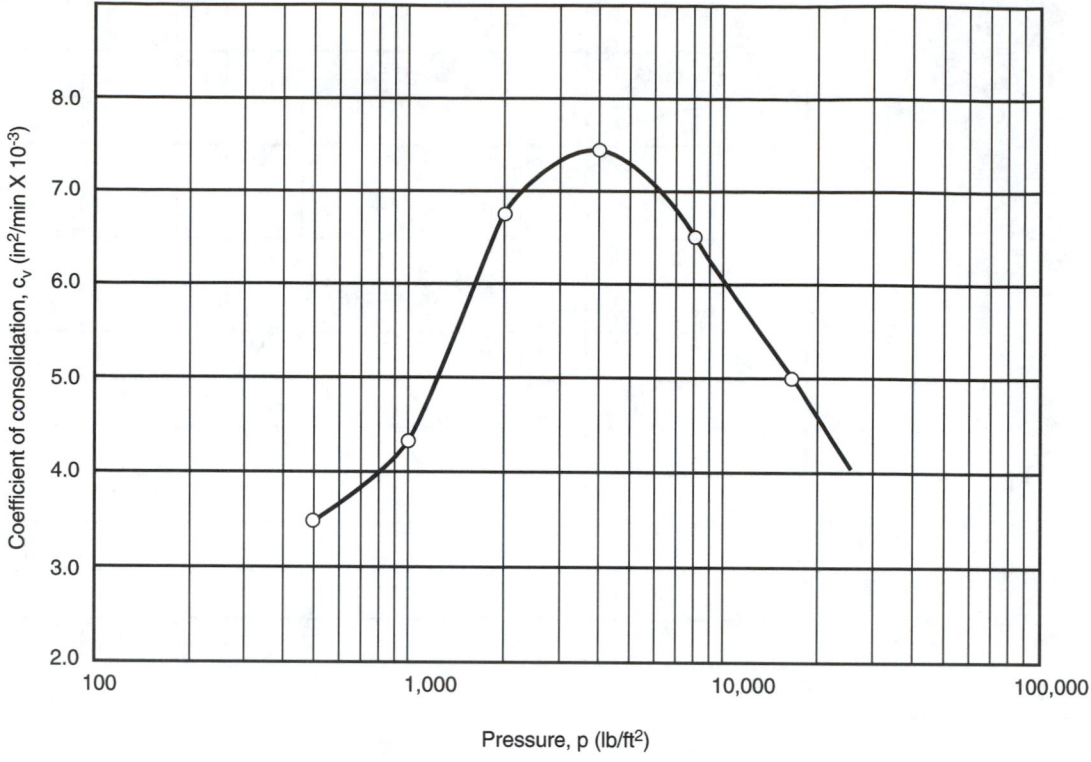

FIGURE 19–6   Coefficient of Consolidation versus Logarithm of Pressure

primary results of a consolidation test. The $c_v - \log p$ curve for this example is shown in Figure 19–6.

CONCLUSIONS   The compressibility of soils is one of the most useful properties that can be obtained from laboratory testing. The primary results of a laboratory consolidation test are (1) $e - \log p$ curve, (2) $c_v - \log p$ curve, and (3) initial void ratio of the soil *in situ*. Using these data, soils engineers can estimate both the rate and magnitude of the consolidation settlement of structures. Estimates of this type are often critical in selecting a type of foundation and then evaluating its adequacy. (For detailed information on such computations, see Liu and Evett, 2001 [3].)

In addition to containing the preceding three primary results of a laboratory consolidation test, the written report would normally include the various time curves (i.e., plots of the logarithm of time versus deformation), values of other parameters determined in the course of the test (e.g., initial and final moisture content, dry mass and initial and final wet unit weight, initial degree of saturation, specific gravity of solids, and Atterberg limits if obtained), and any other pertinent information, such as specimen dimensions and an identification and description of the test sample, including whether the soil is undisturbed, remolded, compacted, or otherwise prepared.

As should be rather obvious, the consolidation test is among the most comprehensive and time-consuming soil tests. It is also among the most expensive tests when performed by a commercial laboratory.

REFERENCES

[1] ASTM, *2001 Annual Book of ASTM Standards,* West Conshohocken, PA, 2001. Copyright, American Society for Testing and Materials, 100 Barr Harbor Drive, West Conshohocken, PA 19428-2959. Reprinted with permission.

[2] Joseph E. Bowles, *Engineering Properties of Soils and Their Measurement,* 2d ed., McGraw-Hill Book Company, New York, 1978.

[3] Cheng Liu and Jack B. Evett, *Soils and Foundations,* 5th ed., Prentice-Hall, Inc., Englewood Cliffs, N.J., 2001.

Soils Testing Laboratory
Consolidation Test

Sample No. _____10_____   Project No. _____I-77-5 (1)_____

Location _____Charlotte, N.C._____   Boring Tube No. _Shelby tube no. 1_

Depth _____24 ft_____   Tested by _____John Doe_____

Description of Soil _____Light brown clay_____

Date of Test _____6/8/02_____

[A] Specimen Data

At Beginning of Test

   (1)   Type of specimen (check one)   ⊠ Undisturbed   ☐ Remolded

   (2)   Diameter of specimen, D __2.50__ in. __6.35__ cm

   (3)   Area of specimen, A __31.68__ cm.^{2}

   (4)   Initial height of specimen, H_0 __0.780__ in. __1.981__ cm

   (5)   Initial volume of specimen, V_0 [i.e., (3) × (4)] __62.76__ cm^3

   (6)   Mass of specimen ring + specimen __208.48__ g

   (7)   Mass of specimen ring __100.50__ g

   (8)   Initial wet mass of specimen, M_{T0} [i.e., (6) − (7)] __107.98__ g

   (9)   Initial wet unit weight, γ_0 $\left[\text{i.e., } \dfrac{(8)}{(5)} \times 62.4\right]$ __107.4__ lb/ft^3

  (10) Initial moisture content, w_0 __42.2__ %

      (a)   Initial wet mass of specimen, M_{T0} [i.e., (8)] __107.98__ g

      (b)   Dry mass of total specimen, M_d [i.e., (21)] __75.94__ g

      (c)   Initial moisture content, w_0 $\left[\text{i.e., } \dfrac{(a) - (b)}{(b)} \times 100\right]$ __42.2%__

  (11) Dry mass of total specimen, M_d, [i.e., (21)] __75.94__ g

  (12) Specific gravity of soil __2.72__

  (13) Volume of solid in soil specimen, V_s $\left[\text{i.e., } \dfrac{(11)}{(12)}\right]$ __27.92__ cm^3

  (14) Height of solid, H_s $\left[\text{i.e., } \dfrac{(13)}{(3)}\right]$ __0.881__ cm

  (15) Initial degree of saturation, S_0 $\left[\text{i.e., } \dfrac{(8) - (11)}{(3)[(4) - (14)]} \times 100\right]$ __91.9__ %

(Note: Density of water = 1g/cm^3)

At End of Test

  (16) Can no. __2-B__

  (17) Mass of can + wet specimen removed from consolidometer __234.54__ g

  (18) Mass of can + oven-dried specimen __203.11__ g

  (19) Mass of can __127.17__ g

(20) Final mass of water in the specimen [i.e., (17) − (18)] __31.43__ g
(21) Dry mass of total specimen [i.e., (18) − (19)] __75.94__ g

(22) Final moisture content, $w_f \left[\text{i.e., } \dfrac{(20)}{(21)} \times 100 \right]$ __41.4__ %

(23) Final degree of saturation __100__ %

Initial Void Ratio

(24) Initial void ratio, $e_0 \left[\text{i.e., } \dfrac{(4) - (14)}{(14)} \right]$ __1.249__

[B] Time-versus-Deformation Data

(1) Pressure increment from __0__ lb/ft^2 to __500__ lb/ft^2

Date	Time	Elapsed Time (min)	Deformation Dial Reading (in.)
6/8/99	9:15 A.M.	0	0
	9:15.1	0.1	0.0067
	9:15.25	0.25	0.0069
	9:15.5	0.5	0.0071
	9:16	1	0.0077
	9:17	2	0.0084
	9:19	4	0.0095
	9:23	8	0.0107
	9:30	15	0.0120
	9:45	30	0.0132
	10:15	60	0.0144
	11:15	120	0.0152
	1:15 P.M.	240	0.0158
	5:15	480	0.0160
6/9/99	8:15 A.M.	1380	0.0162
	11:15	1560	0.0162

[C] Void Ratio

(1) Initial void ratio, e_0 [i.e., from part [A], (24)] __1.249__

(2) Volume of solid in specimen, V_s [i.e., from part [A], (13)] __27.92__ cm^3

(3) Area of specimen, A [i.e., from part [A], (3)] __31.68__ cm^2

(4) Height of solid in specimen, $H_s \left[\text{i.e., } \dfrac{(2)}{(3)} \right]$ __0.881__ cm

Pressure, p (lb/ft^2)	Initial Deformation Dial Reading at Beginning of First Loading (in.)	Deformation Dial Reading Representing 100% Primary Consolidation, d_{100} (in.)	Change in Thickness of Specimen, ΔH (cm)	Change in Void Ratio Δe $\left[\Delta e = \dfrac{\Delta H}{H_s} \right]$	Void Ratio, e $[e = e_0 - \Delta e]$
(5)	(6)	(7)	(8) = [(7) − (6)] × 2.54	$(9) = \dfrac{(8)}{(4)}$	(10) = (1) − (9)
0	0	0	0	0	1.249
500	0	0.0158	0.0401	0.046	1.203
1,000	0	0.0284	0.0721	0.082	1.167
2,000	0	0.0490	0.1245	0.141	1.108
4,000	0	0.0761	0.1933	0.219	1.030
8,000	0	0.1145	0.2908	0.330	0.919
16,000	0	0.1580	0.4013	0.456	0.793

[D] Coefficient of Consolidation

Pressure, p (lb/ft²)	Initial Height of Specimen at Beginning of Test, H_0 (in.)	Deformation Dial Reading at 50% Consolidation (in.)	Thickness of Specimen at 50% Consolidation[a] (in.)	Half-Thickness of Specimen at 50% Consolidation (in.)	Time for 50% Consolidation (min)	Coefficient of Consolidation (in.²/min)
(1)	(2) [from part [A], (4)]	(3) [from dial readings versus log of time curves]	(4) = (2) − (3)	$(5) = \dfrac{(4)}{2}$	(6) [from dial readings versus log of time curves]	$(7) = \dfrac{0.196 \times (5)^2}{(6)}$
0	0.780	—	—	—	—	—
500	0.780	0.0108	0.769	0.385	8.2	3.54×10^{-3}
1,000	0.780	0.0233	0.757	0.378	6.4	4.38×10^{-3}
2,000	0.780	0.0398	0.740	0.370	4.0	6.71×10^{-3}
4,000	0.780	0.0644	0.716	0.358	3.4	7.39×10^{-3}
8,000	0.780	0.0982	0.682	0.341	3.5	6.51×10^{-3}
16,000	0.780	0.1387	0.641	0.320	4.0	5.02×10^{-3}

[a]It is common practice to set the dial to a zero reading at the beginning of the first loading. If this is not done, the actual dial reading at the beginning of the first loading must be subtracted from each entry in this column [(4)] to obtain the correct thickness of the specimen

Soils Testing Laboratory
Consolidation Test

Sample No. _____   Project No. _____

Location _____   Boring Tube No. _____

Depth _____   Tested by _____

Description of Soil _____

Date of Test _____

[A] Specimen Data

At Beginning of Test

    (1)   Type of specimen (check one)   ☐ Undisturbed   ☐ Remolded

    (2)   Diameter of specimen, D _____ in. _____ cm

    (3)   Area of specimen, A _____ cm^2

    (4)   Initial height of specimen, H_0 _____ in. _____ cm

    (5)   Initial volume of specimen, V_0 [i.e., (3) × (4)] = _____ cm^3

    (6)   Mass of specimen ring + specimen _____ g

    (7)   Mass of specimen ring _____ g

    (8)   Initial wet mass of specimen, M_{T0} [i.e., (6) − (7)] _____ g

    (9)   Initial wet unit weight, $\gamma_0 \left[\text{i.e., } \dfrac{(8)}{(5)} \times 62.4 \right]$ _____ lb/ft^3

    (10)   Initial moisture content, w_0 _____ %

          (a) Initial wet mass of specimen, M_{T0} [i.e., (8)] _____ g

          (b) Dry mass of total specimen, M_d [i.e., (21)] _____ g

          (c) Initial moisture content, $w_0 \left[\text{i.e., } \dfrac{(a) - (b)}{(b)} \times 100 \right]$ _____ %

    (11)   Dry mass of total specimen, M_d, [i.e., (21)] _____ g

    (12)   Specific gravity of soil _____

    (13)   Volume of solid in soil specimen, $V_s \left[\text{i.e., } \dfrac{(11)}{(12)} \right]$ _____ cm^3

    (14)   Height of solid, $H_s \left[\text{i.e., } \dfrac{(13)}{(3)} \right]$ _____ cm

    (15)   Initial degree of saturation, $S_0 \left[\text{i.e., } \dfrac{(8) - (11)}{(3)[(4) - (14)]} \times 100 \right]$ _____ %

(Note: Density of water = 1g/cm^3)

At End of Test

    (16)   Can no. _____

    (17)   Mass of can + wet specimen removed from consolidometer _____ g

    (18)   Mass of can + oven-dried specimen _____ g

    (19)   Mass of can _____ g

(20)   Final mass of water in the specimen [i.e., (17) − (18)] _____ g

(21)   Dry mass of total specimen [i.e., (18) − (19)] _____ g

(22)   Final moisture content, w_f $\left[\text{i.e., } \dfrac{(20)}{(21)} \times 100 \right]$ _____ %

(23)   Final degree of saturation _____ %

Initial Void Ratio

(24)   Initial void ratio, e_0 $\left[\text{i.e., } \dfrac{(4) - (14)}{(14)} \right]$ _____

[B] Time-versus-Deformation Data

(1)   Pressure increment from _____ lb/ft^2 to _____ lb/ft^2

Date	Time	Elapsed Time (min)	Deformation Dial Reading (in.)

[C] Void Ratio

(1) Initial void ratio, e_0 [i.e., from part [A], (24)] _____

(2) Volume of solid in specimen, V_s [i.e., from part [A], (13)] _____ cm^3

(3) Area of specimen, A [i.e., from part [A], (3)] _____ cm^2

(4) Height of solid in specimen, $H_s \left[\text{i.e.,} \dfrac{(2)}{(3)} \right]$ _____ cm

Pressure, p (lb/ft^2)	Initial Deformation Dial Reading at Beginning of First Loading (in.)	Deformation Dial Reading Representing 100% Primary Consolidation, d_{100} (in.)	Change in Thickness of Specimen, ΔH (cm)	Change in Void Ratio Δe $\left[\Delta e = \dfrac{\Delta H}{H_s} \right]$	Void Ratio, e $[e = e_0 - \Delta e]$
(5)	(6)	(7)	$(8) = [(7) - (6)] \times 2.54$	$(9) = \dfrac{(8)}{(4)}$	$(10) = (1) - (9)$

[D] Coefficient of Consolidation

Pressure, p (lb/ft²)	Initial Height of Specimen at Beginning of Test, H_0 (in.)	Deformation Dial Reading at 50% Consolidation (in.)	Thickness of Specimen at 50% Consolidation[a] (in.)	Half-Thickness of Specimen at 50% Consolidation (in.)	Time for 50% Consolidation (min)	Coefficient of Consolidation (in.²/min)
(1)	(2) [from part [A]; (4)]	(3) [from dial readings versus log of time curves]	(4) = (2) − (3)	$(5) = \dfrac{(4)}{2}$	(6) [from dial readings versus log of time curves]	$(7) = \dfrac{0.196 \times (5)^2}{(6)}$

[a]It is common practice to set the dial to a zero reading at the beginning of the first loading. If this is not done, the actual dial reading at the beginning of the first loading must be subtracted from each entry in this column [(4)] to obtain the correct thickness of the specimen at 50% consolidation.

Soils Testing Laboratory
Consolidation Test

Sample No. _____   Project No. _____

Location _____   Boring Tube No. _____

Depth _____   Tested by _____

Description of Soil _____

Date of Test _____

[A] Specimen Data

At Beginning of Test

(1) Type of specimen (check one) ☐ Undisturbed ☐ Remolded

(2) Diameter of specimen, D _____ in. _____ cm

(3) Area of specimen, A _____ cm^2

(4) Initial height of specimen, H_0 _____ in. _____ cm

(5) Initial volume of specimen, V_0 [i.e., (3) × (4)] = _____ cm^3

(6) Mass of specimen ring + specimen _____ g

(7) Mass of specimen ring _____ g

(8) Initial wet mass of specimen, M_{T0} [i.e., (6) − (7)] _____ g

(9) Initial wet unit weight, γ_0 $\left[\text{i.e., } \dfrac{(8)}{(5)} \times 62.4 \right]$ _____ lb/ft^3

(10) Initial moisture content, w_0 _____ %

   (a) Initial wet mass of specimen, M_{T0} [i.e., (8)] _____ g

   (b) Dry mass of total specimen, M_d [i.e., (21)] _____ g

   (c) Initial moisture contact, w_0 $\left[\text{i.e., } \dfrac{(a) - (b)}{(b)} \times 100 \right]$ _____ %

(11) Dry mass of total specimen, M_d, [i.e., (21)] _____ g

(12) Specific gravity of soil _____

(13) Volume of solid in soil specimen, V_s $\left[\text{i.e., } \dfrac{(11)}{(12)} \right]$ _____ cm^3

(14) Height of solid, H_s $\left[\text{i.e., } \dfrac{(13)}{(3)} \right]$ _____ cm

(15) Initial degree of saturation, S_0 $\left[\text{i.e., } \dfrac{(8) - (11)}{(3)[(4) - (14)]} \times 100 \right]$ _____ %

(Note: Density of water = 1g/cm^3)

At End of Test

(16) Can no. _____

(17) Mass of can + wet specimen removed from consolidometer _____ g

(18) Mass of can + oven-dried specimen _____ g

(19) Mass of can _____ g

(20) Final mass of water in the specimen [i.e., (17) − (18)] _____ g

(21) Dry mass of total specimen [i.e., (18) − (19)] _____ g

(22) Final moisture content, w_f $\left[\text{i.e., } \dfrac{(20)}{(21)} \times 100\right]$ _____ %

(23) Final degree of saturation _____ %

Initial Void Ratio

(24) Initial void ratio, e_0 $\left[\text{i.e., } \dfrac{(4) - (14)}{(14)}\right]$ _____

[B] Time-versus-Deformation Data

(1) Pressure increment from _____ lb/ft^2 to _____ lb/ft^2

Date	Time	Elapsed Time (min)	Deformation Dial Reading (in.)

[C] Void Ratio

(1) Initial void ratio, e_0 [i.e., from part [A], (24)] _____

(2) Volume of solid in specimen, V_s [i.e., from part [A], (13)] _____ cm^3

(3) Area of specimen, A [i.e., from part [A], (3)] _____ cm^2

(4) Height of solid in specimen, $H_s \left[\text{i.e., } \dfrac{(2)}{(3)} \right]$ _____ cm

Pressure, p (lb/ft^2)	Initial Deformation Dial Reading at Beginning of First Loading (in.)	Deformation Dial Reading Representing 100% Primary Consolidation, d_{100} (in.)	Change in Thickness of Specimen, ΔH (cm)	Change in Void Ratio Δe $\left[\Delta e = \dfrac{\Delta H}{H_s} \right]$	Void Ratio, e $[e = e_0 - \Delta e]$
(5)	(6)	(7)	$(8) = [(7) - (6)] \times 2.54$	$(9) = \dfrac{(8)}{(4)}$	$(10) = (1) - (9)$

[D] Coefficient of Consolidation

Pressure, p (lb/ft²)	Initial Height of Specimen at Beginning of Test, H_0 (in.)	Deformation Dial Reading at 50% Consolidation (in.)	Thickness of Specimen at 50% Consolidation[a] (in.)	Half-Thickness of Specimen at 50% Consolidation (in.)	Time for 50% Consolidation (min)	Coefficient of Consolidation (in.²/min)
(1)	(2) [from part [A], (4)]	(3) [from dial readings versus log of time curves]	(4) = (2) − (3)	$(5) = \dfrac{(4)}{2}$	(6) [from dial readings versus log of time curves]	$(7) = \dfrac{0.196 \times (5)^2}{(6)}$

[a]It is common practice to set the dial to a zero reading at the beginning of the first loading. If this is not done, the actual dial reading at the beginning of the first loading must be subtracted from each entry in this column [(4)] to obtain the correct thickness of the specimen at 50% consolidation.

CHAPTER TWENTY

Determining the Unconfined Compressive Strength of Cohesive Soil

(Referenced Document: ASTM D 2166)

INTRODUCTION

The *unconfined compressive strength of cohesive soil, q_u,* is defined as the load per unit area at which an unconfined prismatic or cylindrical specimen of soil will fail in a simple compression test. It is taken to be the maximum load attained per unit area or the load per unit area at 15% axial strain, whichever is secured first during the performance of a test [1].

The unconfined compression test is perhaps the simplest, easiest, and least expensive test for investigating the approximate shear strength of cohesive soils in terms of total stress. (The reason the test is usually limited to cohesive soils is that there is no lateral support, and the soil sample must be able to stand alone. A cohesionless soil such as sand cannot generally stand alone in this manner without lateral support.) The soil parameter cohesion c is taken to be one-half the unconfined compressive strength (i.e., $c = q_u/2$).

APPARATUS AND SUPPLIES

Compression device (including deflection dial; see Figures 20–1 and 20–2)

Sample ejector, for ejecting the soil core from the sampling tube

Dial comparator or other suitable device, for measuring the specimen's physical dimensions

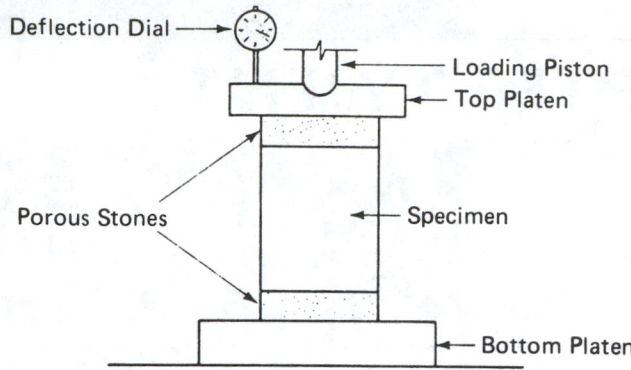

FIGURE 20–1   Unconfined Compression Test Apparatus [2]

FIGURE 20–2   Unconfined Compression
Test Apparatus (Courtesy of Soiltest, Inc.)

Timing device

Drying oven

Balance (with accuracy to 0.01 g)

Miscellaneous equipment: trimming tools, remolding apparatus, moisture content containers, etc.

PREPARATION OF TEST SPECIMENS [1]

(1) *Specimen Size*—Specimens shall have a minimum diameter of 30 mm (1.3 in.), and the largest particle contained within the test specimen shall be smaller than one tenth of the specimen diameter. For specimens having a diameter of 72 mm (2.8 in.) or larger, the largest particle size shall be smaller than one sixth of the specimen diameter. If, after completion of a test on an undisturbed specimen, it is found, based on visual observation, that larger particles than permitted are present, indicate this information in the remarks section of the report of test data (Note 1). The height-to-diameter ratio shall be between 2 and 2.5. Determine the average height and diameter of the test specimen using the dial comparator. Take a minimum of three height measurements (120° apart), and at least three diameter measurements at the quarter points of the height.

> *Note 1*—If large soil particles are found in the sample after testing, a particle-size analysis performed in accordance with ASTM Method D 422 (Chapter 9) may be performed to confirm the visual observation and the results provided with the test report.

(2) *Undisturbed Specimens*—Prepare undisturbed specimens from large undisturbed samples or from samples secured in accordance with ASTM Method D 1587 and preserved and transported in accordance with the practices for Group C samples in ASTM Method D 4220. Tube specimens may be tested without trimming except for the squaring of ends, if conditions of the sample justify this procedure. Handle specimens carefully to prevent disturbance, changes in cross section, or loss of water content. If compression or any type of noticeable disturbance would be caused by the extrusion device, split the sample tube lengthwise or cut it off in small sections to facilitate removal of the specimen without disturbance. Prepare carved specimens without disturbance and, whenever possible, in a humidity-controlled room. Make every effort to prevent any change in water content of the soil. Specimens shall be of uniform circular cross section with ends perpendicular to the longitudinal axis of the specimen. When carving or trimming, remove any small pebbles or shells encountered. Carefully fill voids on the surface of the specimen with remolded soil obtained from the trimmings. When pebbles or crumbling result in excessive irregularity at the ends, cap the specimen with a minimum thickness of plaster of paris, hydrostone, or similar material. When sample condition permits, a vertical lathe that will accommodate the total sample may be used as an aid in carving the specimen to the required diameter. Where prevention of the development of appreciable capillary forces is deemed important, seal the specimen with a rubber membrane, thin plastic coatings, or with a coating of grease or sprayed plastic immediately after preparation and during the entire testing cycle. Determine the mass and dimensions of the test specimen. If the specimen is to be capped, its mass and dimensions

should be determined before capping. If the entire test specimen is not to be used for determination of water content, secure a representative sample of trimmings for this purpose, placing them immediately in a covered container. The water content determination shall be performed in accordance with Method D 2216.

(3) *Remolded Specimens*—Specimens may be prepared either from a failed undisturbed specimen or from a disturbed sample, providing it is representative of the failed undisturbed specimen. In the case of failed undisturbed specimens, wrap the material in a thin rubber membrane and work the material thoroughly with the fingers to assure complete remolding. Avoid entrapping air in the specimen. Exercise care to obtain a uniform density, to remold to the same void ratio as the undisturbed specimen, and to preserve the natural water content of the soil. Form the disturbed material into a mold of circular cross section having dimensions meeting the requirements of (1). After removal from the mold, determine the mass and dimensions of the test specimens.

(4) *Compacted Specimens*—Specimens shall be prepared to the predetermined water content and density prescribed by the individual assigning the test (Note 2). After a specimen is formed, trim the ends perpendicular to the longitudinal axis, remove from the mold, and determine the mass and dimensions of the test specimen.

Note 2—Experience indicates that it is difficult to compact, handle, and obtain valid results with specimens that have a degree of saturation that is greater than 90%.

PROCEDURE The unconfined compression test is quite similar to the test for compressive strength of concrete, where crushing concrete cylinders is carried out solely by measured increases in end loadings. A cohesive soil specimen, prepared as previously described, is placed in a compression device (Figures 20–1 and 20–2) and subjected to an axial load, applied to produce axial strain at a rate of ½ to 2%/min. The resulting stress and strain are then measured. The unconfined compressive strength is taken as the maximum load attained per unit area or the load per unit area at 15% axial strain, whichever is secured first during the test.

The actual step-by-step procedure is as follows (ASTM 2166-00 [1]):

(1) Place the specimen in the loading device so that it is centered on the bottom platen. Adjust the loading device carefully so that the upper platen just makes contact with the specimen. Zero the deformation indicator. Apply the load so as to produce an axial strain at a rate of ½ to 2%/min. Record load, deformation, and time values at sufficient intervals to define the shape of the stress-strain curve (usually 10 to 15 points are sufficient). The rate of strain should be chosen so that the time to failure does not exceed about 15 min (Note 3). Continue loading until the load values decrease with increasing strain, or until 15% strain is reached. The rate of strain

used for testing sealed specimens may be decreased if deemed desirable for better test results. Indicate the rate of strain in the report of the test data. Determine the water content of the test specimen using the entire specimen, unless representative trimmings are obtained for this purpose, as in the case of undisturbed specimens. Indicate on the test report whether the water content sample was obtained before or after the shear test.

> *Note 3*—Softer materials that will exhibit larger deformation at failure should be tested at a higher rate of strain. Conversely, stiff or brittle materials that will exhibit small deformations at failure should be tested at a lower rate of strain.

(2) Make a sketch, or take a photo, of the test specimen at failure showing the slope angle of the failure surface if the angle is measurable.

DATA Data collected in this test should include the following:

[A] Specimen Data

Diameter of specimen, D_0 (in.)

Initial height of specimen, H_0 (in.)

Mass of specimen (g)

Water content data:

Mass of wet soil plus can (g)

Mass of dry soil plus can (g)

Mass of can (g)

[B] Compression Data

Successive load and deformation values as load is applied (loads are determined by multiplying proving ring dial readings by the proving ring calibration factor)

CALCULATIONS [A] Specimen Parameters

With the diameter and initial height of the specimen known, its initial area, volume, and height-to-diameter ratio can be calculated. With the specimen's mass known, its wet unit weight can be found. From the water content data, the water content (see Chapter 3) and then the dry unit weight of the specimen can be determined.

[B] Compression Calculation

For each applied load, axial unit strain, ϵ, can be computed by dividing the specimen's change in height, ΔH, by its initial height, H_0. In equation form,

$$\epsilon = \frac{\Delta H}{H_0} \qquad\qquad (20\text{–}1)$$

The value of ΔH is given by the deformation dial reading, provided that the dial is set to zero initially.

As load is applied to the specimen, its cross-sectional area will increase by a small amount. For each applied load, the cross-sectional area A can be computed by the equation

$$A = \frac{A_0}{1 - \epsilon} \qquad\qquad (20\text{–}2)$$

where A_0 is the initial area of the specimen.

Each applied axial load P can be determined by multiplying the proving ring dial reading by the proving ring calibration factor, and the load per unit area can be computed by dividing the load by the corresponding cross-sectional area.

The largest value of load per unit area or load per unit area at 15% strain, whichever is secured first, is taken to be the unconfined compressive strength q_u, and the cohesion c is taken to be half the unconfined compressive strength.

[C] Graph

A graph of load per unit area (ordinate) versus unit strain (abscissa) should be prepared. From this graph, the unconfined compressive strength may be evaluated as either the maximum value of load per unit area or the load per unit area at 15% strain, whichever occurs first.

NUMERICAL EXAMPLE

An unconfined compression test was performed in the laboratory, and the following data were obtained:

[A] Specimen Data

   Diameter of specimen, D_0 = **2.50 in.**

   Initial height of specimen, H_0 = **5.98 in.**

   Mass of specimen = **991.50 g**

   Water content data:

   Mass of wet soil plus can = **383.41 g**

   Mass of dry soil plus can = **326.78 g**

   Mass of can = **50.56 g**

[B] Compression Data

   Proving ring calibration = **6,000 lb/in.**

Deformation Dial, ΔH (in.)	Proving Ring Dial (in.)
0.000	0.0000
0.025	0.0024
0.050	0.0058
0.075	0.0086
0.100	0.0116
0.125	0.0150
0.150	0.0176
0.175	0.0208
0.200	0.0224
0.225	0.0232
0.240	0.0224
0.260	0.0198

All of these data are shown on forms prepared for recording both collected laboratory data and computed results (see pages 319 and 320). At the end of the chapter, two blank copies of these forms are included for the reader's use.

With the data listed, computations would proceed as follows:

[A] Specimen Parameters

The height-to-diameter ratio is **5.98/2.50,** or 2.4, which is between 2 and 2.5 and, therefore, acceptable. The specimen's initial area, A_0, and volume are easily computed as follows:

$$A_0 = \frac{\pi D_0^2}{4} = \frac{\pi (\mathbf{2.50})^2}{4} = 4.91 \text{ in.}^2$$

$$\text{Volume} = (A_0)(H_0) = (4.91)(\mathbf{5.98}) = 29.36 \text{ in.}^3$$

With the specimen's mass and volume known, its wet unit weight γ_{wet} can be determined:

$$\gamma_{\text{wet}} = \frac{\mathbf{991.50}}{29.36} \times \frac{1,728}{453.6} = 128.6 \text{ lb/ft}^3$$

(The numbers 1,728 and 453.6 are conversion factors: $1,728 \text{ in.}^3 = 1 \text{ ft}^3$; 453.6 g = 1 lb.)

The water content of the specimen can be computed as follows (see Chapter 3):

$$w = \frac{383.41 - 326.78}{326.78 - 50.56} \times 100 = 20.5\%$$

Finally, the dry unit weight γ_{dry} can be computed:

$$\gamma_{\text{dry}} = \frac{\gamma_{\text{wet}}}{w + 100} \times 100 = \frac{128.6}{20.5 + 100} \times 100 = 106.7 \text{ lb/ft}^3$$

[B] Compression Calculations

For the first applied load in this test, the deformation dial ΔH and the proving ring dial readings were 0.025 and 0.0024 in., respectively. The axial strain ϵ for this particular applied load can be computed using Eq. (20–1):

$$\epsilon = \frac{\Delta H}{H_0} \tag{20-1}$$

$$\epsilon = \frac{0.025}{5.98} = 0.004$$

The corresponding cross-sectional area can be computed using Eq. (20–2):

$$A = \frac{A_0}{1 - \epsilon} \tag{20-2}$$

$$A = \frac{4.91}{1 - 0.004} = 4.93 \text{ in.}^2$$

The applied axial load P can be computed by multiplying the proving ring dial reading by the proving ring calibration factor:

$$P = (0.0024)(6{,}000) = 14.4 \text{ lb}$$

The load per unit area for this particular applied load is therefore 14.4/4.93, or 2.92 lb/in.^{2} This converts to 420 lb/ft^2.

    The foregoing data furnish the values to fill in the second row of the form on page 320. Similar computations provide values needed to fill in the remaining rows.

[C] Graph

The required graph showing the relationship between load per unit area and unit strain can be obtained by plotting values of load per unit area (the last column in the form on page 320) as the ordinate and unit strain (the second column in the form on page 320) as the abscissa. The graph for this example is shown in Figure 20–3. From this graph, the unconfined compressive strength q_u, which is the maximum value of load per unit area or the load per unit area at 15% strain, whichever occurs first, is 3,930 lb/ft^2. The cohesion c, which is half the unconfined compressive strength (i.e., $c = q_u/2$), is therefore 3,930/2, or 1,965 lb/ft^2.

Soils Testing Laboratory
Unconfined Compression Test

Sample No. _____18_____   Project No. _____I-85-5(3)_____

Location _____Kannapolis, N.C._____   Boring No. _____14_____

Depth of Sample _____18 ft_____   Date of Test _____6-2-02_____

Description of Soil _____Light brown clay_____

Tested by _____John Doe_____

[A] Specimen Data

(1) Type of specimen (check one)   ☐ Undisturbed   ☒ Remolded
(2) Shape of specimen (check one)   ☒ Cylindrical   ☐ Prismatic
(3) Diameter of specimen, D_0 __2.50__ in.
(4) Initial area of specimen, A_0 __4.91__ in.^{2}
(5) Initial height of specimen, H_0 __5.98__ in.

(6) Height-to-diameter ratio $\left[\text{i.e., } \dfrac{(5)}{(3)} \right]$ __2.4__

(7) Volume of specimen [i.e., (4) × (5)] __29.36__ in.^{3}
(8) Mass of specimen __991.50__ g

(9) Wet unit weight of specimen $\left[\text{i.e., } \dfrac{(8)}{453.6} \times \dfrac{1728}{(7)} \right]$ __128.6__ lb/ft^3

(10) Water content of specimen __20.5__ %
   (a) Can no. __2-D__
   (b) Mass of wet soil + can __383.41__ g
   (c) Mass of dry soil + can __326.78__ g
   (d) Mass of can __50.56__ g
   (e) Mass of water __56.63__ g
   (f) Mass of dry soil __276.22__ g

   (g) Water content $\left[\text{i.e., } \dfrac{(e)}{(f)} \times 100 \right]$ __20.5__ %

(11) Dry unit weight of specimen $\left[\text{i.e., } \dfrac{(9)}{(10) + 100} \times 100 \right]$ __106.7__ lb/ft^3

[B] Compression Data

(1) Proving ring calibration ___6,000___ lb/in.

Deformation Dial, ΔH (in.)	Unit Strain, ϵ (in./in.)	Cross-Sectional Area, A (in.^{2})	Proving Ring Dial (in.)	Applied Axial Load, P (lb)	Load per Unit Area	
					lb/in.^{2}	lb/ft^2
(1)	$(2) = \dfrac{\Delta H}{H_0}$	$(3) = \dfrac{A_0}{1-\epsilon}$	(4)	$(5) = (4) \times$ proving ring calibration	$(6) = \dfrac{(5)}{(3)}$	$(7) = (6) \times 144$
0.000	0.000	4.91	0.0000	0	0	0
0.025	0.004	4.93	0.0024	14.4	2.92	420
0.050	0.008	4.95	0.0058	34.8	7.03	1,012
0.075	0.013	4.97	0.0086	51.6	10.38	1,495
0.100	0.017	4.99	0.0116	69.6	13.95	2,009
0.125	0.021	5.02	0.0150	90.0	17.93	2,582
0.150	0.025	5.04	0.0176	105.6	20.95	3,017
0.175	0.029	5.06	0.0208	124.8	24.66	3,551
0.200	0.033	5.08	0.0224	134.4	26.46	3,810
0.225	0.038	5.10	0.0232	139.2	27.29	3,930
0.240	0.040	5.11	0.0224	134.4	26.30	3,787
0.260	0.043	5.13	0.0198	118.8	23.16	3,335

Unconfined compressive strength, q_u ___3,930___ lb/ft^2

Cohesion, $c \left(\text{i.e., } \dfrac{q_u}{2} \right)$ ___1,965___ lb/ft^2

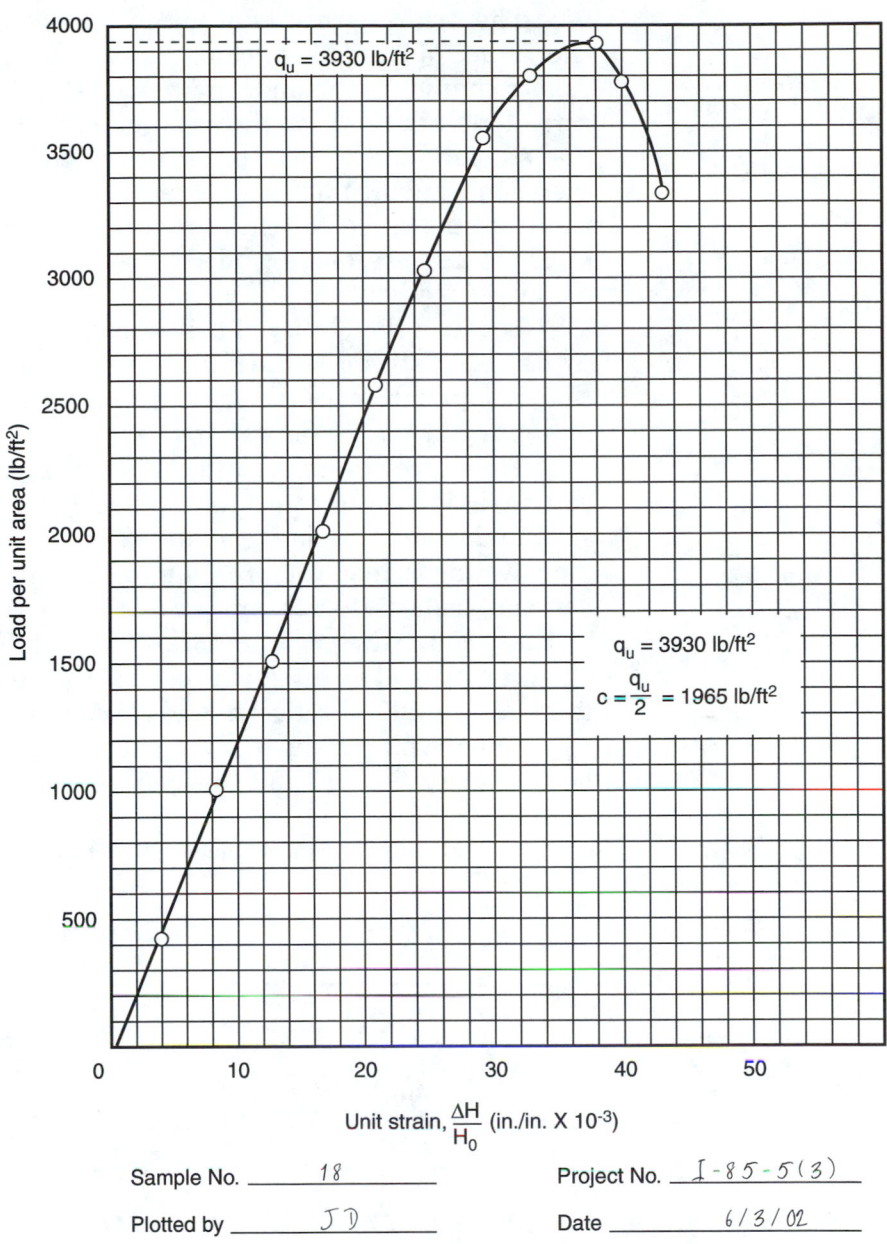

$q_u = 3930$ lb/ft²

$q_u = 3930$ lb/ft²

$c = \dfrac{q_u}{2} = 1965$ lb/ft²

Load per unit area (lb/ft²)

Unit strain, $\dfrac{\Delta H}{H_0}$ (in./in. X 10⁻³)

Sample No. _____18_____   Project No. _I - 8 5 - 5 (3)_

Plotted by _____J D_____   Date _____6 / 3 / 02_____

FIGURE 20–3   Relationship between Load per Unit Area and Unit Strain

CONCLUSIONS

The test described covers the determination of the unconfined compressive strength of cohesive soil in the undisturbed, remolded, or compacted condition, using strain-controlled application of axial load. It provides an approximate value of the strength of cohesive soils in terms of total stresses. The method is applicable only to cohesive materials that will not expel bleed water (water expelled from the soil due to deformation or compaction) during the loading portion of the test and that will retain intrinsic strength after removal of confining pressures, such as clays or cemented soils. Dry and crumbly soils, fissured or varved materials, silts, peats, and sands cannot be tested with this method to obtain valid unconfined compressive strength values. [1]

The primary results of unconfined compressive strength tests are the values of unconfined compressive strength q_u and cohesion c. A test report should also include, however, (1) an identification and visual description of the soil, (2) the type (undisturbed, compacted, or remolded) and shape (cylindrical or prismatic) of the test specimen, and (3) the dimensions, height-to-diameter ratio, initial density, and water content of the test specimen. Normally, a graph of load per unit area versus unit strain is also presented, and sketches of the failed specimen might be included as well.

Because the ability of soil to support imposed loads is determined by its shear strength, the shear strength of soil is important in foundation design, lateral earth pressure calculations, slope stability analysis, and many other considerations.

There are several methods of investigating the shear strength of a soil in a laboratory, including the (1) unconfined compression test, (2) triaxial compression test, and (3) direct shear test. As discussed previously, unconfined compression tests can generally be used only to investigate cohesive soils and provide approximate values of the strength of these soils in terms of total stresses. Triaxial tests, described in Chapter 21, and direct shear tests, discussed in Chapter 22, can be used to investigate both cohesive and cohesionless soils.

Some typical values of the shear strength of cohesive soil (i.e., cohesion) are given in Table 20–1.

Table 20–1   Typical Values of Shear Strength of Cohesive Soil [3]

Consistency of Clay	Shear Strength (Half of Unconfined Compressive Strength) (lb/ft^2)
Very soft	<250
Soft	250–500
Medium	500–1,000
Stiff	1,000–2,000
Very stiff	2,000–4,000
Hard	>4,000

REFERENCES

[1] ASTM *2001 Annual Book of ASTM Standards,* West Conshohocken, PA, 2001. Copyright, American Society for Testing and Materials, 100 Barr Harbor Drive, West Conshohocken, PA 19428-2959. Reprinted with permission.

[2] Joseph E. Bowles, *Engineering Properties of Soils and Their Measurement,* 2d ed., McGraw-Hill Book Company, New York, 1978.

[3] T. William Lambe, *Soil Testing for Engineers,* John Wiley & Sons, Inc., New York, 1951.

Soils Testing Laboratory
Unconfined Compression Test

Sample No. _____   Project No. _____

Location _____   Boring No. _____

Depth of Sample _____   Date of Test _____

Description of Soil _____

Tested by _____

[A] Specimen Data

(1) Type of specimen (check one)   ☐ Undisturbed   ☐ Remolded

(2) Shape of specimen (check one)   ☐ Cylindrical   ☐ Prismatic

(3) Diameter of specimen, D_0 _____ in.

(4) Initial area of specimen, A_0 _____ in.^{2}

(5) Initial height of specimen, H_0 _____ in.

(6) Height-to-diameter ratio $\left[\text{i.e., } \dfrac{(5)}{(3)} \right]$ _____

(7) Volume of specimen [i.e., (4) × (5)] _____ in.^{3}

(8) Mass of specimen _____ g

(9) Wet unit weight of specimen $\left[\text{i.e., } \dfrac{(8)}{453.6} \times \dfrac{1728}{(7)} \right]$ _____ lb/ft^3

(10) Water content of specimen _____ %

    (a) Can no. _____

    (b) Mass of wet soil + can _____ g

    (c) Mass of dry soil + can _____ g

    (d) Mass of can _____ g

    (e) Mass of water _____ g

    (f) Mass of dry soil _____ g

    (g) Water content $\left[\text{i.e., } \dfrac{(e)}{(f)} \times 100 \right]$ _____ %

(11) Dry unit weight of specimen $\left[\text{i.e., } \dfrac{(9)}{(10) + 100} \times 100 \right]$ _____ lb/ft^3

[B] Compression Data

(1) Proving ring calibration _____ lb/in.

Deformation Dial, ΔH (in.)	Unit Strain, ϵ (in./in.)	Cross-Sectional Area, A (in.²)	Proving Ring Dial (in.)	Applied Axial Load, P (lb)	Load per Unit Area	
					lb/in.²	lb/ft²
(1)	$(2) = \dfrac{\Delta H}{H_0}$	$(3) = \dfrac{A_0}{1 - \epsilon}$	(4)	$(5) = (4) \times$ proving ring calibration	$(6) = \dfrac{(5)}{(3)}$	$(7) = (6) \times 144$

Unconfined compressive strength, q_u _____ lb/ft²

Cohesion, $c \left(\text{i.e., } \dfrac{q_u}{2} \right)$ _____ lb/ft²

Soils Testing Laboratory
Unconfined Compression Test

Sample No. _____   *Project No.* _____

Location _____   *Boring No.* _____

Depth of Sample _____   *Date of Test* _____

Description of Soil _____

Tested by _____

[A] Specimen Data

(1)   Type of specimen (check one)   ☐ Undisturbed   ☐ Remolded

(2)   Shape of specimen (check one)   ☐ Cylindrical   ☐ Prismatic

(3)   Diameter of specimen, D_0 _____ in.

(4)   Initial area of specimen, A_0 _____ in.^{2}

(5)   Initial height of specimen, H_0 _____ in.

(6)   Height-to-diameter ratio $\left[\text{i.e., } \dfrac{(5)}{(3)} \right]$ _____

(7)   Volume of specimen [i.e., (4) × (5)] _____ in.^{3}

(8)   Mass of specimen _____ g

(9)   Wet unit weight of specimen $\left[\text{i.e., } \dfrac{(8)}{453.6} \times \dfrac{1728}{(7)} \right]$ _____ lb/ft^3

(10) Water content of specimen _____ %

    (a) Can no. _____

    (b) Mass of wet soil + can _____ g

    (c) Mass of dry soil + can _____ g

    (d) Mass of can _____ g

    (e) Mass of water _____ g

    (f) Mass of dry soil _____ g

    (g) Water content $\left[\text{i.e., } \dfrac{(e)}{(f)} \times 100 \right]$ _____ %

(11) Dry unit weight of specimen $\left[\text{i.e., } \dfrac{(9)}{(10) + 100} \times 100 \right]$ _____ lb/ft^3

[B] Compression Data

(1)   Proving ring calibration _____ lb/in.

Deformation Dial, ΔH (in.)	Unit Strain, ϵ (in./in.)	Cross-Sectional Area, A (in.^{2})	Proving Ring Dial (in.)	Applied Axial Load, P (lb)	Load per Unit Area	
					lb/in.^{2}	lb/ft^2
(1)	$(2) = \dfrac{\Delta H}{H_0}$	$(3) = \dfrac{A_0}{1 - \epsilon}$	(4)	$(5) = (4) \times$ proving ring calibration	$(6) = \dfrac{(5)}{(3)}$	$(7) = (6) \times 144$

Unconfined compressive strength, q_u _____ lb/ft^2

Cohesion, $c \left(\text{i.e., } \dfrac{q_u}{2} \right)$ _____ lb/ft^2

21

Triaxial Compression Test

(Referenced Document: ASTM D 2850)

INTRODUCTION

The triaxial compression test is carried out in a manner somewhat similar to the unconfined compression test (Chapter 20) in that a cylindrical soil specimen is subjected to a vertical (axial) load. The major difference is that, unlike unconfined compression tests, where there is no confining lateral pressure, triaxial tests are performed on cylindrical soil specimens encased in rubber membranes with confining lateral pressure present. The magnitude of the lateral pressure can be chosen and is made possible by placing a specimen within a pressure chamber (see Figures 21–1 and 21–2), into which water or air is then pumped. The soil specimen in the chamber under a chosen lateral pressure is subjected to an increasing axial load until the specimen fails. The procedure is then repeated on additional specimens at other confining pressures.

The lateral pressure, or chamber pressure, which is applied to the ends of the specimen as well as its sides, is called the *minor principal stress*. The externally applied axial load divided by the cross-sectional area of the test specimen is the *unit axial load*. The minor principal stress plus the unit axial load is called the *major principal stress*.

Triaxial compression test results are analyzed by plotting Mohr circles for the stress conditions of each specimen when failure occurs. By evaluating the plotted Mohr circles, the soil's shear strength parameters (cohesion c and angle of internal friction ϕ) can be determined. These parameters are used to evaluate the shear strength of soil.

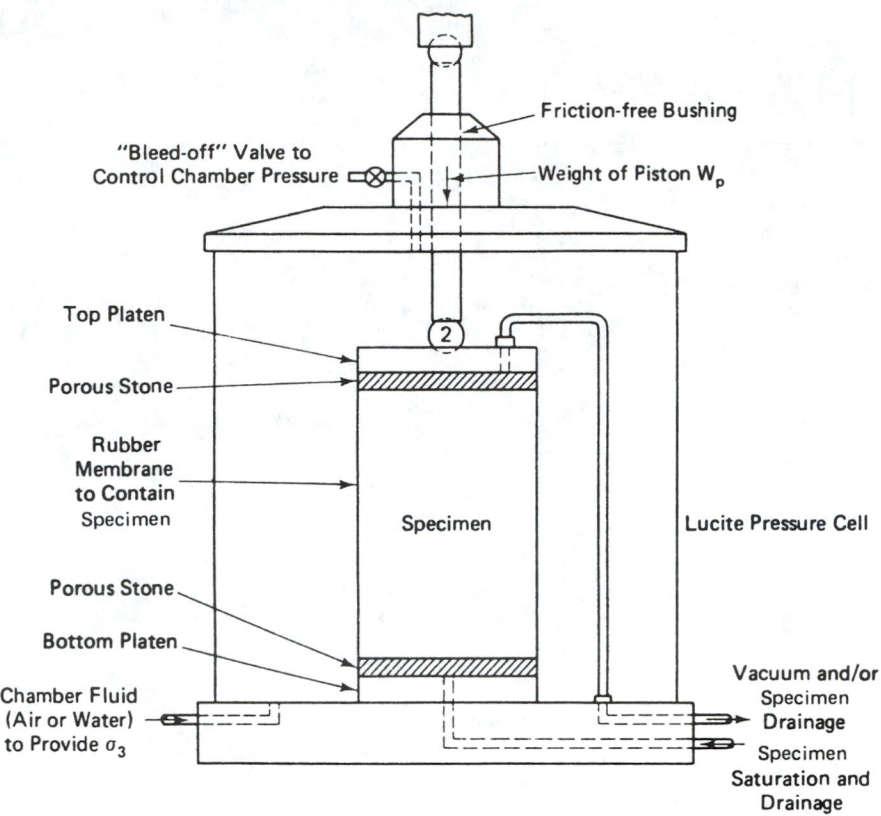

FIGURE 21-1   Schematic Diagram of a Triaxial Chamber [1]

FIGURE 21-2   Triaxial Chamber
(Courtesy of Soiltest, Inc.)

The unconfined compression test (Chapter 20) is a special case of the triaxial compression test in which the confining pressure is zero. Note that triaxial tests can be applied to both cohesive and cohesionless soils, whereas unconfined compression tests are limited more or less to cohesive soils.

There are three basic types of triaxial compression test procedures as determined by sample drainage conditions: unconsolidated undrained, consolidated undrained, and consolidated drained. These can be defined as follows [2].

Unconsolidated undrained (UU) tests are carried out by placing a specimen in the chamber and introducing lateral pressure without allowing the specimen to consolidate (drain) under the confining pressure. Axial load is then applied fairly rapidly without permitting drainage of the specimen. The UU test can be run rather quickly because the specimen is not required to consolidate under the confining pressure or drain during application of the axial load. Because of the short time required to run this test, it is often referred to as the quick, or Q, test.

Consolidated undrained (CU) tests are performed by placing a specimen in the chamber and introducing lateral pressure. The specimen is then allowed to consolidate under the all-around confining pressure by leaving the drain lines open (see Figures 21–1 and 21–2). The drain lines are then closed, and axial stress is induced without allowing further drainage.

Consolidated drained (CD) tests are similar to CU tests, except that the specimen is allowed to drain as axial load is applied, so that high excess pore pressures do not develop. Because the permeability of clayey soils is low, axial load must be added very slowly during CD tests so that pore pressure can be dissipated. CD tests may take a considerable period of time to run because of the time required for both consolidation under the confining pressure and drainage during application of axial load. Inasmuch as the time requirement is long for low-permeability soils, this test is often referred to as the slow, or S, test.

The specific type of triaxial compression test (i.e., UU, CU, or CD) that should be used in any particular case depends on the field conditions to be simulated. The UU and CU tests are covered in this chapter.

APPARATUS AND SUPPLIES

Axial loading device (see Figure 21–3)

Axial load-measuring device

Triaxial chamber (Figures 21–1 and 21–2)

Specimen mold, cap, and base

Rubber membranes, membrane stretcher, rubber binding strips (Figure 21–4)

Deformation indicator

Calipers

Vacuum pump

Sample ejector (for undisturbed, thin-wall-tube sample)

FIGURE 21–3 Axial Loading Device (Courtesy of Soiltest, Inc.)

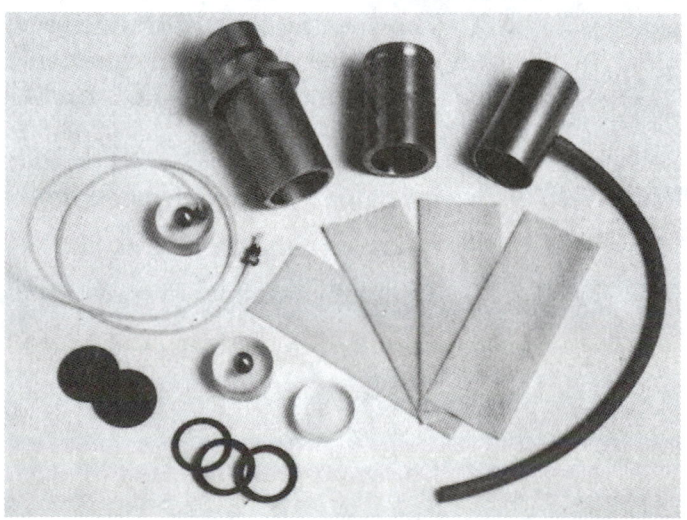

FIGURE 21–4 Triaxial Accessories (Courtesy of Wykeham Farrance, Inc.)

Graduated burette

Pore water pressure measurement device

PREPARATION OF TEST SPECIMENS [3]

(1) *Specimen Size*—Specimens shall be cylindrical and have a minimum diameter of 3.3 cm (1.3 in.). The height-to-diameter ratio shall be between 2 and 2.5. The largest particle size shall be smaller than one-sixth the specimen diameter. If, after completion of a test, it is found based on visual observation that oversize particles are present, indicate this information in the report of test data.

> *Note 1*—If oversize particles are found in the specimen after testing, a particle-size analysis may be performed in accordance with ASTM Test Method D 422 to confirm the visual observation and the results provided with the test report.

(2) *Undisturbed Specimens*—Prepare undisturbed specimens from large undisturbed samples or from samples secured in accordance with ASTM Practice D 1587 or other acceptable undisturbed tube sampling procedures. Samples shall be preserved and transported in accordance with the practices for Group C samples in ASTM Practices D 4220. Specimens obtained by tube sampling may be tested without trimming except for cutting the end surfaces plane and perpendicular to the longitudinal axis of the specimen, provided soil characteristics are such that no significant disturbance results from sampling. Handle specimens carefully to minimize disturbance, changes in cross section, or change in water content. If compression or any type of noticeable disturbance would be caused by the extrusion device, split the sample tube lengthwise or cut the tube in suitable sections to facilitate removal of the specimen with minimum disturbance. Prepare trimmed specimens in an environment such as a controlled high-humidity room where soil water content change is minimized. Where removal of pebbles or crumbling resulting from trimming causes voids on the surface of the specimen, carefully fill the voids with remolded soil obtained from the trimmings. When the sample condition permits, a vertical trimming lathe may be used to reduce the specimen to the required diameter. After obtaining the required diameter, place the specimen in a miter box and cut the specimen to the final height with a wire saw or other suitable device. Trim the surfaces with the steel straightedge. Perform one or more water content determinations on material trimmed from the specimen in accordance with ASTM Test Method D 2216. Determine the mass and dimensions of the specimen. A minimum of three height measurements (120° apart) and at least three diameter measurements at the quarter points of the height shall be made to determine the average height and diameter of the specimen.

(3) *Compacted Specimens*—Soil required for compacted specimens shall be thoroughly mixed with sufficient water to produce

the desired water content. If water is added to the soil, store the material in a covered container for at least 16 h prior to compaction. Compacted specimens may be prepared by compacting material in at least six layers using a split mold of circular cross section having dimensions meeting the requirements enumerated in (1). Specimens may be compacted to the desired density by either: (1) kneading or tamping each layer until the accumulative mass of the soil placed in the mold is compacted to a known volume; or (2) by adjusting the number of layers, the number of tamps per layer, and the force per tamp. The top of each layer shall be scarified prior to the addition of material for the next layer. The tamper used to compact the material shall have diameter equal to or less than one-half the diameter of the mold. After a specimen is formed, with the ends perpendicular to the longitudinal axis, remove the mold and determine the mass and dimensions of the specimen. Perform one or more water content determinations on excess material used to prepare the specimen in accordance with ASTM Test Method D 2216.

> Note 2—It is common for the unit weight of the specimen after removal from the mold to be less than the value based on the volume of the mold. This occurs as a result of the specimen swelling after removal of the lateral confinement due to the mold.

PART I. UNCONSOLIDATED UNDRAINED (UU) TEST

PROCEDURE   To carry out an unconsolidated undrained (UU) triaxial compression test, a cylindrical soil specimen is encased in a rubber membrane and placed in the triaxial chamber, and a specific (and constant) lateral (all-around) pressure is applied by means of water or compressed air within the chamber without allowing the specimen to consolidate (drain) under the lateral pressure. A vertical (axial) load is then applied externally and fairly rapidly without allowing drainage of the specimen and is steadily increased until the specimen fails. The externally applied axial load that causes specimen failure and the lateral pressure are recorded. The soil specimen is then removed and discarded, and another specimen of the same soil sample is placed in the triaxial chamber. The procedure is repeated for the new specimen for a different (either higher or lower) lateral pressure. The axial load at failure and the lateral pressure are recorded for the second test. The entire procedure is usually repeated for two or more other specimens of the same soil, each with a different lateral pressure.

The actual step-by-step procedure is as follows (ASTM D 2850-99 [3]):

(1) Place the membrane on the membrane expander or if it is to be rolled onto the specimen, place the membrane onto the cap or base.

Place the specimen on the base. Place the rubber membrane around the specimen and seal it at the cap and base with O-rings or other positive seals at each end. A thin coating of silicon grease on the vertical surfaces of the cap or base will aid in sealing the membrane.

(2) With the specimen encased in the rubber membrane, which is sealed to the specimen cap and base and positioned in the chamber, assemble the triaxial chamber. Bring the axial load piston into contact with the specimen cap several times to permit proper seating and alignment of the piston with the cap. When the piston is brought into contact the final time, record the reading on the deformation indicator. During this procedure, take care not to apply an axial stress to the specimen exceeding approximately 0.5% of the estimated compressive strength. If the mass of the piston is sufficient to apply an axial stress exceeding approximately 0.5% of the estimated compressive strength, lock the piston in place above the specimen cap after checking the seating and alignment and keep locked until application of the chamber pressure.

(3) Place the chamber in position in the axial loading device. Be careful to align the axial loading device, the axial load-measuring device, and the triaxial chamber to prevent the application of a lateral force to the piston during testing. Attach the pressure-maintaining and measurement device and fill the chamber with the confining liquid. Adjust the pressure-maintaining and measurement device to the desired chamber pressure and apply the pressure to the chamber fluid. Wait approximately 10 min after the application of chamber pressure to allow the specimen to stabilize under the chamber pressure prior to application of the axial load.

> *Note 1*—In some cases the chamber will be filled and the chamber pressure applied before placement in the axial loading device.

> *Note 2*—Make sure the piston is locked or held in place by the axial loading device before applying the chamber pressure.

> *Note 3*—The waiting period may need to be increased for soft or partially saturated soils.

(4) If the axial load-measuring device is located outside of the triaxial chamber, the chamber pressure will produce an upward force on the piston that will react against the axial loading device. In this case, start the test with the piston slightly above the specimen cap, and before the piston comes in contact with the specimen cap, either (1) measure and record the initial piston friction and upward thrust of the piston produced by the chamber pressure and later correct the measured axial load, or (2) adjust the axial load-measuring device to compensate for the friction and thrust. If the axial load-measuring device is located inside the chamber, it will not be necessary to correct or compensate for

the uplift force acting on the axial loading device or for piston friction. In both cases record the initial reading on the deformation indicator when the piston contacts the specimen cap.

(5) Apply the axial load to produce axial strain at a rate of approximately 1%/min for plastic materials and 0.3%/min for brittle materials that achieve maximum deviator stress at approximately 3 to 6% strain. At these rates, the elapsed time to reach maximum deviator stress will be approximately 15 to 20 min. Continue the loading to 15% axial strain, except loading may be stopped when the deviator stress has peaked then dropped 20% or the axial strain has reached 5% beyond the strain at which the peak in deviator stress occurred.

(6) Record load and deformation values at about 0.1, 0.2, 0.3, 0.4, and 0.5% strain; then at increments of about 0.5% strain to 3%; and thereafter at every 1%. Take sufficient readings to define the stress-strain curve; hence more frequent readings may be required in the early stages of the test and as failure is approached.

> *Note 4*—Alternate intervals for the readings may be used, provided sufficient points are obtained to define the stress-strain curve.

(7) After completion of the tests, remove the test specimen from the chamber. Determine the water content of the test specimen in accordance with ASTM Method D 2216 (Chapter 3), using the entire specimen, if possible.

(8) Prior to placing the specimen (or portion thereof) in the oven to dry, sketch a picture or take a photograph of the specimen showing the mode of failure (shear plane, bulging, etc.).

(9) Prepare new specimens, and repeat the entire procedure at other chamber pressures. If disturbed samples are used, the new remolded specimens should have approximately the same unit weight (within 1 to 2 lb/ft^3).

Note: Step (9) does not appear in the ASTM procedure; it has been added by the authors.

DATA Data collected in the unconsolidated undrained triaxial compression test should include the following:

[A] Specimen Data

Diameter or side of specimen, D_0 (in.)

Initial height of specimen, H_0 (in.)

Mass of specimen (g)

Water content data:

Mass of wet soil plus can (g)

Mass of dry soil plus can (g)

Mass of can (g)

[B] Triaxial Compression Data

Chamber pressure on test specimen, σ_3 (psi)

Rate of axial strain (in./min)

Deformation dial readings ΔH and proving ring dial readings (in.)

Note—These data are obtained for each of three or more specimens tested at different chamber pressures.

CALCULATIONS [A] Specimen Parameters

The initial area, volume, height-to-diameter ratio, wet unit weight, water content, dry unit weight, and degree of saturation of the specimen are all calculated by methods described in previous chapters.

[B] Triaxial Compression

For each applied load, the axial strain, ϵ, can be computed by dividing the specimen's change in height, ΔH, as read from the deformation indicator, by its initial height, H_0. In equation form,

$$\epsilon = \frac{\Delta H}{H_0} \qquad (20\text{--}1)$$

Each corresponding cross-sectional area A of the specimen can be computed by the equation

$$A = \frac{A_0}{1 - \epsilon} \qquad (20\text{--}2)$$

where A_0 is the initial area of the specimen and ϵ is the axial strain for the given axial load (expressed as a decimal). Each corresponding applied axial load can be determined by multiplying the proving ring dial reading by the proving ring calibration. Finally, each unit axial load (deviator stress) can be computed by dividing each applied axial load by the corresponding cross-sectional area. These computations must be repeated for each specimen tested.

Note—In the event that the application of the chamber pressure results in a change in the specimen height, A_0 should be corrected to reflect this change in volume. Frequently, this is done by assuming that lateral strains are equal to vertical strains. The diameter after volume change would be given by $D = D_0 (1 - \Delta H/H)$. [3]

[C] Stress-Strain Graph

A stress-strain graph should be prepared by plotting unit axial load (deviator stress) (ordinate) versus axial strain (abscissa). From this graph, the unit axial load at failure can be determined by taking the maximum unit axial load or the unit axial load at 15% axial strain, whichever occurs first. The unit axial load (deviator stress) at failure is denoted by Δp.

[D] Shear Diagram (Mohr Circles for Triaxial Compression)

The minor and major principal stresses must be determined in order to plot the required shear diagram (Mohr circle for triaxial compression), from which values of the shear strength parameters (cohesion c and angle of internal friction ϕ) can be obtained.

The minor principal stress is equal to the chamber pressure and is denoted by σ_3. The major principal stress at failure, denoted by σ_1, is equal to the sum of unit axial load at failure, Δp and minor principal stress. That is,

$$\sigma_1 = \Delta p + \sigma_3 \qquad (21\text{--}1)$$

After the minor and major principal stresses and unit axial loads at failure have been determined for each specimen tested, the required shear diagram may be prepared with shear stresses plotted along the ordinate and normal stresses on the abscissa. From results of one of the triaxial tests, a point is located along the abscissa at a distance σ_3 from the origin. This point is denoted by A in Figure 21–5 and is indicated as being located along the abscissa at a distance $(\sigma_3)_1$ from the origin. It is also necessary to locate another point along the abscissa at a distance σ_1 from the origin, by measuring either the distance σ_1 from the origin or the distance Δp from point A (the point located at a distance σ_3 from the origin). This point is denoted by B in Figure 21–5 and is noted as being located along the abscissa at a distance $(\Delta p)_1$ from point A. With AB as a diameter, a semicircle known as the *Mohr circle* is then constructed. This entire procedure is repeated using data obtained from the triaxial test of another specimen of the same soil sample at a different chamber pressure. In such manner, point C is located along the abscissa at a distance $(\sigma_3)_2$ from the origin and point D at a distance $(\Delta p)_2$ from point C. With CD as a diameter, another semicircle is then constructed. The next step is to draw a straight line tangent to the two semicircles, as shown in Figure 21–5. The angle between this straight line and a horizontal line (ϕ in the figure) gives the angle of internal friction, and the value of stress where the straight line intersects the ordinate (c in the figure) is the cohesion. The same scale must be used along abscissa and ordinate.

In theory, it is adequate to have only two Mohr circles to define the straight-line relationship of Figure 21–5. In practice, however, it is better to have three (or more) such semicircles that can be used to draw the best straight line. That is why the test procedure calls for three or more separate tests to be performed on three or more specimens from the same soil sample.

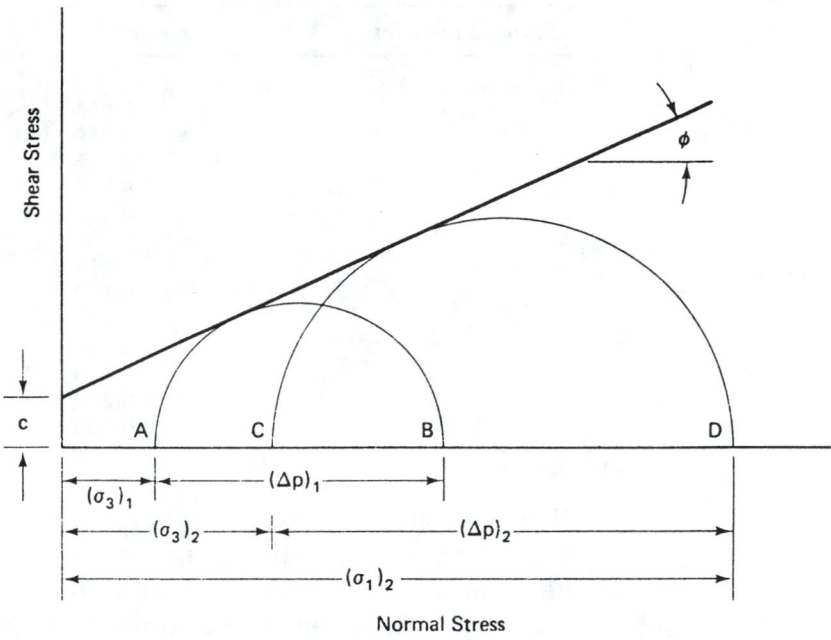

FIGURE 21–5 Shear Diagram for Triaxial Compression Test

NUMERICAL EXAMPLE An unconsolidated undrained triaxial compression test was performed on a soil specimen in the laboratory according to the procedure described. The following data were obtained:

[A] Specimen Data

Diameter of specimen, $D_0 = $ **2.50 in.**

Initial height of specimen, $H_0 = $ **5.82 in.**

Mass of specimen = **920.20 g**

Water content data:

Mass of wet soil plus can = **939.92 g**

Mass of dry soil plus can = **811.07 g**

Mass of can = **48.62 g**

The specific gravity of the soil is known to be **2.78**

[B] Triaxial Compression Data

Chamber pressure on test specimen, $\sigma_3 = $ **10.0 psi**

Rate of axial strain = **0.02 in./min**

Proving ring calibration = **6,000 lb/in.**

Elapsed Time (min)	Deformation Dial, ΔH (in.)	Proving Ring Dial (in.)
0	0	0
	0.005	0.0012
	0.010	0.0025
	0.015	0.0037
	0.020	0.0053
	0.025	0.0066
	0.050	0.0140
	0.075	0.0201
	0.100	0.0256
	0.125	0.0294
	0.150	0.0321
	0.175	0.0337
	0.200	0.0331
11.25	0.225	0.0305

[It is, of course, necessary to perform triaxial tests on two (preferably, even three or more) specimens from the same sample at different chamber pressures in order to evaluate the shear strength parameters. In this example, however, actual data are shown for only one specimen tested.]

With the preceding data known, necessary computations can be made as follows:

[A] Specimen Parameters

The height-to-diameter ratio is **5.82/2.50,** or 2.33, which is between 2 and 2.5 and therefore acceptable. The initial area A_0 and volume V_0 of the specimen are easily computed as follows:

$$A_0 = \frac{\pi D_0}{4} = \frac{\pi (2.50)^2}{4} = 4.91 \text{ in.}^2$$

$$V_0 = (H_0)(A_0) = (5.82)(4.91) = 28.58 \text{ in.}^3$$

With the mass and volume of the specimen known, its wet unit weight γ_{wet} can be determined:

$$\gamma_{wet} = \left(\frac{920.20}{28.58}\right)\left(\frac{1,728}{453.6}\right) = 122.7 \text{ lb/ft}^3$$

(The numbers 1,728 and 453.6 are conversion factors: $1,728 \text{ in.}^3 = 1 \text{ ft}^3$; $453.6 \text{ g} = 1 \text{ lb.}$)

The water content of the specimen may be computed as follows (see Chapter 3):

$$w = \frac{939.92 - 811.07}{811.07 - 48.62} \times 100 = 16.9\%$$

With water content known, dry unit weight γ_{dry} can be determined next:

$$\gamma_{dry} = \frac{\gamma_{wet}}{w + 100} \times 100 = \frac{122.7}{16.9 + 100} \times 100 = 105.0 \text{ lb/ft}^3$$

In order to find the degree of saturation, a single cubic foot of soil specimen may be isolated as follows:

Weight of water in 1 ft^3 of soil specimen

$$= \gamma_{wet} - \gamma_{dry} = 122.7 - 105.0 = 17.7 \text{ lb}$$

Volume of water in 1 ft^3 of soil specimen

$$= \frac{\text{weight of water}}{\text{unit weight of water}} = \frac{17.7}{62.4} = 0.284 \text{ ft}^3$$

Weight of solid in 1 ft^3 of soil specimen $= \gamma_{dry} = 105.0 \text{ lb}$

Volume of solid in 1 ft^3 of soil specimen

$$= \frac{\text{weight of solid}}{\text{specific gravity of solid} \times 62.4} = \frac{105.0}{(\mathbf{2.78})(62.4)} = 0.605 \text{ ft}^3$$

Volume of void in 1 ft^3 of soil specimen

$$= 1 - \text{volume of solid} = 1 - 0.605 = 0.395 \text{ ft}^3$$

$$\text{Degree of saturation} = \frac{\text{volume of water}}{\text{volume of void}} \times 100$$

$$= \frac{0.284}{0.395} \times 100 = 71.9\%$$

These data, both given and computed, are shown on the form on page 349. At the end of the chapter, two blank copies of this form are included for the reader's use.

[B] Triaxial Compression

The first deformation dial reading ΔH is **0.005 in.** The axial strain ϵ for this particular applied load can be computed using Eq. (20–1):

$$\epsilon = \frac{\Delta H}{H_0} \tag{20–1}$$

$$\epsilon = \frac{\mathbf{0.005}}{\mathbf{5.82}} = 0.0009$$

The corresponding cross-sectional area of the specimen can be computed using Eq. (20–2):

$$A = \frac{A_0}{1 - \epsilon} \qquad\qquad\qquad \textbf{(20–2)}$$

$$A = \frac{4.91}{1 - 0.0009} = 4.91 \text{ in}^2$$

Multiplying the proving ring dial reading (**0.0012 in.**) by the proving ring calibration (**6,000 lb/in.**) gives a corresponding applied axial load of 7.2 lb. Finally, dividing the applied axial load (7.2 lb) by the corresponding cross-sectional area (4.91 in.^{2}) gives a unit axial load of 1.5 lb/in.^{2} (or psi).

The values above fill in the second row of the form on **page 350**. Similar calculations furnish values for succeeding given deformation dial readings and proving ring dial readings needed to fill in the remaining rows on the form.

It should be emphasized that one or two additional evaluations are required for additional triaxial compression tests on other specimens of the same sample at different chamber pressures. Such evaluations are not included here.

[C] Stress-Strain Curve

The required stress-strain curve is obtained by plotting axial strain (third column on page 350) along the abscissa versus unit axial load (seventh column on page 350) along the ordinate. The curve for this example is the lower one in Figure 21–6. As indicated in the figure, the values of σ_3 (given), Δp (determined from the curve), and σ_1 ($\Delta p + \sigma_3$) are 10.0, 40.0, and 50.0 psi, respectively.

It will be noted that two additional stress-strain curves for other triaxial compression tests on different specimens, along with associated values of σ_3, Δp, and σ_1, are presented in Figure 21–6. Neither the test data nor required subsequent calculations for defining these curves are given in this example. They are provided at this point in order to complete the evaluation of the shear strength parameters.

[D] Shear Diagram (Mohr Circles for Triaxial Compression)

From results of the triaxial test for which data are given in this example, a point is located along the abscissa at a distance σ_3, or 10.0 psi, from the origin (see Figure 21–7). Another point is located along the abscissa at a distance σ_1, or 50.0 psi, from the origin. With the straight line connecting these points as a diameter, a semicircle is then constructed as shown in Figure 21–7. By repeating the process for the test for the second specimen ($\sigma_3 = 20.0$ psi, $\sigma_1 = 67.6$ psi), points are located along the abscissa at $\sigma_3 = 20.0$ psi and $\sigma_1 = 67.6$ psi from the origin, and a semicircle is drawn. Another semicircle is then constructed in the same manner for the test for the third specimen ($\sigma_3 = 30.0$ psi, $\sigma_1 = 85.5$ psi).

With the three semicircles thus constructed, a straight line tangent to the semicircles is drawn as shown in Figure 21–7. The angle between

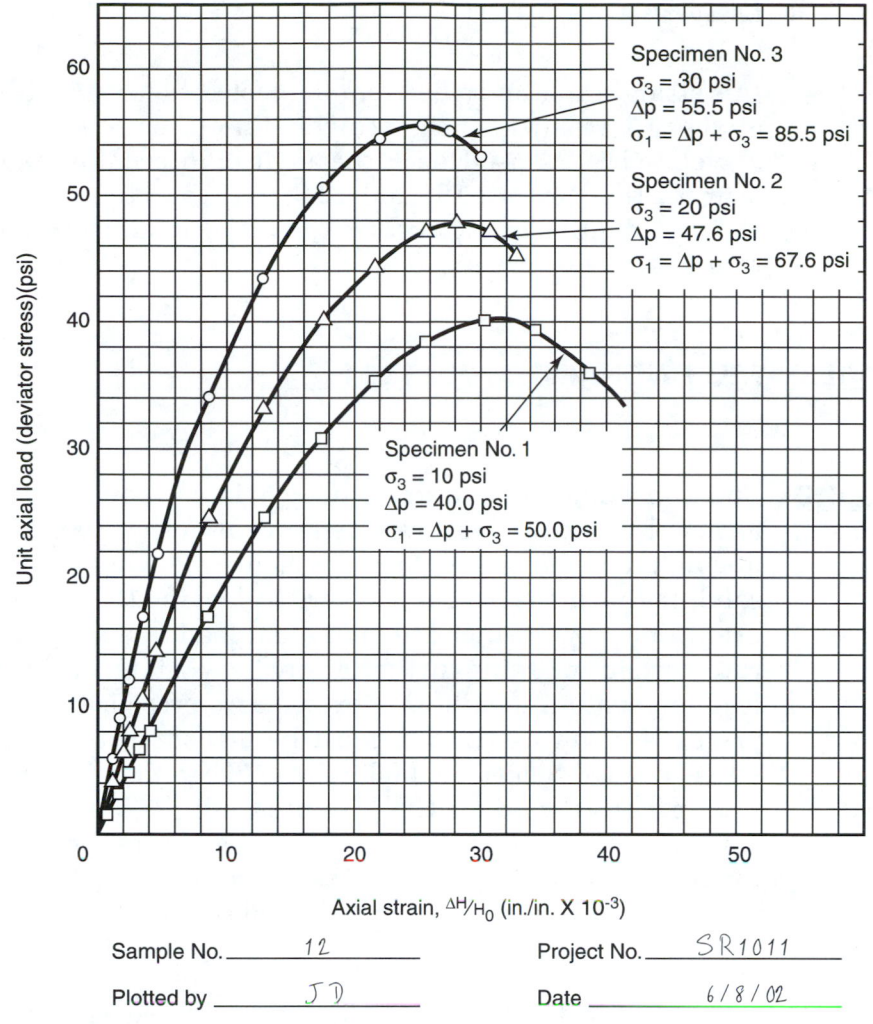

FIGURE 21–6   Unit Axial Load versus Axial Strain Curves

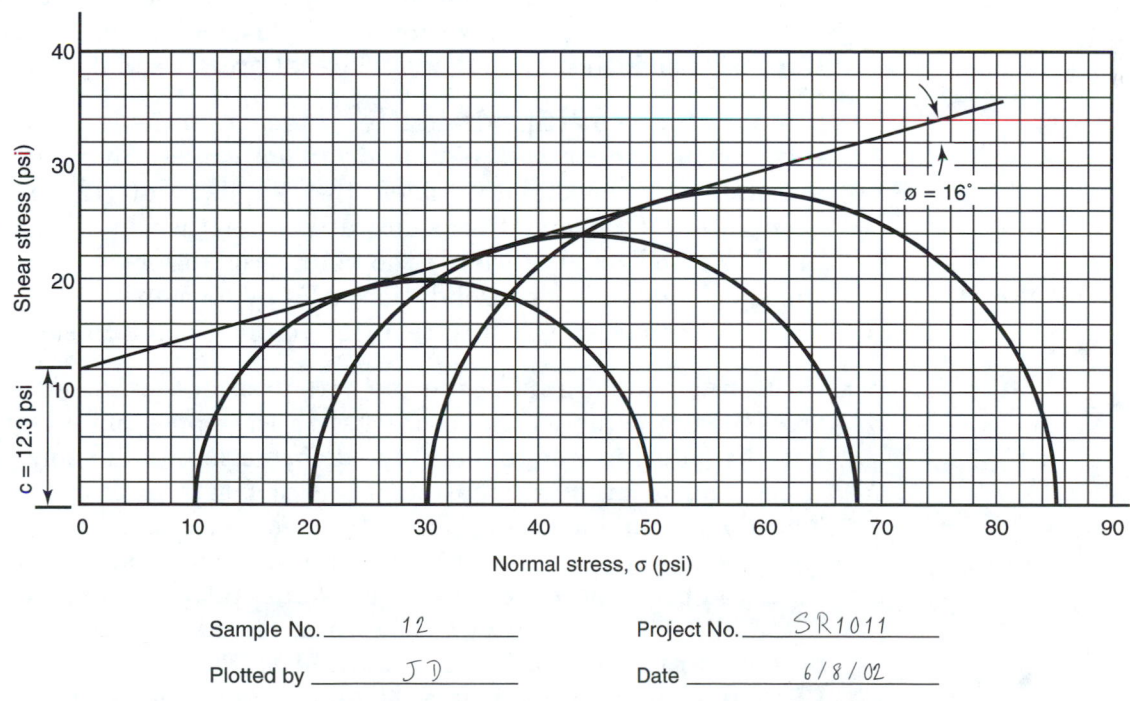

FIGURE 21–7   Mohr Circles for Triaxial Compression

this straight line and a horizontal line is measured as 16°, and the value of shear stress where the straight line intersects the ordinate is observed to be 12.3 psi. Hence, the angle of internal friction and the cohesion for this soil are 16° and 12.3 psi, respectively. These values (the shear strength parameters) are the major results of a triaxial compression test.

PART II. CONSOLIDATED UNDRAINED (CU) TEST

PROCEDURE   The consolidated undrained (CU) triaxial compression test is performed by placing a saturated specimen in the triaxial chamber, introducing lateral (confining) pressure, and allowing the specimen to consolidate under the lateral pressure by leaving the drain lines open. The drain lines are then closed, and axial load is applied at a fairly rapid rate without allowing further drainage. With no drainage during axial load application, a buildup of excess pore pressure will result. The pore pressure μ during the test must be measured to obtain the effective stress needed to plot the Mohr circle. (The effective stress $\bar{\sigma}$ equals the total pressure σ minus the pore pressure μ—that is, $\bar{\sigma} = \sigma - \mu$.) The pore pressure can be determined using a pressure-measuring device connected to the drain lines at each end of the specimen.

A step-by-step procedure is as follows:

(1) With the saturated specimen encased in a rubber membrane and inside the triaxial chamber, assemble the triaxial chamber in the axial loading device. Carefully adjust the piston of the cell so that it just makes contact with the top platen of the specimen. Fill the chamber with fluid and apply lateral pressure (σ_3).

(2) With the connecting drain lines attached to the graduated burette, open both top and bottom drainage valves. This allows the specimen to consolidate under the applied lateral pressure. Completion of consolidation is indicated when the water level in the burette stabilizes. Determine the volume of water extruded from the specimen during consolidation (ΔV) by measuring the volume of water accrued in the burette at the completion of consolidation.

(3) Connect the drain lines to the pore pressure measurement device. Bring the piston into contact with the upper platen, and set both the proving ring dial and the deformation dial to zero. Apply axial load to produce axial strain at the rate of approximately 1%/min. As axial load is applied, the specimen's pore pressure will increase. Record proving ring dial readings, deformation dial readings, and pore water pressure measurements at about 0.1, 0.2, 0.3, 0.4, and 0.5% strain. After 0.5% strain, record at every increment of 0.5% strain up to 3% strain and thereafter at every 1% strain. Continue loading to 15% axial strain, except that loading may be

stopped when the deviator stress has peaked and then dropped 20%, or when the axial strain has reached 5% beyond the strain at which the peak in deviator stress occurred.

(4) Upon completion of the test, stop the compression and release the axial load. Remove the test specimen from the chamber, and make a sketch or take a photograph of the test specimen at failure. Record the angle of failure surface. Use the entire specimen to determine its water content (see Chapter 3).

(5) Prepare new specimens, and repeat the entire procedure at other chamber pressures. If disturbed samples are used, the new remolded specimens should have approximately the same unit weight (within 1 to 2 lb/ft^3).

DATA

Data collected in the consolidated undrained triaxial compression test should include the following:

[A] Specimen Data

Diameter or side of specimen, D_0 (in.)

Initial height of specimen, H_0 (in.)

Mass of specimen (g)

Water content data:

Mass of wet soil plus can (g)

Mass of dry soil plus can (g)

Mass of can (g)

[B] Triaxial Compression Data

Chamber pressure on test specimen, σ_3 (psi)

Volume of water extruded from the specimen during consolidation, ΔV (i.e., volume of water accrued in burette) (in.^{3})

Rate of axial strain (in./min)

Deformation dial readings, ΔH, and proving ring dial readings (in.)

Pore water pressure, μ (psi)

Note—These data are obtained for each of three or more specimens tested at different chamber pressures.

CALCULATIONS

[A] Specimen Parameters

The specimen's initial area, volume, height-to-diameter ratio, wet unit weight, water content, dry unit weight, and degree of saturation are all calculated by methods described in previous chapters.

[B] Triaxial Compression (After Consolidation in Triaxial Chamber)

The specimen's volume after consolidation (V_c) can be obtained by subtracting the volume of drainage out of the specimen during consolidation (i.e., the volume of water accrued in the burette) (ΔV) from the specimen's original volume (V_0). In equation form,

$$V_c = V_0 - \Delta V \tag{21-2}$$

If the specimen's height and diameter are assumed to decrease in the same proportions during consolidation, area of the specimen after consolidation (A_c) can be derived from its initial area A_0 using the equation

$$A_c = A_0\left(\frac{V_c}{V_0}\right)^{2/3} \tag{21-3}$$

and the height of the specimen after consolidation H_c can be computed by

$$H_c = H_0\left(\frac{V_c}{V_0}\right)^{1/3} \tag{21-4}$$

For each applied load, the axial unit strain ϵ can be computed by dividing the change in height of the specimen, ΔH, by its height after consolidation, H_c. In equation form,

$$\epsilon = \frac{\Delta H}{H_c} \tag{21-5}$$

Each corresponding cross-sectional area of the specimen, A, can be computed by the equation

$$A = \frac{A_c}{1 - \epsilon} \tag{21-6}$$

where A_c is the initial area of the specimen after consolidation. Each corresponding applied axial load can be determined by multiplying the proving ring dial reading by the proving ring calibration. Finally, each unit axial load can be computed by dividing each applied axial load by the corresponding cross-sectional area. These computations must be repeated for each specimen tested.

[C] Stress-Strain and Pore Pressure–Strain Graphs

A stress-strain graph should be prepared by plotting unit axial load on the ordinate versus axial strain on the abscissa. From this graph, the unit axial load at failure can be determined by taking the maximum unit axial load or the unit axial load at 15% axial strain, whichever occurs first. The unit axial load at failure is denoted by Δp.

A graph of pore pressure μ versus axial strain ϵ should also be prepared by plotting pore pressure on the ordinate versus axial strain on the abscissa. The same scale for axial strain in the stress-strain graph should be used in this graph. The pore pressure corresponding to the unit axial load at failure can be determined from the graph.

[D] Shear Diagram (Mohr Circles for Triaxial Compression)

Results of CU tests are commonly presented with Mohr circles plotted in terms of effective stress $\bar{\sigma}$. The effective lateral pressure $\bar{\sigma}_3$ and major effective principal stress $\bar{\sigma}_1$ can be computed as follows:

$$\bar{\sigma}_3 = \sigma_3 - \mu \tag{21-7}$$

$$\bar{\sigma}_1 = \sigma_1 - \mu \tag{21-8}$$

(It should be noted that μ is the pore pressure corresponding to the unit axial load at failure and $\sigma_1 = \sigma_3 + \Delta p$.) The strength envelope in this case is referred to as the *effective stress strength envelope*.

Results of CU tests are also commonly presented with Mohr circles plotted in terms of total stress σ. The strength envelope in this case is referred to as the *total stress strength envelope*.

Both the effective stress strength envelope and the total stress strength envelope obtained from a CU test on normally consolidated clay are shown in Figure 21–8. It will be noted that the Mohr circle has equal diameters for total stresses and effective stresses but the Mohr circle for effective stresses is displaced leftward by an amount equal to the pore pressure at failure μ_f.

If several CU tests are performed on the same normally consolidated clay initially consolidated under different lateral pressures σ_3, the total stress strength envelope is approximately a straight line passing through the origin (see Figure 21–8). Hence, results of CU triaxial tests on normally consolidated clays can be expressed by Coulomb's equation,

$$s = \sigma \tan \phi_{CU} \tag{21-9}$$

where ϕ_{CU} is known as the consolidated undrained angle of internal friction.

NUMERICAL EXAMPLE

A consolidated undrained triaxial compression test was performed on a soil sample in the laboratory according to the procedure described. The data obtained in the laboratory and subsequent computed data are

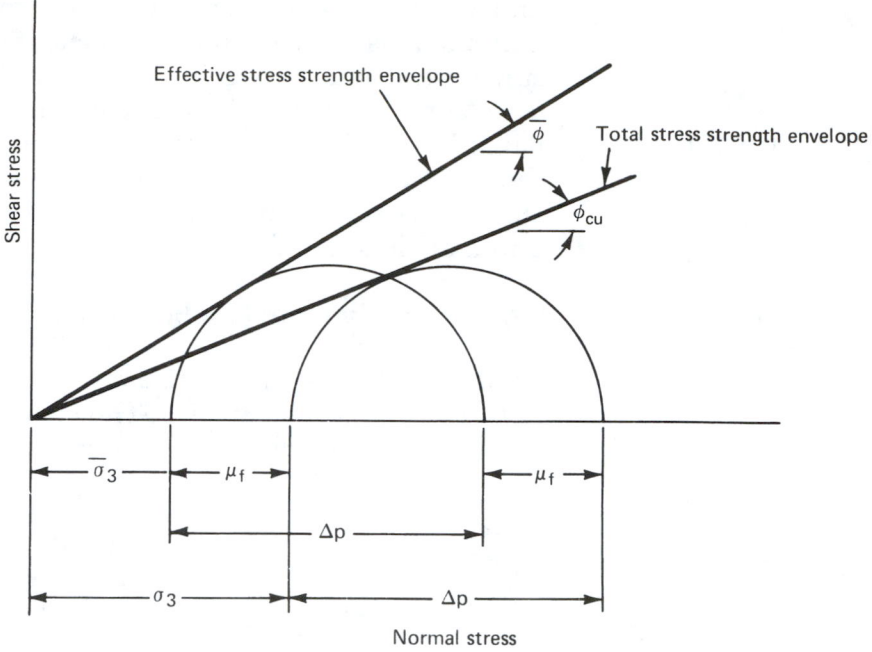

FIGURE 21–8   Results of Consolidated Undrained Triaxial Tests on Normally Consolidated Clay

given in the form on page 351 and 352. (At the end of the chapter, two blank copies of this form are included for the reader's use.) The required graphs of stress versus strain and pore pressure versus strain are presented in Figures 21–9 and 21–10, respectively. Both the effective stress strength envelope and the total stress strength envelope are shown in Figure 21–11.

CONCLUSIONS

As indicated previously, values of the angle of internal friction and cohesion are the major results of a triaxial test. In addition to these parameters, however, reports should include values of the initial unit weight, moisture content, and degree of saturation, as well as the height-to-diameter ratio of the specimen tested. The type of test performed (UU, CU, CD) and the type (undisturbed, remolded, compacted) and shape (cylindrical, prismatic) of specimen should also be reported. Of course, a visual description of the soil and any unusual conditions should be noted. The average rate of axial strain to failure, the axial strain at failure, and whether strain control or stress control was used should also be indicated. The stress-strain curves, pore pressure–strain curve (if the CU test is performed), and shear diagram (Mohr circles) should be presented in the report. Sketches of the failed specimen might also be included.

For any dry soil, about the same shear strength parameters would be obtained from any of the three triaxial tests (UU, CU, or CD). For a saturated or partially saturated cohesionless soil, the CD test will yield about the same ϕ angle as for a dry soil, unless the material is very fine grained (low permeability) and/or the test is performed at an extremely rapid rate of strain. For any saturated cohesive soil, results are dependent on which

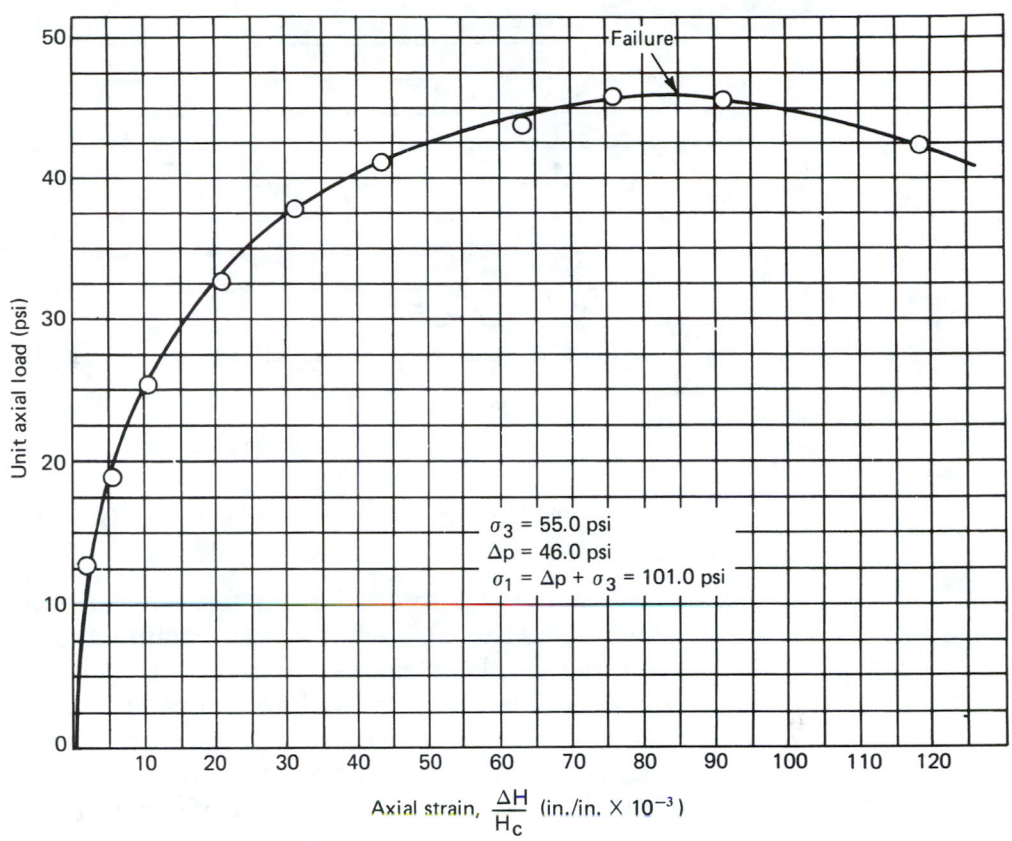

FIGURE 21-9 Curve of Unit Axial Load Versus Axial Strain

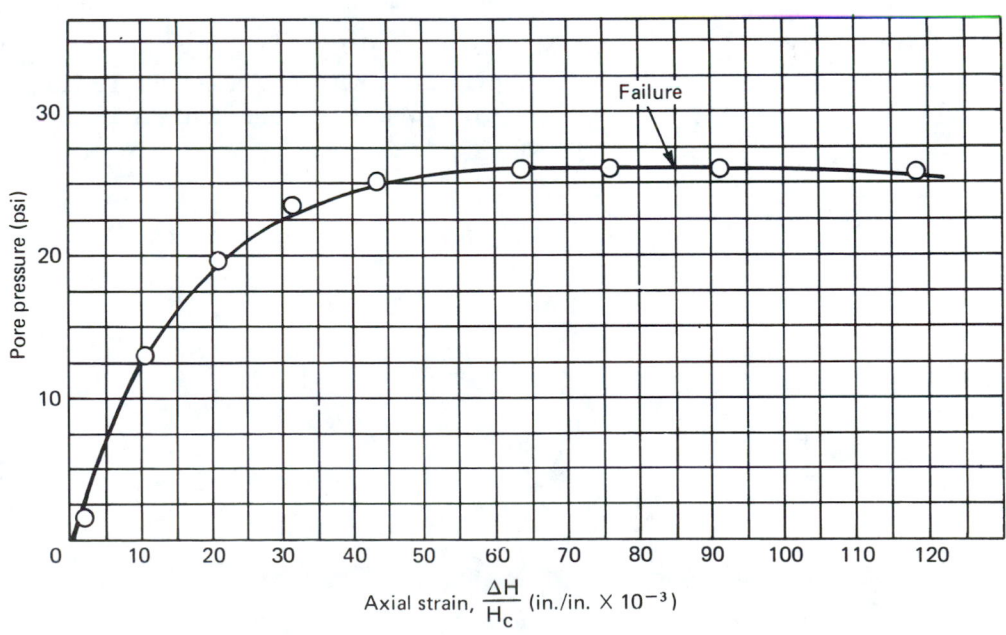

FIGURE 21-10 Curve of Pore Pressure Versus Axial Strain

347

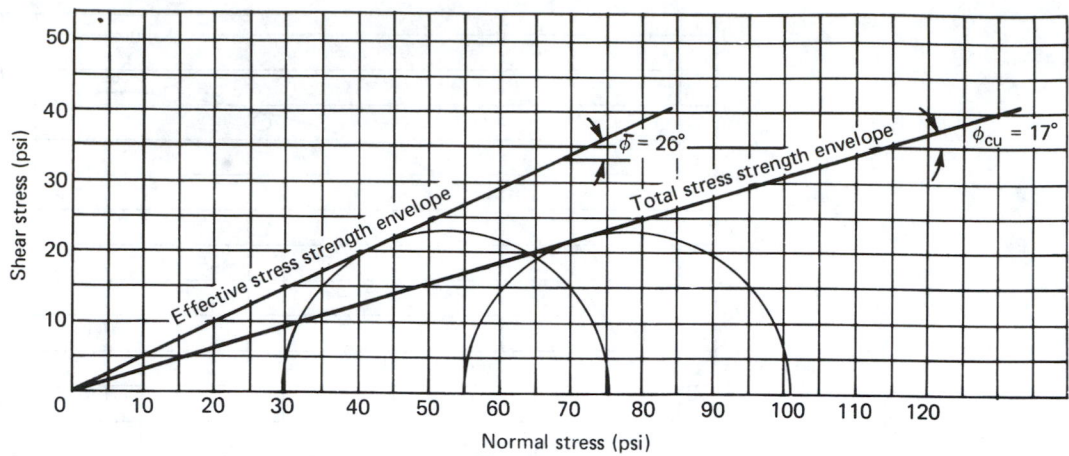

FIGURE 21–11 Results of CU Test for Given Numerical Example

of the three tests is used. Parameters will range from $\phi \cong 0$ and $c =$ some value using the UU test to $\phi =$ true value and $c = 0$ using the CD test. Results will also depend on whether the soil is normally consolidated, overconsolidated, or a remolded sample. For any partially saturated cohesive soil, results depend on both degree of saturation and type of drained test performed. Results from an undrained test will be highly dependent on the sample's degree of saturation, ranging from $\phi = 0$ for $S = 100\%$ to $\phi =$ true value for $S = 0$ [1].

REFERENCES

[1] Joseph E. Bowles, *Engineering Properties of Soils and Their Measurement*, 2d ed., McGraw-Hill Book Company, New York, 1978.

[2] Irving S. Dunn, Loren R. Anderson, and Fred W. Kiefer, *Fundamentals of Geotechnical Analysis*, John Wiley & Sons, Inc., New York, 1980.

[3] ASTM, *2001 Annual Book of ASTM Standards*, West Conshohocken, PA, 2001. Copyright, American Society for Testing and Materials, 100 Barr Harbor Drive, West Conshohocken, PA 19428-2959. Reprinted with permission.

Soils Testing Laboratory
Unconsolidated Undrained (UU)
Triaxial Compression Test

Sample No. _____ 12 _____   Project No. _____ SR 1011 _____

Location _____ Charlotte, N.C. _____   Boring No. _____ 2 _____

Depth _____ 11 ft _____   Date of Test _____ 6/5/02 _____

Description of Soil _____ Brown silty clay _____

Tested by _____ John Doe _____

[A] Specimen Data   (Specimen no. __1__)

(1)   Type of test performed __unconsolidated undrained, without pore pressure measurement__

(2)   Type of specimen (check one)   ☐ Undisturbed   ☒ Remolded

(3)   Diameter of specimen, D_0 __2.50__ in.

(4)   Initial area of specimen, A_0 __4.91__ in.^{2}

(5)   Initial height of specimen, H_0 __5.82__ in.

(6)   Height-to-diameter ratio $\left[\text{i.e., } \dfrac{(5)}{(3)}\right]$ __2.33__

(7)   Volume of specimen, V_0 [i.e., (4) × (5)] __28.58__ in.^{3}

(8)   Mass of specimen __920.20__ g

(9)   Wet unit weight of specimen $\left[\text{i.e., } \dfrac{(8)}{453.6} \times \dfrac{1{,}728}{(7)}\right]$ __122.7__ lb/ft^3

(10)   Water content of specimen __16.9__ %

    (a)   Can no. __2-B__

    (b)   Mass of wet soil + can __939.92__ g

    (c)   Mass of dry soil + can __811.07__ g

    (d)   Mass of can __48.62__ g

    (e)   Mass of water [i.e., (b) − (c)] __128.85__ g

    (f)   Mass of dry soil [i.e., (c) − (d)] __762.45__ g

    (g)   Water content $\left[\text{i.e., } \dfrac{(e)}{(f)} \times 100\right]$ __16.9__ %

(11)   Dry unit weight of specimen $\left[\text{i.e., } \dfrac{(9)}{(10) + 100} \times 100\right]$ __105.0__ lb/ft^3

(12)   Specific gravity of soil __2.78__

(13)   Weight of water in 1 ft^3 of soil specimen [i.e., (9) − (11)] __17.7__ lb

(14)   Volume of water in 1 ft^3 of soil specimen $\left[\text{i.e., } \dfrac{(13)}{62.4}\right]$ __0.284__ ft^3

(15)   Weight of solid in 1 ft^3 of soil specimen [i.e., (11)] __105.0__ lb

(16)   Volume of solid in 1 ft^3 of soil specimen $\left[\text{i.e., } \dfrac{(15)}{(12) \times 62.4}\right]$ __0.605__ ft^3

(17)   Volume of void in 1 ft^3 of soil specimen [i.e., 1 − (16)] __0.395__ ft^3

(18)   Degree of saturation $\left[\text{i.e., } \dfrac{(14)}{(17)} \times 100\right]$ __71.9__ %

[B] Triaxial Compression Data   (Specimen no. __1__)

(1) Chamber pressure on test specimen, σ_3 __10.0__ psi
(2) Rate of axial strain __0.02__ in./min
(3) Initial height of specimen, H_0 __5.82__ in.
(4) Proving ring calibration __6,000__ lb/in.

Elapsed Time (min)	Deformation Dial, ΔH (in.)	Axial Strain, ϵ (in./in.)	Cross-Sectional Area, A (in.^{2})	Proving Ring Dial (in.)	Applied Axial Load (lb)	Unit Axial Load (Deviator Stress) (psi)
(1)	(2)	$(3) = \dfrac{\Delta H}{H_0}$	$(4) = \dfrac{A_0}{1 - \epsilon}$	(5)	$(6) = (5) \times$ proving ring calibration	$(7) = \dfrac{(6)}{(4)}$
0	0	0	4.91	0	0	0
	0.005	0.0009	4.91	0.0012	7.2	1.5
	0.010	0.0017	4.92	0.0025	15.0	3.0
	0.015	0.0026	4.92	0.0037	22.2	4.5
	0.020	0.0034	4.93	0.0053	31.8	6.5
	0.025	0.0043	4.93	0.0066	39.6	8.0
	0.050	0.0086	4.95	0.0140	84.0	17.0
	0.075	0.0129	4.97	0.0201	120.6	24.3
	0.100	0.0172	5.00	0.0256	153.6	30.7
	0.125	0.0215	5.02	0.0294	176.4	35.1
	0.150	0.0258	5.04	0.0321	192.6	38.2
	0.175	0.0301	5.06	0.0337	202.2	40.0
	0.200	0.0344	5.08	0.0331	198.6	39.1
11.25	0.225	0.0387	5.11	0.0305	183.0	35.8

Minor principal stress (i.e., chamber pressure, σ_3) __10.0__ psi
Unit axial load at failure, Δp __40.0__ psi
Major principal stress at failure (i.e., $\sigma_1 = \sigma_3 + \Delta p$) __50.0__ psi

Soils Testing Laboratory
Consolidated Undrained (CU)
Triaxial Compression Test

Sample No. _____10_____   Project No. _____SR 1012_____

Location _____Newell, N.C._____   Boring No. _____1_____

Depth _____20 ft_____   Date of Test _____6/6/02_____

Description of Soil _____Brown clay_____

Tested by _____John Doe_____

[A] Specimen Data   (before consolidation) (Specimen no. ___1___)

(1) Type of test performed __consolidated undrained, with pore__ __pressure measurement_____

(2) Type of specimen (check one)   ☐ Undisturbed   ☒ Remolded
(3) Diameter of specimen, D_0 __2.50__ in.
(4) Initial area of specimen, A_0 __4.91__ in.^{2}
(5) Initial height of specimen, H_0 __6.10__ in.

(6) Height-to-diameter ratio $\left[\text{i.e., } \frac{(5)}{(3)} \right]$ __2.44__

(7) Volume of specimen, V_0 [i.e., (4) × (5)] __29.95__ in.^{3}
(8) Mass of specimen __910.12__ g

(9) Wet unit weight of specimen $\left[\text{i.e., } \frac{(8)}{453.6} \times \frac{1,728}{(7)} \right]$ __115.8__ lb/ft^3

(10) Water content of specimen __36.07__ %
   (a) Can no. __1-A__
   (b) Mass of wet soil + can __525.21__ g
   (c) Mass of dry soil + can __398.40__ g
   (d) Mass of can __46.83__ g
   (e) Mass of water [i.e., (b) − (c)] __126.81__ g
   (f) Mass of dry soil [i.e., (c) − (d)] __351.57__ g

   (g) Water content $\left[\text{i.e., } \frac{(e)}{(f)} \times 100 \right]$ __36.07__ %

(11) Dry unit weight of specimen $\left[\text{i.e., } \frac{(9)}{(10) + 100} \times 100 \right]$ __85.1__ lb/ft^3

[B] Specimen Data   (after consolidation)

(1) Chamber consolidation pressure __55.0__ psi
(2) Volume of water extruded from specimen during consolidation [i.e., volume of water accrued in burette], ΔV __12.8__ cm^3 or __0.78__ in.^{3}
(3) Volume of specimen after consolidation, V_c [i.e., $V_0 - \Delta V$] __29.17__ in.^{3}
(4) Area of specimen after consolidation, A_c [i.e., $A_0(V_c/V_0)^{2/3}$] __4.82__ in.^{2}
(5) Height of specimen after consolidation, H_c [i.e., $H_0(V_c/V_0)^{1/3}$] __6.05__ in.

[C] Triaxial Compression Data   (Specimen no. __1__)

(1) Consolidation pressure on test specimen, σ_3 __55.0__ psi
(2) Rate of axial strain __0.05__ in./min
(3) Height of specimen after consolidation, H_c __6.05__ in.
(4) Proving ring calibration __6,000__ lb/in.

Elapsed Time (min)	Deformation Dial, ΔH (in.)	Axial Strain, ϵ (in./in.)	Cross-Sectional Area, A (in.^{2})	Proving Ring Dial (in.)	Applied Axial Load (lb)	Unit Axial Load (Deviator Stress) (psi)	Pore Pressure, μ (psi)
(1)	(2)	$(3) = \dfrac{\Delta H}{H_0}$	$(4) = \dfrac{A_0}{1-\epsilon}$	(5)	$(6) = (5) \times$ proving ring calibration	$(7) = \dfrac{(6)}{(4)}$	(8)
0	0	0	4.82	0	0	0	0
	0.012	0.0020	4.83	0.0021	12.6	2.6	1.5
	0.033	0.0055	4.85	0.0152	91.2	18.8	7.1
	0.065	0.0107	4.87	0.0205	123.0	25.3	12.9
	0.127	0.0210	4.92	0.0267	160.2	32.6	19.5
	0.191	0.0316	4.98	0.0314	188.4	37.8	23.4
	0.265	0.0438	5.04	0.0345	207.0	41.1	25.0
	0.384	0.0635	5.15	0.0374	224.4	43.6	25.8
	0.460	0.0760	5.22	0.0398	238.8	45.7	25.8
	0.551	0.0911	5.30	0.0402	241.2	45.5	25.8
14.3	0.717	0.1185	5.47	0.0386	231.6	42.3	25.6

Minor principal stress (i.e., chamber pressure, σ_3) __55.0__ psi
Unit axial load at failure, Δp __46.0__ psi
Major principal stress at failure (i.e., $\sigma_1 = \sigma_3 + \Delta p$) __101.0__ psi
Pore pressure corresponding to unit axial load at failure, μ_f __25.8__ psi
Effective minor principal stress, $\bar{\sigma}_3$ [i.e., $\bar{\sigma}_3 - \mu_f$] __29.2__ psi
Effective major principal stress at failure, $\bar{\sigma}_1$ [i.e., $\sigma_1 - \mu_f$] __75.2__ psi

Soils Testing Laboratory
Unconsolidated Undrained (UU)
Triaxial Compression Test

Sample No. _____ Project No. _____

Location _____ Boring No. _____

Depth _____ Date of Test _____

Description of Soil _____

Tested by _____

[A] Specimen Data (Specimen no. _____)

(1) Type of test performed _____

_____

(2) Type of specimen (check one) ☐ Undisturbed ☐ Remolded

(3) Diameter of specimen, D_0 _____ in.

(4) Initial area of specimen, A_0 _____ in.^{2}

(5) Initial height of specimen, H_0 _____ in.

(6) Height-to-diameter ratio $\left[\text{i.e., } \dfrac{(5)}{(3)}\right]$ _____

(7) Volume of specimen, V_0 [i.e., (4) × (5)] _____ in.^{3}

(8) Mass of specimen _____ g

(9) Wet unit weight of specimen $\left[\text{i.e., } \dfrac{(8)}{453.6} \times \dfrac{1,728}{(7)}\right]$ _____ lb/ft^3

(10) Water content of specimen _____ %

   (a) Can no. _____

   (b) Mass of wet soil + can _____ g

   (c) Mass of dry soil + can _____ g

   (d) Mass of can _____ g

   (e) Mass of water [i.e., (b) − (c)] _____ g

   (f) Mass of dry soil [i.e., (c) − (d)] _____ g

   (g) Water content $\left[\text{i.e., } \dfrac{(e)}{(f)} \times 100\right]$ _____ %

(11) Dry unit weight of specimen $\left[\text{i.e., } \dfrac{(9)}{(10) + 100} \times 100\right]$ _____ lb/ft^3

(12) Specific gravity of soil _____

(13) Weight of water in 1 ft^3 of soil specimen [i.e., (9) − (11)] _____ lb

(14) Volume of water in 1 ft^3 of soil specimen $\left[\text{i.e., } \dfrac{(13)}{62.4}\right]$ _____ ft^3

(15) Weight of solid in 1 ft^3 of soil specimen [i.e., (11)] _____ lb

(16) Volume of solid in 1 ft^3 of soil specimen $\left[\text{i.e., } \dfrac{(15)}{(12) \times 62.4}\right]$ _____ ft^3

(17) Volume of void in 1 ft^3 of soil specimen [i.e., 1 − (16)] _____ ft^3

(18) Degree of saturation $\left[\text{i.e., } \dfrac{(14)}{(17)} \times 100\right]$ _____ %

[B] Triaxial Compression Data   (Specimen no. _____)

(1)   Chamber pressure on test specimen, σ_3 _____ psi
(2)   Rate of axial strain _____ in./min
(3)   Initial height of specimen, H_0 _____ in.
(4)   Proving ring calibration _____ lb/in.

Elapsed Time (min)	Deformation Dial, ΔH (in.)	Axial Strain, ϵ (in./in.)	Cross-Sectional Area, A (in.^{2})	Proving Ring Dial (in.)	Applied Axial Load (lb)	Unit Axial Load (Deviator Stress) (psi)
(1)	(2)	$(3) = \dfrac{\Delta H}{H_0}$	$(4) = \dfrac{A_0}{1 - \epsilon}$	(5)	$(6) = (5) \times$ proving ring calibration	$(7) = \dfrac{(6)}{(4)}$

Minor principal stress (i.e., chamber pressure, σ_3) _____ psi
Unit axial load at failure, Δp _____ psi
Major principal stress at failure (i.e., $\sigma_1 = \sigma_3 + \Delta p$) _____ psi

Soils Testing Laboratory
Unconsolidated Undrained (UU)
Triaxial Compression Test

Sample No. _____   Project No. _____

Location _____   Boring No. _____

Depth _____   Date of Test _____

Description of Soil _____

Tested by _____

[A] Specimen Data   (Specimen no. _____)

    (1)   Type of test performed _____
_____

    (2)   Type of specimen (check one)   ☐ Undisturbed   ☐ Remolded

    (3)   Diameter of specimen, D_0 _____ in.

    (4)   Initial area of specimen, A_0 _____ in.^{2}

    (5)   Initial height of specimen, H_0 _____ in.

    (6)   Height-to-diameter ratio $\left[\text{i.e., } \dfrac{(5)}{(3)} \right]$ _____

    (7)   Volume of specimen, V_0 [i.e., (4) × (5)] _____ in.^{3}

    (8)   Mass of specimen _____ g

    (9)   Wet unit weight of specimen $\left[\text{i.e., } \dfrac{(8)}{453.6} \times \dfrac{1,728}{(7)} \right]$ _____ lb/ft^3

    (10) Water content of specimen _____ %

        (a)   Can no. _____

        (b)   Mass of wet soil + can _____ g

        (c)   Mass of dry soil + can _____ g

        (d)   Mass of can _____ g

        (e)   Mass of water [i.e., (b) − (c)] _____ g

        (f)   Mass of dry soil [i.e., (c) − (d)] _____ g

        (g)   Water content $\left[\text{i.e., } \dfrac{(e)}{(f)} \times 100 \right]$ _____ %

    (11) Dry unit weight of specimen $\left[\text{i.e., } \dfrac{(9)}{(10) + 100} \times 100 \right]$ _____ lb/ft^3

    (12) Specific gravity of soil _____

    (13) Weight of water in 1 ft^3 of soil specimen [i.e., (9) − (11)] _____ lb

    (14) Volume of water in 1 ft^3 of soil specimen $\left[\text{i.e., } \dfrac{(13)}{62.4} \right]$ _____ ft^3

    (15) Weight of solid in 1 ft^3 of soil specimen [i.e., (11)] _____ lb

    (16) Volume of solid in 1 ft^3 of soil specimen $\left[\text{i.e., } \dfrac{(15)}{(12) \times 62.4} \right]$ _____ ft^3

    (17) Volume of void in 1 ft^3 of soil specimen [i.e., 1 − (16)] _____ ft^3

    (18) Degree of saturation $\left[\text{i.e., } \dfrac{(14)}{(17)} \times 100 \right]$ _____ %

[B] Triaxial Compression Data   (Specimen no. _____)

(1)   Chamber pressure on test specimen, σ_3 _____ psi
(2)   Rate of axial strain _____ in./min
(3)   Initial height of specimen, H_0 _____ in.
(4)   Proving ring calibration _____ lb/in.

Elapsed Time (min)	Deformation Dial, ΔH (in.)	Axial Strain, ϵ (in./in.)	Cross-Sectional Area, A (in.^{2})	Proving Ring Dial (in.)	Applied Axial Load (lb)	Unit Axial Load (Deviator Stress) (psi)
(1)	(2)	$(3) = \dfrac{\Delta H}{H_0}$	$(4) = \dfrac{A_0}{1-\epsilon}$	(5)	$(6) = (5) \times$ proving ring calibration	$(7) = \dfrac{(6)}{(4)}$

Minor principal stress (i.e., chamber pressure, σ_3) _____ psi
Unit axial load at failure, Δp _____ psi
Major principal stress at failure (i.e., $\sigma_1 = \sigma_3 + \Delta p$) _____ psi

Soils Testing Laboratory
Consolidated Undrained (CU)
Triaxial Compression Test

Sample No. _____   Project No. _____

Location _____   Boring No. _____

Depth _____   Date of Test _____

Description of Soil _____

Tested by _____

[A] Specimen Data   (before consolidation) (Specimen no. _____)

(1)   Type of test performed _____
_____

(2)   Type of specimen (check one)   ☐ Undisturbed   ☐ Remolded

(3)   Diameter of specimen, D_0 _____ in.

(4)   Initial area of specimen, A_0 _____ in.^{2}

(5)   Initial height of specimen, H_0 _____ in.

(6)   Height-to-diameter ratio $\left[\text{i.e., } \dfrac{(5)}{(3)} \right]$ _____

(7)   Volume of specimen, V_0 [i.e., (4) × (5)] _____ in.^{3}

(8)   Mass of specimen _____ g

(9)   Wet unit weight of specimen $\left[\text{i.e., } \dfrac{(8)}{453.6} \times \dfrac{1,728}{(7)} \right]$ _____ lb/ft^3

(10)   Water content of specimen _____ %

    (a)   Can no. _____

    (b)   Mass of wet soil + can _____ g

    (c)   Mass of dry soil + can _____ g

    (d)   Mass of can _____ g

    (e)   Mass of water [i.e., (b) − (c)] _____ g

    (f)   Mass of dry soil [i.e., (c) − (d)] _____ g

    (g)   Water content $\left[\text{i.e., } \dfrac{(e)}{(f)} \times 100 \right]$ _____ %

(11)   Dry unit weight of specimen $\left[\text{i.e., } \dfrac{(9)}{(10) + 100} \times 100 \right]$ _____ lb/ft^3

[B] Specimen Data   (after consolidation)

(1)   Chamber consolidation pressure _____ psi

(2)   Volume of water extruded from specimen during consolidation [i.e., volume of water accrued in burette], ΔV _____ cm^3 or _____ in.^{3}

(3)   Volume of specimen after consolidation, V_c [i.e., $V_0 - \Delta V$] _____ in.^{3}

(4)   Area of specimen after consolidation, A_c [i.e., $A_0(V_c/V_0)^{2/3}$] _____ in.^{2}

(5)   Height of specimen after consolidation, H_c [i.e., $H_0(V_c/V_0)^{1/3}$] _____ in.

[C] Triaxial Compression Data   (Specimen no. _____)

(1) Consolidation pressure on test specimen, σ_3 _____ psi
(2) Rate of axial strain _____ in./min
(3) Height of specimen after consolidation, H_c _____ in.
(4) Proving ring calibration _____ lb/in.

Elapsed Time (min)	Deformation Dial, ΔH (in.)	Axial Strain, ϵ (in./in.)	Cross Sectional Area, A (in.²)	Proving Ring Dial (in.)	Applied Axial Load (lb)	Unit Axial Load (Deviator Stress) (psi)	Pore Pressure, μ (psi)
(1)	(2)	$(3) = \dfrac{\Delta H}{H_0}$	$(4) = \dfrac{A_0}{1 - \epsilon}$	(5)	(6) = (5) × proving ring calibration	$(7) = \dfrac{(6)}{(4)}$	(8)

Minor principal stress (i.e., chamber pressure, σ_3) _____ psi
Unit axial load at failure, Δp _____ psi
Major principal stress at failure (i.e., $\sigma_1 = \sigma_3 + \Delta p$) _____ psi
Pore pressure corresponding to unit axial load at failure, μ_f _____ psi
Effective minor principal stress, $\bar{\sigma}_3$ [i.e., $\sigma_3 - \mu_f$] _____ psi
Effective major principal stress at failure, $\bar{\sigma}_1$ [i.e., $\sigma_1 - \mu_f$] _____ psi

Soils Testing Laboratory
Consolidated Undrained (CU)
Triaxial Compression Test

Sample No. _____   Project No. _____

Location _____   Boring No. _____

Depth _____   Date of Test _____

Description of Soil _____

Tested by _____

[A] Specimen Data   (before consolidation) (Specimen no. _____)

(1) Type of test performed _____

_____

(2) Type of specimen (check one) ☐ Undisturbed ☐ Remolded

(3) Diameter of specimen, D_0 _____ in.

(4) Initial area of specimen, A_0 _____ in.^{2}

(5) Initial height of specimen, H_0 _____ in.

(6) Height-to-diameter ratio $\left[\text{i.e., } \dfrac{(5)}{(3)}\right]$ _____

(7) Volume of specimen, V_0 [i.e., (4) × (5)] _____ in.^{3}

(8) Mass of specimen _____ g

(9) Wet unit weight of specimen $\left[\text{i.e., } \dfrac{(8)}{453.6} \times \dfrac{1{,}728}{(7)}\right]$ _____ lb/ft^3

(10) Water content of specimen _____ %

    (a) Can no. _____

    (b) Mass of wet soil + can _____ g

    (c) Mass of dry soil + can _____ g

    (d) Mass of can _____ g

    (e) Mass of water [i.e., (b) − (c)] _____ g

    (f) Mass of dry soil [i.e., (c) − (d)] _____ g

    (g) Water content $\left[\text{i.e., } \dfrac{(e)}{(f)} \times 100\right]$ _____ %

(11) Dry unit weight of specimen $\left[\text{i.e., } \dfrac{(9)}{(10) + 100} \times 100\right]$ _____ lb/ft^3

[B] Specimen Data   (after consolidation)

(1) Chamber consolidation pressure _____ psi

(2) Volume of water extruded from specimen during consolidation [i.e., volume of water accrued in burette], ΔV _____ cm^3 or _____ in.^{3}

(3) Volume of specimen after consolidation, V_c [i.e., $V_0 - \Delta V$] _____ in.^{3}

(4) Area of specimen after consolidation, A_c [i.e., $A_0(V_c/V_0)^{2/3}$] _____ in.^{2}

(5) Height of specimen after consolidation, H_c [i.e., $H_0(V_c/V_0)^{1/3}$] _____ in.

[C] Triaxial Compression Data   (Specimen no. _____)

(1) Consolidation pressure on test specimen, σ_3 _____ psi
(2) Rate of axial strain _____ in./min
(3) Height of specimen after consolidation, H_c _____ in.
(4) Proving ring calibration _____ lb/in.

Elapsed Time (min)	Deformation Dial, ΔH (in.)	Axial Strain, ϵ (in./in.)	Cross-Sectional Area, A (in.^{2})	Proving Ring Dial (in.)	Applied Axial Load (lb)	Unit Axial Load (Deviator Stress) (psi)	Pore Pressure, μ (psi)
(1)	(2)	$(3) = \dfrac{\Delta H}{H_0}$	$(4) = \dfrac{A_0}{1 - \epsilon}$	(5)	$(6) = (5) \times$ proving ring calibration	$(7) = \dfrac{(6)}{(4)}$	(8)

Minor principal stress (i.e., chamber pressure, σ_3) _____ psi
Unit axial load at failure, Δp _____ psi
Major principal stress at failure (i.e., $\sigma_1 = \sigma_3 + \Delta p$) _____ psi
Pore pressure corresponding to unit axial load at failure, μ_f _____ psi
Effective minor principal stress, $\bar{\sigma}_3$ [i.e., $\sigma_3 - \mu_f$] _____ psi
Effective major principal stress at failure, $\bar{\sigma}_3$ [i.e., $\sigma_1 - \mu_f$] _____ psi

22

CHAPTER TWENTY TWO

Direct Shear Test

(Referenced Document: ASTM D 3080)

INTRODUCTION This chapter presents a third method for investigating the shear strength of soil in a laboratory. The unconfined compression test was discussed in Chapter 20, and the triaxial compression test in Chapter 21. Chapter 22 presents the *direct shear test*. It differs somewhat from the other two in that they are "compression" tests (i.e., shear failure is effected by a compression force), whereas it is a "shear" test [i.e., shear failure is caused by a shear force along a predetermined horizontal surface (surface *A* in Figure 22–1)].

Like the triaxial test, the direct shear test can be performed on both cohesive and cohesionless soils, and it evaluates both cohesion *c* and angle of internal friction ø. These parameters are used to evaluate a soil's shear strength.

As in the triaxial test, there are three basic types of direct shear test procedures, determined by sample drainage conditions. In an unconsolidated undrained (UU) test, shear is started before the soil sample is consolidated under the applied normal load. This test is analogous to the UU triaxial test. For a consolidated undrained (CU) test, shear is not started (i.e., the shearing force is not applied) until after settlement resulting from the applied normal load stops. This test is somewhere between the CD and CU triaxial tests. For a consolidated drained (CD) test, shear is not started until after settlement resulting from the applied normal load stops; the shearing force is then applied so slowly that

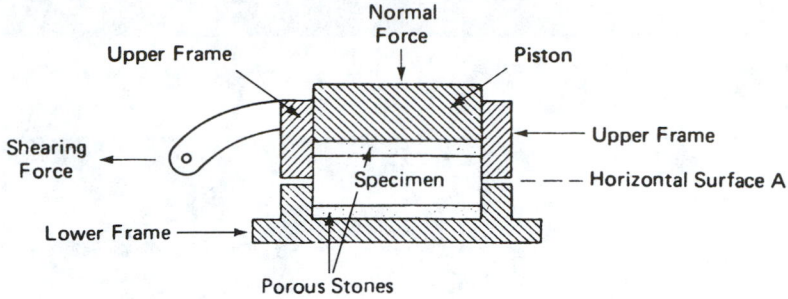

FIGURE 22–1   Typical Direct Shear Box for Single Shear [1]

no pore pressures develop in the sample. This test is analogous to the CD triaxial test [2].

APPARATUS AND SUPPLIES

Direct shear box (see Figures 22–1 and 22–2)

Direct shear apparatus (see Figures 22–3 and 22–4)

Porous stones

Axial-loading device (see Figure 22–4)

Axial load-measuring device (see Figure 22–4)

Shear-loading device (see Figure 22–4)

Shear load-measuring device (see Figure 22–4)

Tools for preparing specimens: cutting ring, wire saw, knife, etc.

Displacement indicators

Equipment for remolding or compacting specimens

FIGURE 22–2   Direct Shear Box (Courtesy of Soiltest, Inc.)

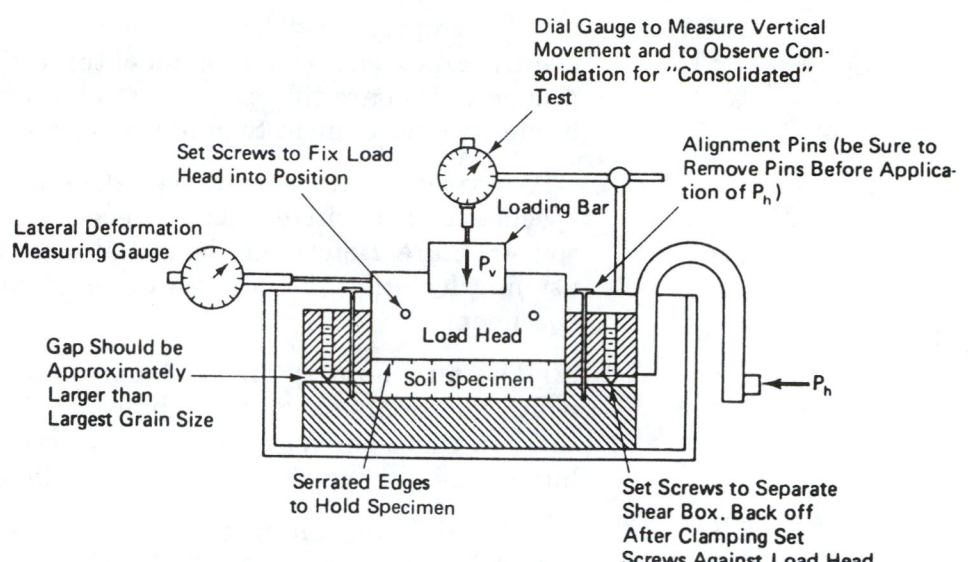

FIGURE 22–3   Direct Shear Apparatus [2]

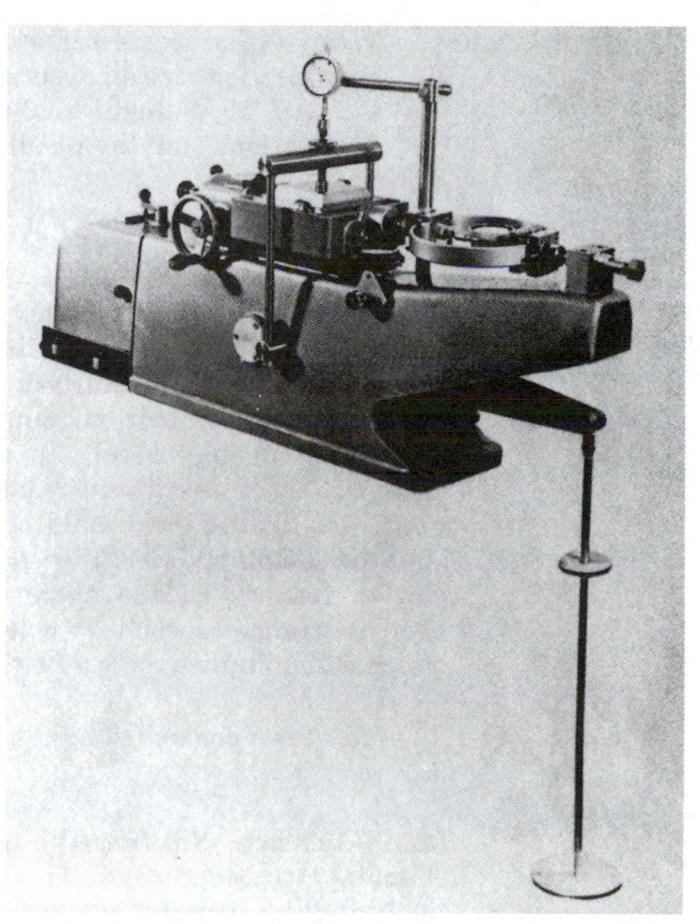

FIGURE 22–4   Direct Shear Apparatus (Courtesy of Wykeham Farrance, Inc.)

TEST SPECIMEN [1]

(1) The sample used for specimen preparation should be sufficiently large so that a minimum of three similar specimens can be prepared. Prepare the specimens in a controlled temperature and humidity environment to minimize moisture loss or gain.

(1.1) Extreme care should be taken in preparing undisturbed specimens of sensitive soils to prevent disturbance to the natural soil structure. Determine the initial mass of the wet specimen for use in calculating the initial water content and unit weight of the specimen.

(2) The minimum specimen diameter for circular specimens, or width for square specimens, shall be 2.0 in. (50 mm), or not less than 10 times the maximum particle size diameter, whichever is larger, and conform to the width to thickness ratio specified in (4).

(3) The minimum initial specimen thickness shall be 0.5 in. (12 mm), but not less than six times the maximum particle diameter.

(4) The minimum specimen diameter to thickness or width to thickness ratio shall be 2:1.

> *Note 1*—If large soil particles are found in the soil after testing, a particle size analysis should be performed in accordance with ASTM Method D 422 (Chapter 9) to confirm the visual observations, and the result should be provided with the test report.

SPECIMEN PREPARATION [1]

(1.1) *Undisturbed Specimens*—Prepare undisturbed specimens from large undisturbed samples or from samples secured in accordance with ASTM Practice D 1587 or other undisturbed tube sampling procedures. Undisturbed samples shall be preserved and transported as outlined for Group C or D samples in Practice D 4220. Handle specimens carefully to minimize disturbance, changes in cross section, or loss of water content. If compression or any type of noticeable disturbance would be caused by the extrusion device, split the sample tube lengthwise or cut it off in small sections to facilitate removal of the specimen with minimum disturbance. Prepare trimmed specimens, whenever possible, in an environment which will minimize the gain or loss of specimen moisture.

> *Note 2*—A controlled high-humidity room is desirable for this purpose.

(1.2) *Compacted Specimens*—Specimens shall be prepared using the compaction method, water content, and unit weight prescribed by the individual assigning the test. Assemble and secure the shear box. Place a moist porous insert in the bottom of the shear box. Specimens may be molded by either kneading or tamping each layer until the accumulative mass of the soil placed in the shear box is compacted to a known volume, or by adjusting the number of layers, the number of tamps per layer, and the force per tamp. The top of each layer shall be scarified prior to the addition of material for the next layer. The compacted layer boundaries should be positioned so they are not coincident with the shear plane defined by the shear box halves, unless this is the stated purpose for a particular test. The tamper used to compact the material shall have an area in contact with the soil equal to or less than half the area of the mold. Determine the mass of wet soil required for a single compacted lift and place it in the shear box. Compact the soil until the desired unit weight is obtained. Continue placing the compacting soil until the entire specimen is compacted.

> *Note 3*—A light coating of grease applied to the inside of the shear box may be used to reduce friction between the specimen and shear box during consolidation. However, the upper ring in some shear devices requires friction to support the ring after the shear plates have been gapped. A light coating of grease applied between the halves of the shear box may be used to reduce friction between the halves of the shear box during shear. TFE-fluorocarbon coating may also be used on these surfaces instead of grease to reduce friction.

> *Note 4*—The required thickness of the compacted lift may be determined by directly measuring the thickness of the lift or from the marks on the tamping rod which correspond to the thickness of the lift being placed.

> *Note 5*—The decision to dampen the porous inserts by inundating the shear box before applying the normal force depends on the problem under study. For undisturbed samples obtained below the water table, the porous inserts are usually dampened. For swelling soils, the sequence of consolidation, wetting, and shearing should model field conditions. Determine the compacted mass of the specimen from either the measured mass placed and compacted in the mold, or the difference between the mass of the shear box and compacted specimen and the tare mass of the shear box.

(2) Material required for the specimen shall be batched by thoroughly mixing soil with sufficient water to produce the desired water content. Allow the specimen to stand prior to compaction in accordance with the following guide:

Classification ASTM D 2487	Minimum Standing Time, h
SW, SP	No Requirement
SM	3
SC, ML, CL	18
MH, CH	36

(3) Compacted specimens may also be prepared by compacting soil using the procedures and equipment used to determine moisture-density relationships of soils (ASTM Test Methods D 698 or D 1557) and trimming the direct shear test specimen from the larger test specimen as though it were an undisturbed specimen.

PROCEDURE   To carry out a direct shear test, a soil specimen is prepared and placed in a direct shear box (see Figures 22–1 and 22–2), which may be round or square. A normal load of specific (and constant) magnitude is applied. The box is "split" into two parts horizontally (see Figures 22–1 and 22–2), and if the lower half is held stationary while the upper half is pushed with increasing force, the soil will ultimately experience shear failure along horizontal surface A. This procedure is carried out in the direct shear apparatus (Figures 22–3 and 22–4), and the particular normal load and shear stress that produced shear failure are recorded. The soil specimen is then removed from the shear box and discarded, and another specimen of the same soil is placed in the shear box. A normal load either higher or lower than that used in the first test is applied to the second specimen, and a shear force is again applied with sufficient magnitude to cause shear failure. The normal load and shear stress that produced shear failure are recorded for the second test. The entire procedure may be repeated for another specimen and another different normal load.

The actual step-by-step procedure is as follows:

(1) Measure the diameter (or side), height, and mass of the specimen. Assemble the shear box with the frames aligned and locked in position. Carefully insert the test specimen. Place the loading block in place and connect the loading device. Position the vertical displacement indicator and apply the appropriate normal load. (The normal load includes both the weight of the loading block and the externally applied normal force.)

Note 1—The decision to dampen the porous stones before insertion of the specimen and before application of the normal force depends upon the problem under study. For undisturbed samples from below the water table, the porous stones are usually dampened. For swelling soils, wetting should probably follow application of the normal force to prevent swell not representative of field conditions.

(2) For a consolidated test, consolidate the test specimen under the appropriate normal force. As soon as possible after applying the initial normal force, fill the water reservoir to a point above the top of the specimen. Maintain this water level during the consolidation and subsequent shear phases so that the specimen is at all times effectively submerged. Allow the specimen to drain and consolidate under the desired normal force or increments thereof prior to shearing. During the consolidation process, record the normal displace-

ment readings before each increment of normal force is applied and at appropriate times [see ASTM Method D 2435 (Chapter 19)]. Plot the normal displacement readings against elapsed time. Allow each increment of normal force to remain until primary consolidation is complete. The final increment should equal the previous normal force developed and should produce the specified normal stress.

> *Note 2*—The normal force used for each of the three or more specimens will depend upon the information required. Application of the normal force in one increment may be appropriate for relatively firm soils. For relatively soft soils, however, several increments may be necessary to prevent damage to the specimen. The initial increment will depend upon the strength and sensitivity of the soil. This force should not be so large as to squeeze the soil out of the device.

(3) Separate the upper and lower halves of the shear box frame by a gap of approximately 0.025 in. (0.64 mm) so the specimen can be sheared. Position the shear-deformation (horizontal displacement) indicator and set both the vertical and the horizontal displacement indicators to zero. Fill the shear box with water for saturated tests. Apply the shearing force and shear the specimen. After reaching failure, stop the test apparatus. This displacement may range from 10 to 20% of the specimen's original diameter or length. For all tests except those under consolidated drained conditions (in a controlled-displacement case), the rate of shear (i.e., the rate of horizontal displacement) should be on the order of 0.05 in./min.

Obtain data readings of time, vertical and horizontal displacement, and shear force at desired interval of displacement. Data readings should be taken at displacement intervals equal to 2% of the specimen diameter or width to accurately define a shear stress-displacement curve.

> *Note 3*—Additional readings may be helpful in identifying the value of peak shear stress of overconsolidated or brittle material.

> *Note 4*—It may be necessary to stop the test and re-gap the shear box halves to maintain clearance between the shear box halves.

(4) For a consolidated drained test, apply the shearing force and shear the specimen slowly to ensure complete dissipation of excess pore pressure. The following guide for total elapsed time to failure may be useful in determining rate of loading

$$\text{Time to failure } (t_f) = 50t_{50}$$

where:

t_{50} = time required for the specimen to achieve 50% consolidation under the normal force.

Note 5—If the material exhibits a tendency to swell, the soil must be inundated with water and must be permitted to achieve equilibrium under an increment of normal stress large enough to counteract the swell tendency before the minimum time to failure can be determined. The time-consolidation curve for subsequent normal stress increments is then valid for use in determining t_f.

Note 6—Some soils, such as dense sands and overconsolidated clays, may not exhibit well-defined time-settlement curves. Consequently, the calculation of t_f may produce an inappropriate estimate of the time required to fail the specimen under drained conditions. For overconsolidated clays which are tested under normal stresses less than the soil's preconsolidation pressure, it is suggested that a time to failure be estimated using a value of t_{50} equivalent to one obtained from normal consolidation time-settlement behavior. For clean dense sands which drain quickly, a value of 10 min may be used for t_f. For dense sands with more than 5% fines, a value of 60 min may be used for t_f. If an alternative value of t_f is selected, the rationale for the selection shall be explained with the test results.

Determine the appropriate displacement from the following equation:

$$d_r = d_f/t_f$$

where:
d_r = displacement rate (in./min, mm/min)
d_f = estimated horizontal displacement at failure (in., mm)
t_f = total estimated elapsed time to failure (min)

Note 7—The magnitude of the estimated displacement at failure is dependent on many factors including the type and the stress history of the soil. As a guide, use d_f = 0.5 in. (12 mm) if the material is normally or lightly overconsolidated fine-grained soil; otherwise use d_f = 0.2 in. (5 mm).

(5) At the completion of the test, remove the normal force from the specimen by removing the mass from the lever and hanger or by releasing the pressure.

For cohesive test specimens, separate the shear box halves with a sliding motion along the failure plane. Do not pull the shear box halves apart perpendicularly to the failure surface, since it would damage the specimen. Photograph, sketch, or describe in writing the failure surface. This procedure is not applicable to cohesionless specimens.

Remove the specimen from the shear box and determine its water content according to ASTM Test Method D 2216 (Chapter 3).

(6) Repeat the entire procedure for two or more specimens at different normal loads.

Note—The test procedure described can be used for each of the three basic types of test. For an unconsolidated undrained test, follow steps (1), (3), (5), and (6). For a consolidated undrained test, follow steps (1) to (3), (5), and (6). For a consolidated drained test, follow steps (1) to (6).
Note—The foregoing procedure is adapted from ASTM D 3080-98 [1].

DATA

Data collected in the direct shear test should include the following:

[A] Specimen Data

Diameter or side of specimen (in.)

Initial height of specimen (in.)

Mass of specimen at beginning of test (g)

Initial water content data:

Mass of wet soil sample at beginning of test (g)

Mass of oven-dried soil sample (at end of test) plus can (g)

Mass of can (g)

Final water content data:

Mass of wet soil specimen plus can at end of test (g)

Mass of oven-dried soil specimen (at end of test) plus can (g)

Mass of can (g)

[B] Shear Stress Data

Normal load on test specimen (lb)

Rate of shear (rate of horizontal displacement) (in./min)

Vertical dial readings, horizontal displacement dial readings, and proving ring dial readings (in.)

Note—These data are obtained for each of three specimens tested at different normal loads.

CALCULATIONS

[A] Specimen Parameters

The initial area, volume, wet unit weight, water content, dry unit weight of the specimen, and final water content are all calculated by methods described in previous chapters.

[B] Shear Stress

Normal stress can be determined by dividing normal load (which includes the weight of the loading block and externally applied normal force) by the initial area of the specimen. Horizontal shearing forces can be computed by multiplying each dial reading by the proving ring calibration. Each corresponding shear stress can be calculated by dividing each horizontal shearing force by the initial area of the specimen. These computations must be completed for each specimen tested at different normal loads.

[C] Curves of Shear Stress and Specimen Thickness Change versus Shear Displacement

For each test specimen, graphs of shear stress versus shear (horizontal) displacement and of specimen thickness change (vertical dial readings) versus shear (horizontal) displacement should be prepared. Both graphs can be placed on the same graph sheet using shear (horizontal) displacement as a common abscissa.

With these graphs completed, one can determine the maximum shear stress for each specimen tested by evaluating the graph of shear stress versus shear (horizontal) displacement. The maximum stresses are taken to be either the peak shear stress on the graph or the shear stress at a shear (horizontal) displacement of 10% of the original diameter, whichever is obtained first during the test.

[D] Curve of Maximum Shear Stress versus Normal Stress

In order to evaluate the shear strength parameters (cohesion and angle of internal friction), it is necessary to prepare a graph of maximum shear stress (ordinate) versus normal stress (abscissa) for each specimen tested (see Figure 22–5). The same scale must be used along both

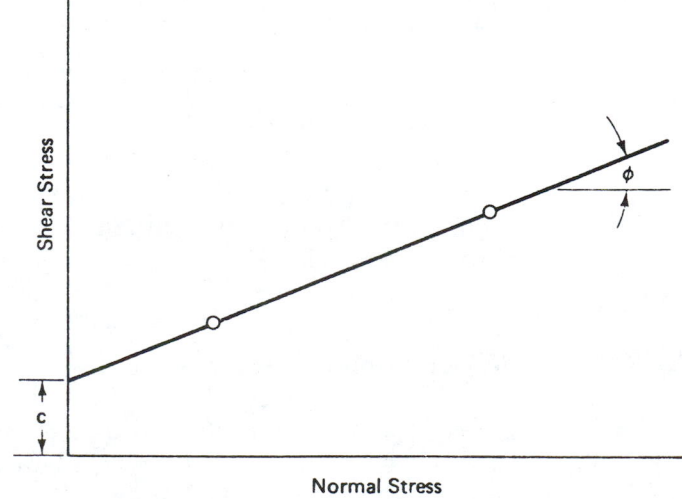

FIGURE 22–5   Shear Diagram for Direct Shear Test

abscissa and ordinate. A straight line is drawn through the plotted points and extended to intersect the ordinate. The angle between the straight line and a horizontal line (ø in the figure) gives the angle of internal friction, and the value of shear stress where the straight line intersects the ordinate (c in the figure) gives the cohesion.

It is adequate, in theory, to have only two points to define the straight-line relationship shown in the figure. In practice, however, it is better to have three (or more) such points through which the best straight line can be drawn. That is why the test procedure calls for three or more separate tests to be performed on three or more specimens from the same soil sample at different normal stresses.

NUMERICAL EXAMPLE

An unconsolidated undrained direct shear test was performed in the laboratory according to the procedure described in this chapter. The following data were obtained:

[A] Specimen Data

Diameter of specimen, D_0 = **2.50 in.**

Initial height of specimen, H_0 = **1.00 in.**

Mass of specimen at beginning of test = **161.52 g**

Initial water content data:

Mass of wet soil sample plus can = **248.43 g**

Mass of oven-dried soil sample plus can = **216.72 g**

Mass of can = **45.30 g**

Final water content data:

Mass of wet soil specimen plus can at end of test = **226.20 g**

Mass of oven-dried soil specimen (at end of test) plus can = **182.58 g**

Mass of can = **46.28 g**

[B] Shear Stress Data

Normal load on test specimen = **20.6 lb**

Rate of shear = **0.05 in./min**

Proving ring calibration = **3,125 lb/in.**

Vertical Dial Reading (in.)	Shear (Horizontal) Displacement Reading (in.)	Proving Ring Dial Reading (in.)
0	0	0
0.0030	0.025	0.0030
0.0050	0.050	0.0063
0.0070	0.075	0.0102
0.0080	0.100	0.0124

(continued)

Vertical Dial Reading (in.)	Shear (Horizontal) Displacement Reading (in.)	Proving Ring Dial Reading (in.)
0.0090	0.125	0.0142
0.0095	0.150	0.0155
0.0100	0.175	0.0163
0.0100	0.200	0.0166
0.0110	0.220	0.0163
0.0115	0.240	0.0159
0.0115	0.260	0.0150

It is, of course, necessary to perform a direct shear test on two (preferably three) specimens from the same sample at different normal stresses in order to evaluate the shear strength parameters. In this example, however, actual data are shown for only one such specimen tested.

All of the preceding data are shown on the form on pages 377 and 378. At the end of the chapter, two copies of this form are included for the reader's use.

With these data known, necessary computations can be made as follows:

[A] Specimen Parameters

The initial area A_0 and volume V_0 of the specimen are easily computed as follows:

$$A_0 = \frac{\pi D_0^2}{4} = \frac{\pi (2.50)^2}{4} = 4.91 \text{ in.}^2$$

$$V_0 = (H_0)(A_0) = (1.00)(4.91) = 4.91 \text{ in.}^3$$

The wet unit weight γ_{wet} of the soil specimen at the beginning of the test can be determined next:

$$\gamma_{\text{wet}} = \left(\frac{161.52}{4.91}\right)\left(\frac{1,728}{453.6}\right) = 125.3 \text{ lb/ft}^3$$

The initial water content may be computed as follows (see Chapter 3):

$$w = \frac{248.43 - 216.72}{216.72 - 45.30} \times 100 = 18.5\%$$

With water content known, dry unit weight γ_{dry} can be determined next:

$$\gamma_{\text{dry}} = \frac{\gamma_{\text{wet}}}{w + 100} \times 100 = \frac{125.3}{18.5 + 100} \times 100 = 105.7 \text{ lb/ft}^3$$

The final water content may also be computed:

$$w = \frac{226.20 - 182.58}{182.58 - 46.28} \times 100 = 32.0\%$$

[B] Shear Stress

The normal load (**20.6 lb**) divided by the initial area of the specimen (4.91 in.^{2}) gives a normal stress of 4.20 lb/in.^{2}, or 604 lb/ft^2. The first proving ring dial reading (**0.0030 in.**) multiplied by the proving ring calibration (**3,125 lb/in.**) gives a corresponding horizontal shear force of 9.4 lb. Dividing this horizontal shear force by the initial area of the specimen (4.91 in.^{2}) and multiplying by 144 to convert square inches to square feet gives a corresponding shear stress of 276 lb/ft^2.

The foregoing values fill in the second row of the form on page 378. Similar calculations furnish required values for succeeding given dial readings and fill in the remaining rows on the form.

It should be emphasized that one or two additional evaluations would be required for testing additional specimens at different normal loads. Such evaluations are not included here.

[C] Curves of Shear Stress and Specimen Thickness Change versus Shear Displacement

Curves of shear stress and specimen thickness change versus shear displacement are shown in Figure 22–6. The maximum shear stresses for the specimens tested are determined from the graph of shear stress versus shear (horizontal) displacement as follows:

For specimen no. 1, with a normal stress = 604 lb/ft^2, maximum shear stress = 1,522 lb/ft^2

For specimen no. 2, with a normal stress = 926 lb/ft^2, maximum shear stress = 1,605 lb/ft^2

For specimen no. 3, with a normal stress = 1,248 lb/ft^2, maximum shear stress = 1,720 lb/ft^2

It should be noted that all input data and necessary computations for determining normal and maximum shear stress for specimen 1 are included in this sample problem. Input data and necessary computations for specimens 2 and 3 are not included; they are presented at this point (see Figure 22–6) merely to complete determination of the shear strength parameters.

[D] Curve of Maximum Shear Stress versus Normal Stress

Using results of the direct shear test for which data are given in this example (i.e., a normal stress of 604 lb/ft^2 and maximum shear stress of 1,522 lb/ft^2), an initial point on the curve of shear stress versus normal stress can be plotted (see Figure 22–7). Using similar results of two additional direct shear tests performed on other specimens at different normal loads for which data are not given in this example (but for which results are listed), two additional points on the curve can be plotted and the best straight line drawn through these points. The angle between this straight line and a horizontal line is measured to be 17°, and the value of shear stress where the straight line intersects the ordinate is

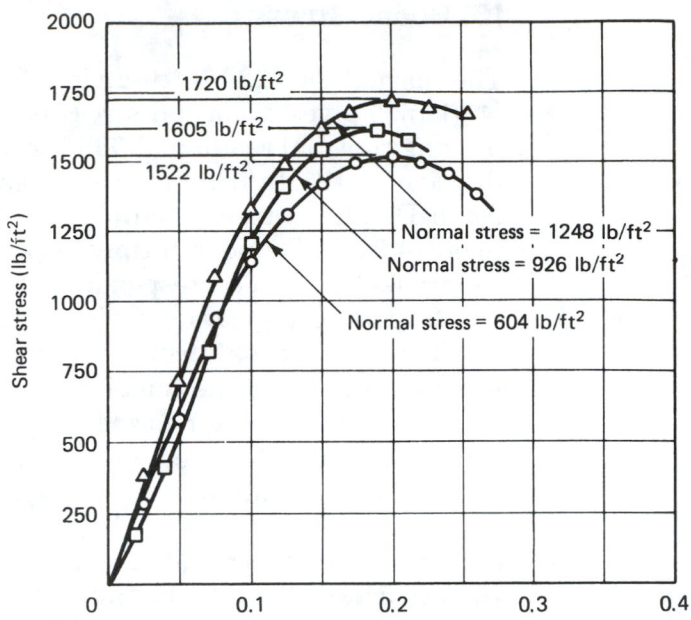

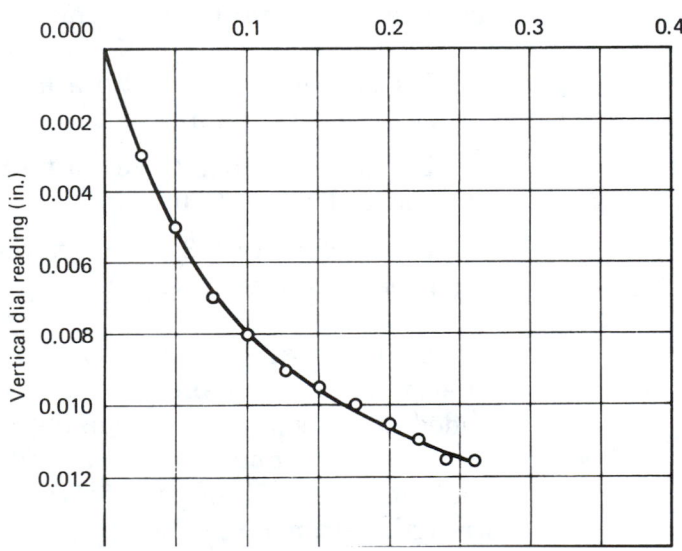

FIGURE 22–6 Curves of Shear Stress and Specimen Thickness Change versus Shear Displacement

found to be 1,340 lb/ft². Hence, the angle of internal friction ø and cohesion c for this soil are 17° and 1,340 lb/ft², respectively. These values (the shear strength parameters) are the major results of a direct shear test.

CONCLUSIONS   As indicated, values of the angle of internal friction and cohesion are the major results of a direct shear test. In addition to these parameters, however, reports should include values of the initial water content and

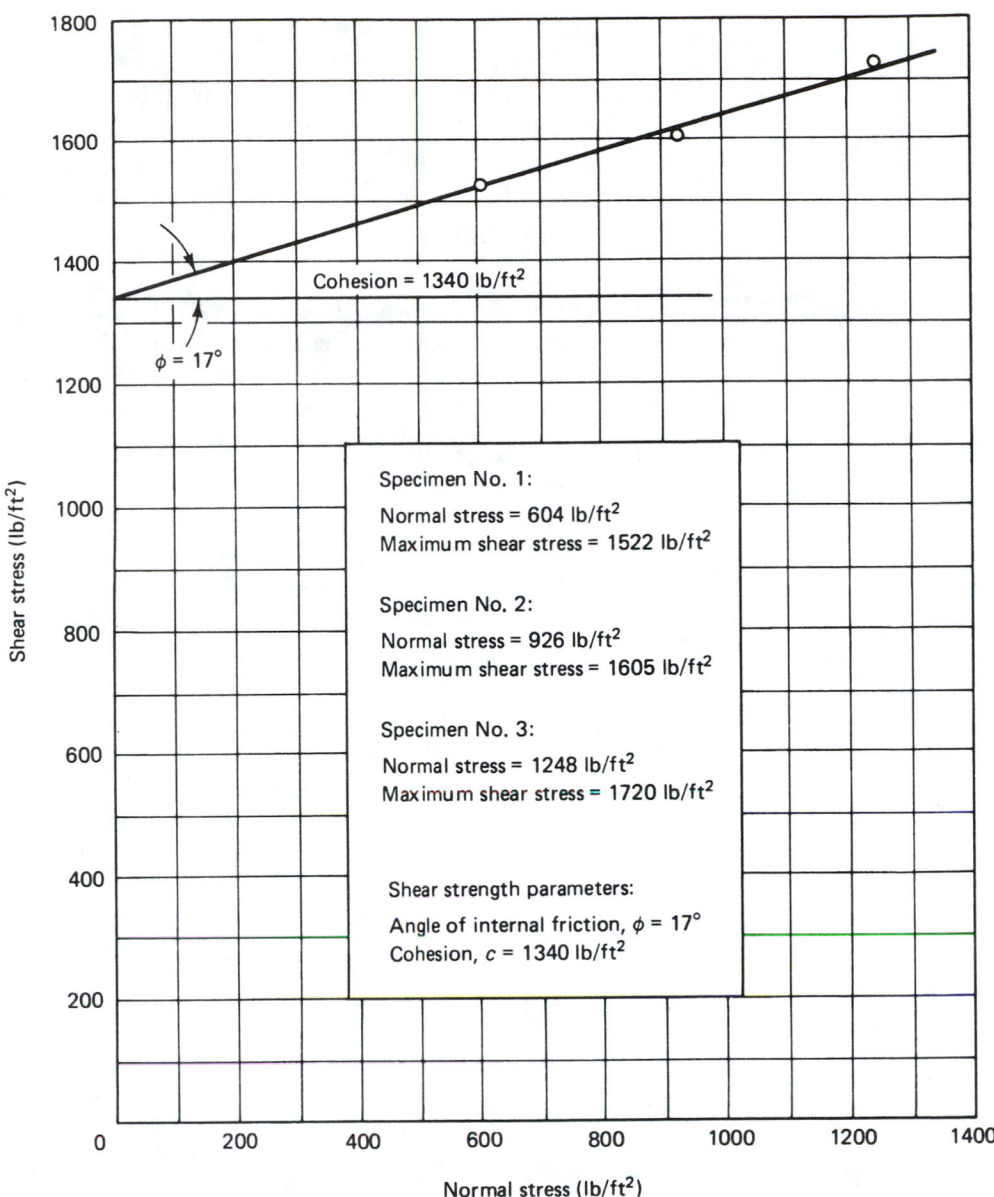

FIGURE 22–7 Curve of Maximum Shear Stress versus Normal Stress

the specimen's size and initial thickness. The type of test performed (UU, CU, CD) and the type (undisturbed, remolded) and shape (cylindrical, prismatic) of specimens should be reported as well. Of course, a visual description of the soil and any unusual conditions should be noted. Finally, graphs of shear stress versus shear (horizontal) displacement for each specimen tested should be included, as well as a graph of maximum shear stress versus normal stress.

For cohesionless soils, if the shearing rate is not extremely rapid, all of the three direct shear tests (UU, CU, and CD) will give about the same results whether the sample is saturated or unsaturated. For cohesive soils, shear strength parameters are influenced significantly by the test method, degree of saturation, and whether the soil is normally consolidated or overconsolidated. Generally, two sets of shear strength parameters are obtained for overconsolidated soils: one for tests using normal loads less than the preconsolidation pressure and a second set for normal loads greater than the preconsolidation pressure. Where an overconsolidated soil is suspected, it may be necessary to perform six or more tests to ensure that the appropriate shear strength parameters are obtained [2].

REFERENCES

[1]   ASTM, *2001 Annual Book of ASTM Standards,* West Conshohocken, PA, 2001. Copyright, American Society for Testing and Materials, 100 Barr Harbor Drive, West Conshohocken, PA 19428-2959. Reprinted with permission.

[2]   Joseph E. Bowles, *Engineering Properties of Soils and Their Measurement,* 2d ed., McGraw-Hill Book Company, 1978.

Sample No. _____2_____   Project No. _____SR 1180_____

Sample Depth _____2.8 ft_____   Boring No. _____5_____

Location _____Charlotte, N.C._____   Date _____6/23/02_____

Tested by _____John Doe_____

Description of Soil _____Light brown silty clay_____

[A] Specimen Data   (Specimen no. __1__)

(1) Type of test performed __unconsolidated undrained__

(2) Type of specimen (check one)   ☒ Undisturbed   ☐ Remolded

(3) Diameter or side of specimen __(diam.) 2.50__ in.

(4) Initial area of specimen __4.91__ in.^{2}

(5) Initial height of specimen __1.00__ in.

(6) Volume of specimen [i.e., (4) × (5)] __4.91__ in.^{3}

(7) Mass of soil specimen (at beginning of test) __161.52__ g

(8) Wet unit weight of specimen $\left[\text{i.e., } \dfrac{(7)}{453.6} \times \dfrac{1728}{(6)} \right]$ __125.3__ lb/ft^3

(9) Initial water content of specimen __18.5__ %

  (a) Can no. __2-A__

  (b) Mass of wet soil sample + can __248.43__ g

  (c) Mass of oven-dried soil sample + can __216.72__ g

  (d) Mass of can __45.30__ g

  (e) Mass of oven-dried soil sample [i.e., (c) − (d)] __171.42__ g

  (f) Mass of water [i.e., (b) − (c)] __31.71__ g

  (g) Water content $\left[\text{i.e., } \dfrac{(f)}{(e)} \times 100 \right]$ __18.5__ %

(10) Dry unit weight of specimen $\left[\text{i.e., } \dfrac{(8)}{(9) + 100} \times 100 \right]$ __105.7__ lb/ft^3

(11) Final water content of specimen __32.0__ %

  (a) Can no. __1-A__

  (b) Mass of entire wet soil specimen (at end of test) + can __226.20__ g

  (c) Mass of entire oven-dried soil specimen (at end of test) + can __182.58__ g

  (d) Mass of can __46.28__ g

  (e) Mass of entire oven-dried soil specimen [i.e., (c) − (d)] __136.30__ g

  (f) Mass of water [i.e., (b) − (c)] __43.62__ g

  (g) Water content $\left[\text{i.e., } \dfrac{(f)}{(e)} \times 100 \right]$ __32.0__ %

[B] Shear Stress Data (Specimen no. ___1___)

    (1)   Normal load ___20.6___ lb

    (2)   Normal stress $\left[\text{i.e., } \dfrac{\text{normal load}}{\text{initial area of specimen [i.e., part [A] (4)]}} \times 144\right]$

        ___604___ lb/ft^2

    (3)   Proving ring calibration ___3,125___ lb/in.

    (4)   Rate of shear ___0.05___ in./min

Vertical Dial Reading (in.)	Horizontal Displacement Dial Reading (in.)	Proving Ring Dial Reading (in.)	Horizontal Shear Force (lb)	Cross-Sectional Area of Specimen (in.^{2})	Shear Stress (lb/ft^2)
(1)	(2)	(3)	(4) = (3) × proving ring calibration	(5) = initial area of specimen	$(6) = \dfrac{(4)}{(5)} \times 144$
0	0	0	0	4.91	0
0.0030	0.025	0.0030	9.4	4.91	276
0.0050	0.050	0.0063	19.7	4.91	578
0.0070	0.075	0.0102	31.9	4.91	936
0.0080	0.100	0.0124	38.8	4.91	1,138
0.0090	0.125	0.0142	44.4	4.91	1,302
0.0095	0.150	0.0155	48.4	4.91	1,419
0.0100	0.175	0.0163	50.9	4.91	1,493
0.0100	0.200	0.0166	51.9	4.91	1,522
0.0110	0.220	0.0163	50.9	4.91	1,493
0.0115	0.240	0.0159	49.7	4.91	1,458
0.0115	0.260	0.0150	46.9	4.91	1,375

Soils Testing Laboratory
Direct Shear Test

Sample No. _____   Project No. _____

Sample Depth _____   Boring No. _____

Location _____   Date _____

Tested by _____

Description of Soil _____

[A] Specimen Data   (Specimen no. _____)

(1)   Type of test performed _____

(2)   Type of specimen (check one)   ☐ Undisturbed   ☐ Remolded

(3)   Diameter or side of specimen _____ in.

(4)   Initial area of specimen _____ in.^{2}

(5)   Initial height of specimen _____ in.

(6)   Volume of specimen [i.e., (4) × (5)] _____ in.^{3}

(7)   Mass of soil specimen (at beginning of test) _____ g

(8)   Wet unit weight of specimen $\left[\text{i.e., } \dfrac{(7)}{453.6} \times \dfrac{1728}{(6)}\right]$ _____ lb/ft^3

(9)   Initial water content of specimen _____ %

    (a)   Can no. _____

    (b)   Mass of wet soil sample + can _____ g

    (c)   Mass of oven-dried soil sample + can _____ g

    (d)   Mass of can _____ g

    (e)   Mass of oven-dried soil sample [i.e., (c) − (d)] _____ g

    (f)   Mass of water [i.e., (b) − (c)] _____ g

    (g)   Water content $\left[\text{i.e., } \dfrac{(f)}{(e)} \times 100\right]$ _____ %

(10)   Dry unit weight of specimen $\left[\text{i.e., } \dfrac{(8)}{(9) + 100} \times 100\right]$ _____ lb/ft^3

(11)   Final water content of specimen _____ %

    (a)   Can no. _____

    (b)   Mass of entire wet soil specimen (at end of test) + can _____ g

    (c)   Mass of entire oven-dried soil specimen (at end of test) + can

        _____ g

    (d)   Mass of can _____ g

    (e)   Mass of entire oven-dried soil specimen [i.e., (c) − (d)] _____ g

    (f)   Mass of water [i.e., (b) − (c)] _____ g

    (g)   Water content $\left[\text{i.e., } \dfrac{(f)}{(e)} \times 100\right]$ _____ %

[B] Shear Stress Data (Specimen no. _____)

    (1)   Normal load _____ lb

    (2)   Normal stress $\left[\text{i.e., } \dfrac{\text{normal load}}{\text{initial area of specimen [i.e., part [A] (4)]}} \times 144\right]$

        _____ lb/ft^2

    (3)   Proving ring calibration _____ lb/in.

    (4)   Rate of shear _____ in./min

Vertical Dial Reading (in.)	Horizontal Displacement Dial Reading (in.)	Proving Ring Dial Reading (in.)	Horizontal Shear Force (lb)	Cross-Sectional Area of Specimen (in.^{2})	Shear Stress (lb/ft^2)
(1)	(2)	(3)	(4) = (3) × proving ring calibration	(5) = initial area of specimen	(6) = $\dfrac{(4)}{(5)} \times 144$

Soils Testing Laboratory
Direct Shear Test

Sample No. _____   Project No. _____

Sample Depth _____   Boring No. _____

Location _____   Date _____

Tested by _____

Description of Soil _____

[A] Specimen Data   (Specimen no. _____)

(1) Type of test performed _____

(2) Type of specimen (check one)   ☐ Undisturbed   ☐ Remolded

(3) Diameter or side of specimen _____ in.

(4) Initial area of specimen _____ in.^{2}

(5) Initial height of specimen _____ in.

(6) Volume of specimen [i.e., (4) × (5)] _____ in.^{3}

(7) Mass of soil specimen (at beginning of test) _____ g

(8) Wet unit weight of specimen $\left[\text{i.e., } \dfrac{(7)}{453.6} \times \dfrac{1728}{(6)} \right]$ _____ lb/ft^3

(9) Initial water content of specimen _____ %

  (a) Can no. _____

  (b) Mass of wet soil sample + can _____ g

  (c) Mass of oven-dried soil sample + can _____ g

  (d) Mass of can _____ g

  (e) Mass of oven-dried soil sample [i.e., (c) − (d)] _____ g

  (f) Mass of water [i.e., (b) − (c)] _____ g

  (g) Water content $\left[\text{i.e., } \dfrac{(f)}{(e)} \times 100 \right]$ _____ %

(10) Dry unit weight of specimen $\left[\text{i.e., } \dfrac{(8)}{(9) + 100} \times 100 \right]$ _____ lb/ft^3

(11) Final water content of specimen _____ %

  (a) Can no. _____

  (b) Mass of entire wet soil specimen (at end of test) + can _____ g

  (c) Mass of entire oven-dried soil specimen (at end of test) + can _____ g

  (d) Mass of can _____ g

  (e) Mass of entire oven-dried soil specimen [i.e., (c) − (d)] _____ g

  (f) Mass of water [i.e., (b) − (c)] _____ g

  (g) Water content $\left[\text{i.e., } \dfrac{(f)}{(e)} \times 100 \right]$ _____ %

[B] Shear Stress Data (Specimen no. _____)

    (1)   Normal load _____ lb

    (2)   Normal stress $\left[\text{i.e., } \dfrac{\text{normal load}}{\text{initial area of specimen [i.e., part [A] (4)]}} \times 144 \right]$

          _____ lb/ft^2

    (3)   Proving ring calibration _____ lb/in.

    (4)   Rate of shear _____ in./min

Vertical Dial Reading (in.)	Horizontal Displacement Dial Reading (in.)	Proving Ring Dial Reading (in.)	Horizontal Shear Force (lb)	Cross-Sectional Area of Specimen (in.^{2})	Shear Stress (lb/ft^2)
(1)	(2)	(3)	(4) = (3) × proving ring calibration	(5) = initial area of specimen	(6) = $\dfrac{(4)}{(5)} \times 144$

CHAPTER TWENTY THREE

California Bearing Ratio Test

(Referenced Document: ASTM D 1883)

INTRODUCTION The California Bearing Ratio (CBR) test is a relatively simple test that is commonly used to obtain an indication of the strength of a subgrade soil, sub-base, and base course material for use in road and airfield pavements. The test is used primarily to determine empirically required thicknesses of flexible pavements for highways and airfield pavements.

CBR tests are normally performed on remolded (compacted) specimens, although they may be conducted on undisturbed soils or on soil *in situ*. Remolded specimens may be compacted to their maximum unit weights at their optimum moisture contents (see Chapter 11 and ASTM Methods D 698 and D 1557) if the CBR is desired at 100% maximum dry unit weight and optimum moisture content. CBR tests may be performed, however, over the ranges of unit weights and moisture contents that are expected during construction. Soil specimens may be tested unsoaked or soaked—the latter by immersing them in water for a certain period of time in order to simulate very poor soil conditions.

The CBR for a soil is the ratio (expressed as a percentage) obtained by dividing the penetration stress required to cause a 3-in.^{2} area (hence, a 1.95-in. diameter) piston to penetrate 0.10 in. into the soil by a standard penetration stress of 1,000 psi. This standard penetration stress is roughly what is required to cause the same piston to penetrate 0.10 in. into a mass of crushed rock. The CBR may be thought of, therefore, as an indication of the strength of the soil relative to that of crushed rock.

The CBR may be expressed in equation form as

$$\text{CBR} = \frac{\text{penetration stress (psi) required to penetrate 0.10 in.}}{1,000 \text{ psi}} \times 100$$

(23–1)

Note—The 1,000 psi in the denominator is the standard penetration stress for 0.10-in. penetration.

On occasion, the bearing ratio based on a penetration stress required to penetrate 0.20 in. with a corresponding standard penetration stress of 1,500 psi may be greater than the one for a 0.10-in. penetration. When this occurs, the test should be run again, and if the retest result is similar, the ratio based on the 0.20-in. penetration should be reported as the CBR.

APPARATUS AND SUPPLIES

CBR test apparatus: compaction mold (6-in. diameter and 7-in. height), collar, spacer disk (5 15/16-in. diameter and 2.416-in. height), adjustable stem and perforated plate, weights, penetration piston (3 in.^{2} in area) (for more details, including additional dimensions, weights, etc., see Figure 23–1)

Loading (compression) machine: with load capacity of at least 10,000 lb and penetration rate of 0.05 in./min (see Figure 23–2)

Expansion measuring apparatus

Two dial gages (with accuracy to 0.001 in.)

Standard compaction hammer

Miscellaneous equipment: mixing bowl, scales, soaking tank, oven

SAMPLE [1]

(1) The sample shall be handled and specimen(s) for compaction shall be prepared in accordance with the procedures given in ASTM Test Methods D 698 or D 1557 for compaction in a 6-in. (152.4-mm) mold except as follows:

(1.1) If all material passes a ¾-in. (19-mm) sieve, the entire gradation shall be used for preparing specimens for compaction without modification. If there is material retained on the ¾-in. (19-mm) sieve, the material retained on the ¾-in. (19-mm) sieve shall be removed and replaced by an equal amount of material passing the ¾-in. (19-mm) sieve and retained on the No. 4 sieve obtained by separation from portions of the sample not otherwise used for testing.

PREPARATION OF TEST SPECIMEN [1]

(1) *Bearing Ratio at Optimum Water Content Only*—Using material prepared as described above, conduct a control compaction test with a sufficient number of test specimens to definitely establish the optimum water content for the soil using the compaction method specified, either ASTM Test Methods D 698 or D 1557. A

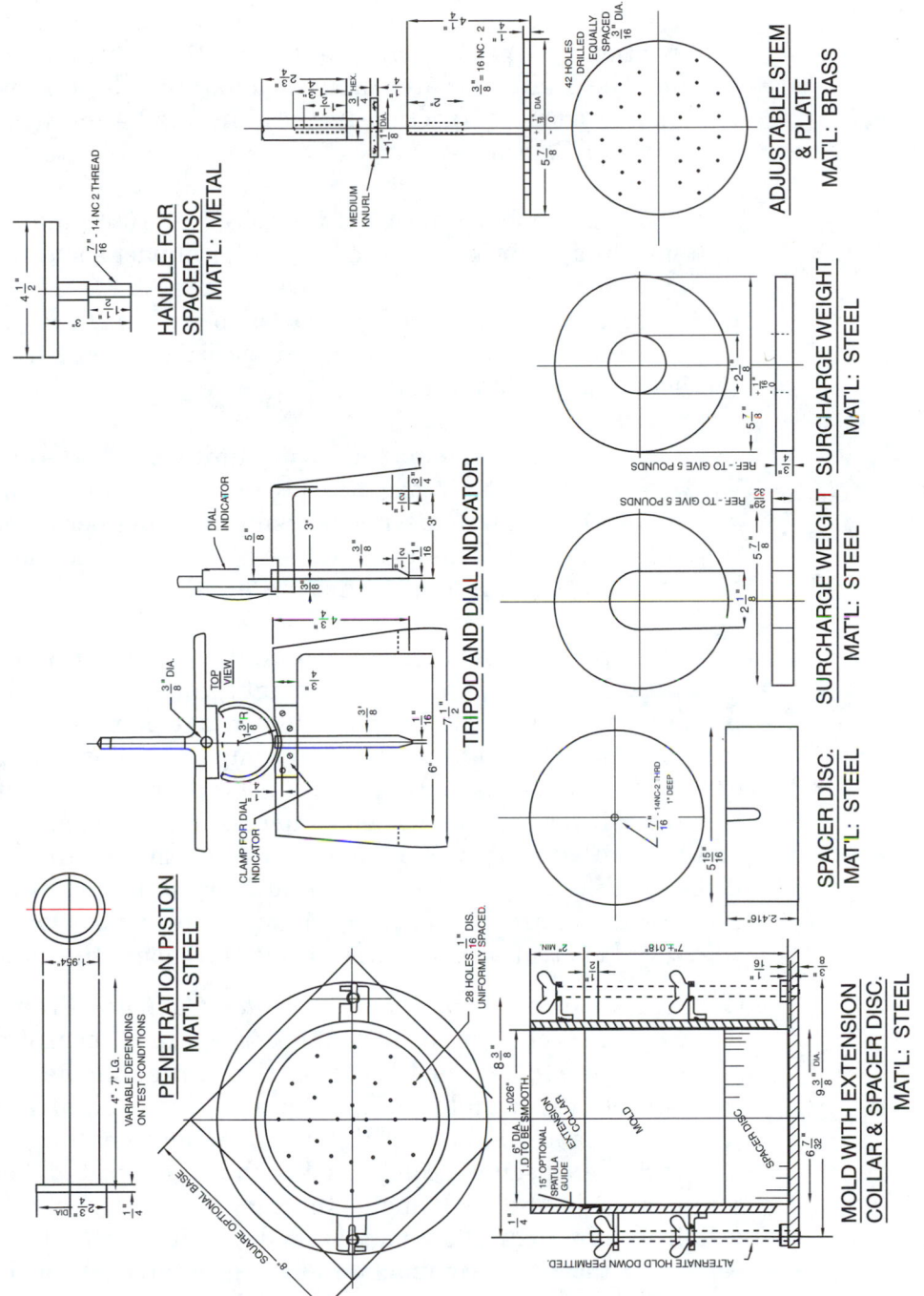

FIGURE 23–1 Bearing Ratio Test Apparatus [1]

385

previously performed compaction test on the same material may be substituted for the compaction test just described, provided that if the sample contains material retained on the ¾-in. (19-mm) sieve, soil prepared as described above is used (Note 1).

> *Note 1*—Maximum dry unit weight obtained from a compaction test performed in a 4-in. (101.6-mm) diameter mold may be slightly greater than the maximum dry unit weight obtained from compaction in the 6-in. (152.4-mm) compaction mold or CBR mold.

(1.1) For cases where the CBR is desired at 100% maximum dry unit weight and optimum water content, compact a specimen using the specified compaction procedure, either ASTM Test Methods D 698 or D 1557, from soil prepared to within ±0.5 percentage point of optimum water content in accordance with ASTM Test Method D 2216 (Chapter 3).

> *Note 2*—Where the maximum dry unit weight was determined from compaction in the 4-in. (101.6-mm) mold, it may be necessary to compact specimens, using 75 blows per layer or some other value sufficient to produce a specimen having a density equal to or greater than that required.

(1.2) Where the CBR is desired at optimum water content and some percentage of maximum dry unit weight, compact three specimens from soil prepared to within ±0.5 percentage point of optimum water content and using the specified compaction but using a different number of blows per layer for each specimen. The number of blows per layer shall be varied as necessary to prepare specimens having unit weights above and below the desired value. Typically, if the CBR for soil at 95% of maximum dry unit weight is desired, specimens compacted using 56, 25, and 10 blows per layer are satisfactory. Penetration shall be performed on each of these specimens.

(2) *Bearing Ratio for a Range of Water Content*—Prepare specimens in a manner similar to that described in (1) except that each specimen used to develop the compaction curve shall be penetrated. In addition, the complete water content–unit weight relation for the 25-blow and 10-blow per layer compactions shall be developed, and each test specimen compacted shall be penetrated. Perform all compaction in the CBR mold. In cases where the specified unit weight is at or near 100% maximum dry unit weight, it will be necessary to include a compactive effort greater than 56 blows per layer (Note 3).

> *Note 3*—A semilog plot of dry unit weight versus compactive effort usually gives a straight-line relation when compactive effort in ft-lb/ft^3 is plotted on the log scale. This type of plot is useful in establishing the compactive effort and number of

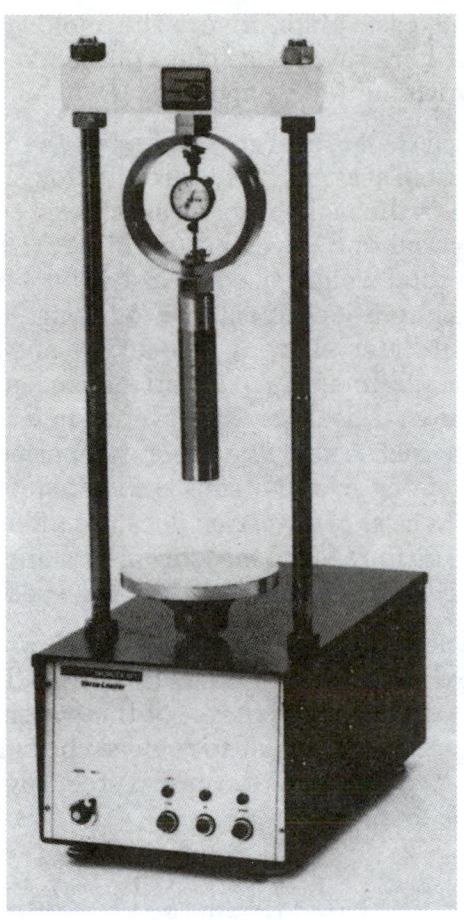

FIGURE 23–2 Loading Machine
(Courtesy of Soiltest, Inc.)

blows per layer needed to bracket the specified dry unit weight and water content range.

(2.1) If the sample is to be soaked, take a representative sample of the material, for the determination of moisture, at the beginning of compaction and another sample of the remaining material after compaction. Use ASTM Method D 2216 to determine the moisture content. If the sample is not to be soaked, take a moisture content sample in accordance with ASTM Test Methods D 698 or D 1557 if the average moisture content is desired.

(2.2) Clamp the mold (with extension collar attached) to the base plate with the hole for the extraction handle facing down. Insert the spacer disk over the base plate and place a disk of filter paper on top of the spacer disk. Compact the soil-water mixture into the mold in accordance with (1), (1.1), or (1.2).

(2.3) Remove the extension collar and carefully trim the compacted soil even with the top of the mold by means of a straightedge. Patch with smaller size material any holes that may have developed in the surface by the removal of coarse material. Remove the perforated base plate and spacer disk, weigh and record the mass of the mold plus compacted soil. Place a disk of coarse filter paper on the perforated base plate, invert the mold and compacted soil, and clamp the perforated base plate to the mold with compacted soil in contact with the filter paper.

(2.4) Place the surcharge weights on the perforated plate and adjustable stem assembly and carefully lower onto the compacted soil specimen in the mold. Apply a surcharge equal to the weight of the base material and pavement within 2.27 kg (5 lb), but in no case shall the total weight used be less than 4.54 kg (10 lb). If no pavement weight is specified, use 4.54 kg. Immerse the mold and weights in water allowing free access of water to the top and bottom of the specimen. Take initial measurements for swell and allow the specimen to soak for 96 h. Maintain a constant water level during this period. A shorter immersion period is permissible for fine-grained soils or granular soils that take up moisture readily, if tests show that the shorter period does not affect the results. At the end of 96 h, take final swell measurements and calculate the swell as a percentage of the initial height of the specimen.

(2.5) Remove the free water and allow the specimen to drain downward for 15 min. Take care not to disturb the surface of the specimen during the removal of the water. It may be necessary to tilt the specimen in order to remove the surface water. Remove the weights, perforated plate, and filter paper, and determine and record the mass.

PROCEDURE The CBR test is designed to simulate conditions that will exist at the surface of the subgrade. A surcharge (weight) is placed on the surface of the compacted specimen to represent the weight of pavement above the subgrade. Furthermore, the specimen is soaked to approximate the poorest field conditions. After soaking, the force required to push a standard piston into the soil a specified amount is determined and is used to evaluate the CBR.

The actual step-by-step procedure is as follows (ASTM D 1883-99 [1]):

(1) Place a surcharge of weights on the specimen sufficient to produce an intensity of loading equal to the weight of the base material. If no pavement weight is specified, use 4.54 kg mass. If the specimen has been soaked previously, the surcharge shall be equal to that used during the soaking period. To prevent upheaval of soil into the hole of the surcharge weights, place the 2.27-kg annular weight on the soil surface prior to seating the penetration piston, after which place the remainder of the surcharge weights.

(2) Seat the penetration piston with the smallest possible load, but in no case in excess of 10 lb (44 N). Set both the stress and penetration gages to zero. This initial load is required to ensure satisfactory seating of the piston and shall be considered as the zero load when determining the load penetration relation. Anchor the strain gage to the load measuring device, if possible; in no case attach it to the testing machine's support bars (legs).

Note 4—At high loads the supports may torque and affect the reading of the penetration gage. Checking the depth of piston penetration is one means of checking for erroneous strain indications.

(3) Apply the load on the penetration piston so that the rate of penetration is approximately 0.05 in. (1.27 mm)/min. Record the load readings at penetrations of 0.025 in. (0.64 mm), 0.050 in. (1.27 mm), 0.075 in. (1.91 mm), 0.100 in. (2.54 mm), 0.125 in. (3.18 mm), 0.150 in. (3.81 mm), 0.175 in. (4.45 mm), 0.200 in. (5.08 mm), 0.300 in. (7.62 mm), 0.400 in. (10.16 mm) and 0.500 in. (12.70 mm). Note the maximum load and penetration if it occurs for a penetration of less than 0.500 in. (12.70 mm). With manually operated loading devices, it may be necessary to take load readings at closer intervals to control the rate of penetration. Measure the depth of piston penetration into the soil by putting a ruler into the indentation and measuring the difference from the top of the soil to the bottom of the indentation. If the depth does not closely match the depth of penetration gage, determine the cause and test a new sample.

(4) Remove the soil from the mold and determine the moisture content of the top 1-in. (25.4-mm) layer. Take a moisture content sample in accordance with ASTM Test Methods D 698 or D 1557 if the average moisture content is desired. Each moisture content sample shall weigh not less than 100 g for fine-grained soils nor less than 500 g for granular soils.

Note 5—The load readings at penetrations of more than 0.300 in. (7.6 mm) may be omitted if the testing machine's capacity has been reached.

DATA The following data should be collected during a CBR test:

[A] Moisture Content Determination

Mass of wet soil plus can (g)

Mass of dry soil plus can (g)

Mass of can (g)

Note—These data must be obtained for each of four specimens: (1) before compaction, (2) after compaction, (3) top 1 in. after soaking, (4) average moisture content after soaking.

[B] Density Determination

Mass of mold plus compacted soil specimen (g)

Mass of mold (g)

Diameter of mold (in.)

Height of soil specimen (in.)

Note—These data must be obtained for each of two specimens: (1) before soaking, (2) after soaking.

[C] Swell Data

> Surcharge weight (lb)
>
> Time and date
>
> Dial reading

Note—These data must be obtained for the initial swell measurement and again for the final swell measurement.

[D] Bearing Ratio Data

> Weight of surcharge (lb)
>
> Area of piston (in.^{2})
>
> Successive proving ring dial readings for each particular penetration specified in the section "Procedure"

CALCULATIONS   [A] Moisture Content Determination

The four moisture contents are determined by the method described in Chapter 3.

[B] Density Determination

The two densities are determined by methods described in previous chapters.

[C] Swell Data

The swell percentage is computed by dividing the final (swell) dial reading by the initial height of the soil specimen. (This assumes an initial dial reading of zero; if the initial reading is not zero, it should be subtracted from the final reading and the difference divided by the height.)

[D] Bearing Ratio

Respective piston loads can be determined by multiplying each proving ring dial reading by the proving ring calibration. Then each penetration stress in pounds per square inch can be obtained by dividing piston loads in pounds by the area of the piston (which is 3 in.^{2}). A curve of penetration stress (psi) versus penetration (in.) should be prepared by plotting values of penetration stress on the piston (ordinate) versus corresponding values of penetration (abscissa), both on an arithmetic scale. In some cases, the curve of penetration stress versus penetration may be concave upward initially, because of surface irregularities or other causes. In such cases, the zero point should be adjusted as shown in Figure 23–3.

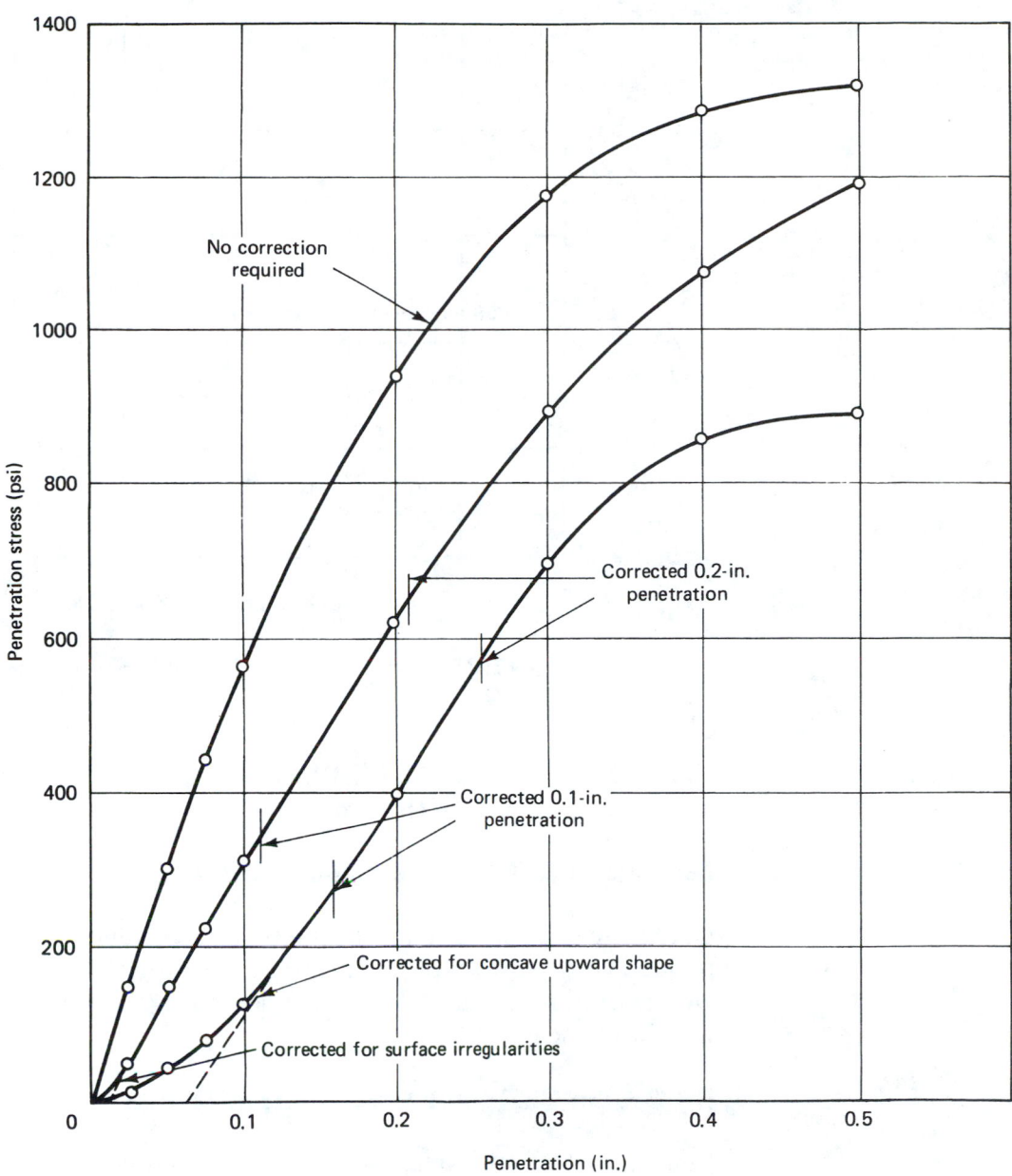

FIGURE 23–3 Correction of Curves of Penetration Stress versus Penetration [1]

The bearing ratio at 0.10-in. penetration is determined by dividing the corrected penetration stress (psi) on the piston for 0.10-in. penetration (from the curve of penetration stress versus penetration) by the standard penetration stress of 1,000 psi and multiplying by 100 (to express the answer as a percentage). [This is the application of Eq. (23–1).] The bearing ratio at 0.20-in. penetration is determined similarly, using the corrected penetration stress (psi) on the piston for 0.20-in. penetration and a standard penetration stress of 1,500 psi. If the bearing ratio based on 0.20-in. penetration is larger, the test should be rerun. If the retest gives a similar result, the ratio based on the 0.20-in. penetration

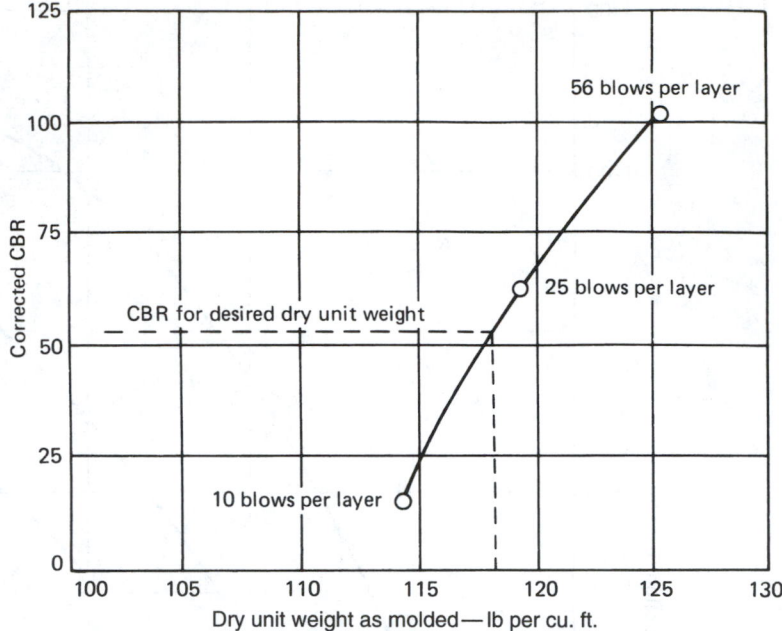

FIGURE 23–4   Dry Unit Weight Versus CBR [1]

should be reported as the CBR. Otherwise, the ratio based on 0.10-in. penetration should be the reported CBR value.

[E] Design CBR for One Water Content Only

Using the data obtained from the three specimens, plot the CBR versus molded dry unit weight relation as illustrated in Figure 23–4. Determine the design CBR at the percentage of the maximum dry unit weight requested. [1]

[F] Design CBR for Water Content Range

Plot the data from the tests at the three compactive efforts as shown in Figure 23–5. The data plotted as shown represent the response of the soil over the range of water content specified. Select the CBR for reporting as the lowest CBR within the specified water content range having a dry unit weight between the specified minimum and the dry unit weight produced by compaction within the water content range. [1]

NUMERICAL EXAMPLE   A CBR test was performed according to the procedure described. The CBR was desired at 100% maximum dry unit weight and optimum moisture content. The following data were obtained:

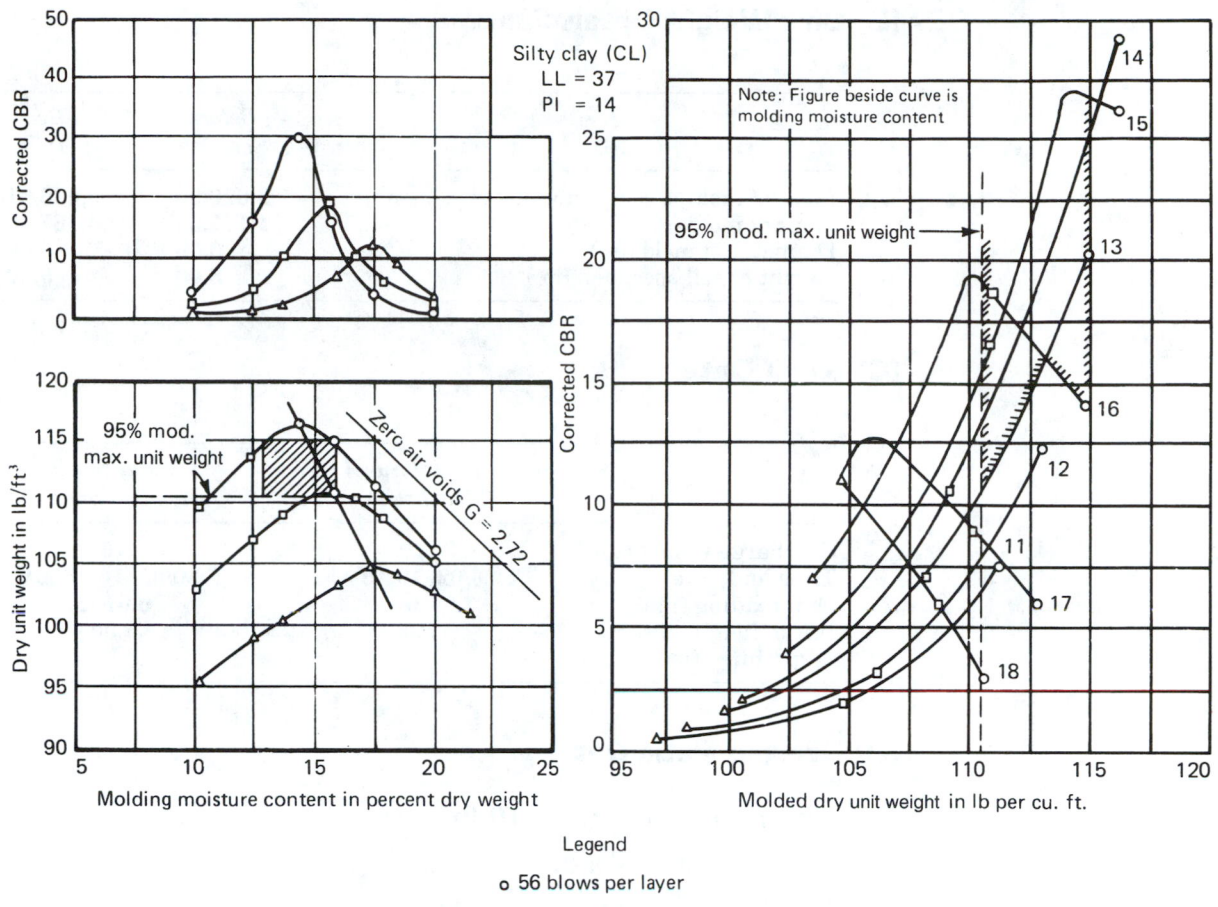

FIGURE 23–5   Determining CBR for Water Content Range and Minimum Dry Unit Weight [1]

[A] Moisture Content Determination

	Before Compaction	After Compaction	Top 1-in. Layer after Soaking	Average Moisture Content after Soaking
Can no.	1-A	1-B	1-C	1-D
Mass of wet soil + can (g)	315.94	326.01	304.71	356.37
Mass of dry soil + can (g)	273.69	283.37	261.53	305.82
Mass of can (g)	45.23	47.82	43.44	46.59

[B] Unit Weight Determination

	Before Soaking	After Soaking
Mass of mold + compacted soil specimen (g)	9,020.90	9,036.81
Mass of mold (g)	4,167.50	4,167.50
Diameter of mold (in.)	6.00	6.00
Height of soil specimen (in.)	5.00	5.00

[C] Swell Data

	Initial Swell Measurement	Final Swell Measurement
Surcharge weight (lb)	10	10
Time and date	5/26/02, 10:16 A.M.	5/30/02, 10:16 A.M.
Dial reading (in.)	0	0.0135
Initial height of soil specimen (in.)	5.00	5.00

[D] Bearing Ratio Data

Weight of surcharge = **10 lb**

Area of piston = **3.00 in.^{2}**

Proving ring calibration = **74,000 lb/in.**

Penetration (in.)	Proving Ring Dial Reading (in.)
0.000	0
0.025	0.0004
0.050	0.0008
0.075	0.0013
0.100	0.0016
0.125	0.0019
0.150	0.0020
0.175	0.0022
0.200	0.0023
0.300	0.0026
0.400	0.0030
0.500	0.0032

With these data known, necessary computations can be made as follows:

[A] Moisture Content Determinations

The four required moisture contents may be computed as follows (see Chapter 3):

Before compaction:

$$w = \frac{315.94 - 273.69}{273.69 - 45.23} \times 100 = 18.5\%$$

After compaction:

$$w = \frac{326.01 - 283.37}{283.37 - 47.82} \times 100 = 18.1\%$$

Top 1-in. layer after soaking:

$$w = \frac{304.71 - 261.53}{261.53 - 43.44} \times 100 = 19.8\%$$

Average moisture content after soaking:

$$w = \frac{356.37 - 305.82}{305.82 - 46.59} \times 100 = 19.5\%$$

These data, both given and computed, are shown on the form on page 397. At the end of the chapter, two blank copies of this form are included for the reader's use.

[B] Unit Weight Determination

The area and volume of the specimen are easily computed as follows:

$$\text{Area} = \frac{\pi D^2}{4} = \frac{\pi (6.00)^2}{4} = 28.27 \text{ in.}^2$$

$$\text{Volume} = \text{height} \times \text{area} = (5.00)(28.27) = 141.35 \text{ in.}^3$$

The mass of the compacted soil specimen can be determined by subtracting the mass of the mold from the mass of the mold plus compacted soil specimen:

Mass of specimen = (mass of mold + specimen) − (mass of mold)

Mass of specimen before soaking = **9,020.90** − **4,167.50** = 4,853.40 g

Mass of specimen after soaking = **9,036.81** − **4,167.50** = 4,869.31 g

Wet unit weights γ_{wet} can now be determined:

$$\gamma_{\text{wet}} \text{ before soaking} = \left(\frac{4,853.40}{141.35}\right)\left(\frac{1,728}{453.6}\right) = 130.8 \text{ lb/ft}^3$$

$$\gamma_{wet} \text{ after soaking} = \left(\frac{4{,}869.31}{141.35}\right)\left(\frac{1{,}728}{453.6}\right) = 131.2 \text{ lb/ft}^3$$

(The numbers 1,728 and 453.6 are conversion factors: 1,728 in.^{3} = 1 ft^3; 453.6 g = 1 lb.) With the water contents having been computed previously, dry unit weights γ_{dry} can be determined next:

$$\gamma_{dry} \text{ before soaking} = \frac{130.8}{18.3 + 100} \times 100 = 110.6 \text{ lb/ft}^3$$

$$\gamma_{dry} \text{ after soaking} = \frac{131.2}{19.5 + 100} \times 100 = 109.8 \text{ lb/ft}^3$$

[C] Swell Data

The swell percentage is computed by dividing the final (swell) dial reading by the initial height of the specimen:

$$\text{Swell} = \frac{\text{final (swell) dial reading}}{\text{initial height of specimen}} \times 100 = \frac{\mathbf{0.0135}}{\mathbf{5.00}} \times 100 = 0.27\%$$

[D] Bearing Ratio

Respective piston loads can be determined by multiplying each proving ring dial reading by the proving ring calibration. Then each penetration stress can be obtained by dividing the piston load by the area of the piston (which is 3 in.^{2}). Hence, for the penetration of **0.025 in.,** the piston load is **0.0004 × 74,000,** or 29.6 lb, and the penetration stress is 29.6/**3.00,** or 9.9 psi.

   These values fill in the second row of the form on page 398. Similar calculations furnish required values for succeeding piston loads and penetration stresses for completing the remaining rows on the form.

   The curve of penetration stress versus penetration is plotted and shown in Figure 23–6. Since the curve is not concave upward initially, no zero point correction is required.

   The bearing ratio at 0.10-in. penetration is determined by dividing the penetration stress for 0.10-in. penetration (from the curve of penetration stress versus penetration) by the standard penetration stress of 1,000 psi and multiplying by 100:

$$\text{CBR at 0.10-in. penetration} = \frac{39.5}{1{,}000} \times 100 = 4.0\%$$

The bearing ratio at 0.20-in. penetration is determined similarly using the penetration stress for the 0.20-in. penetration and a standard penetration stress of 1,500 psi:

$$\text{CBR at 0.20-in. penetration} = \frac{56.7}{1{,}500} \times 100 = 3.8\%$$

Soils Testing Laboratory
California Bearing Ratio Test

Sample No. _____18_____   Project No. _____SR 2128_____

Location ___Indian Trail, N.C.___   Boring No. _____5_____

Depth of Sample _____3 ft_____ Date of Test _____5/26/02_____

Description of Soil _____Brown silty clay_____

Tested by _____John Doe_____

Method Used for Preparation and Compaction of Specimen ___ASTM D 698, Method D___

[A] Moisture Content Determination

	Before Compaction	After Compaction	Top 1-in. Layer after Soaking	Average Moisture Content after Soaking
Can no.	1-A	1-B	1-C	1-D
Mass of wet soil + can (g)	315.94	326.01	304.71	356.37
Mass of dry soil + can (g)	273.69	283.37	261.53	305.82
Mass of can (g)	45.23	47.82	43.44	46.59
Mass of water (g)	42.25	42.64	43.18	50.55
Mass of dry soil (g)	228.46	235.55	218.09	259.23
Moisture content (%)	18.5	18.1	19.8	19.5

Average moisture content before soaking ___18.3___ %
Average moisture content after soaking ___19.5___ %

[B] Unit Weight Determination

		Before Soaking	After Soaking
(1)	Mass of mold + compacted soil specimen (g)	9,020.90	9,036.81
(2)	Mass of mold (g)	4,167.50	4,167.50
(3)	Mass of compacted soil specimen (g) $[(3) = (1) - (2)]$	4,853.40	4,869.31
(4)	Diameter of mold (in.)	6.00	6.00
(5)	Area of soil specimen (in.^{2})	28.27	28.27
(6)	Height of soil specimen (in.)	5.00	5.00
(7)	Volume of soil specimen (in.^{3}) $[(7) = (5) \times (6)]$	141.35	141.35
(8)	Wet unit weight (lb/ft^3) $\left[(8) = \dfrac{(3)}{453.6} \times \dfrac{1728}{(7)}\right]$	130.8	131.2
(9)	Moisture content (%) (from part [A])	18.3	19.5
(10)	Dry unit weight (lb/ft^3) $\left[(10) = \dfrac{(8)}{(9) + 100} \times 100\right]$	110.6	109.8

[C] Swell Data

		Initial Swell Measurement	Final Swell Measurement
(1)	Surcharge weight (lb)	10	10
(2)	Time and date	5/26/02 10:16 A.M.	5/30/02 10:16 A.M.
(3)	Elapsed time (h)	0	96
(4)	Dial reading (in.)	0	0.0135
(5)	Initial height of soil specimen (in.) [from part [B], (6)]	5.00	5.00
(6)	Swell (% of initial height) $\left[(6) = \dfrac{(4)}{(5)} \times 100 \right]$	0	0.27

[D] Bearing Ratio Data

(1) ☒ Soaked ☐ Unsoaked (check one)

(2) Weight of surcharge __10__ lb

(3) Proving ring calibration __74,000__ lb/in.

Penetration (in.)	Proving Ring Dial Reading (in.)	Piston Load (lb)	Area of Piston (in.^{2})	Penetration Stress (psi)
(1)	(2)	(3) = (2) × proving ring calibration	(4)	(5) = $\dfrac{(3)}{(4)}$
0.000	0	0	3.00	0
0.025	0.0004	29.6	3.00	9.9
0.050	0.0008	59.2	3.00	19.7
0.075	0.0013	96.2	3.00	32.1
0.100	0.0016	118.4	3.00	39.5
0.125	0.0019	140.6	3.00	46.9
0.150	0.0020	148.0	3.00	49.3
0.175	0.0022	162.8	3.00	54.3
0.200	0.0023	170.2	3.00	56.7
0.300	0.0026	192.4	3.00	64.1
0.400	0.0030	222.0	3.00	74.0
0.500	0.0032	236.8	3.00	78.9

$$\text{CBR at 0.10-in. penetration} = \frac{\left(\begin{array}{l}\text{corrected penetration stress} \\ \text{for 0.10-in. penetration (from} \\ \text{curve of penetration stress} \\ \text{versus penetration)}\end{array}\right)}{1{,}000} \times 100 = \underline{\quad 4.0 \quad} \%$$

$$\text{CBR at 0.20-in. penetration} = \frac{\left(\begin{array}{l}\text{corrected penetration stress} \\ \text{for 0.20-in. penetration (from} \\ \text{curve of penetration stress} \\ \text{versus penetration)}\end{array}\right)}{1{,}500} \times 100 = \underline{\quad 3.8 \quad} \%$$

Design CBR = $\underline{\quad 4.0 \quad}$ %

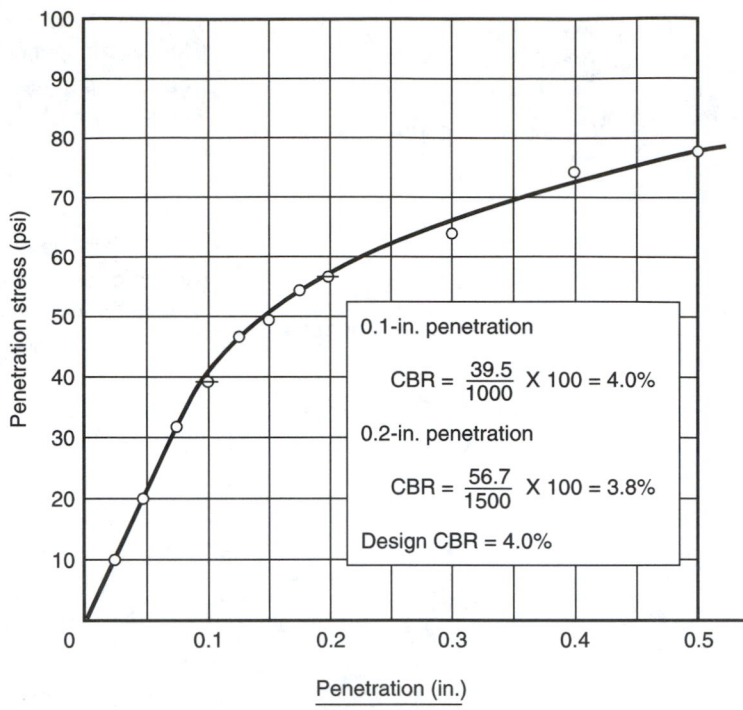

FIGURE 23–6   Curve of Penetration Stress versus Penetration

Inasmuch as the bearing ratio at 0.10-in. penetration is greater than that at 0.20-in. penetration, the ratio at 0.10-in. penetration should be reported as the design CBR.

Hence, in this example, the CBR is 4.0%.

CONCLUSIONS   The major result of a CBR test is, of course, the CBR value. Reports of this test should also include the values of other parameters determined during the test, including moisture contents of the soil sample (before and after compaction, of the top 1-in. layer after soaking, and the average after soaking), dry unit weights of the soil specimen before and after soaking, and the swell. The condition of the sample (soaked or unsoaked), the method used for preparation and compaction of the specimen [either ASTM Method D 698 (Chapter 11) or D 1557], and the curve of penetration stress versus penetration should also be reported.

ASTM D 1883-94 permits the omission of load readings at penetrations of 0.40 in. and 0.50 in. if the capacity of the testing machine has been reached. If, however, bearing ratio values are desired for penetrations of 0.30 in., 0.40 in., and 0.50 in., corrected penetration stress values for these penetrations should be divided by the standard penetration stresses of 1,900, 2,300 and 2,600 psi, respectively (and then multiplied by 100 to express the answer as a percentage).

As mentioned in the introduction, the CBR of a soil gives an indication of its strength as subgrade, sub-base, and base course material. The CBR is used primarily to determine empirically the required thickness of flexible pavements for highways and airfield pavements. Some

Table 23–1 Soil Ratings for Roads and Runways [2, 3]

CBR No.	General Rating	Uses	Classification System	
			Unified	*AASHTO*
0–3	Very poor	Subgrade	OH, CH, MH, OL	A5, A6, A7
3–7	Poor to fair	Subgrade	OH, CH, MH, OL	A4, A5, A6, A7
7–20	Fair	Sub-base	OL, CL, ML, SC, SM, SP	A2, A4, A6, A7
20–50	Good	Base, subbase	GM, GC, SW, SM, SP, GP	A1b, A2–5, A3, A2–6
>50	Excellent	Base	GW, GM	A1a, A2–4, A3

general ratings of soils to be used as subgrade, sub-base, and base courses of roads and airfield runways corresponding to various ranges of CBR values are given in Table 23–1.

REFERENCES

[1] ASTM, *2001 Annual Book of ASTM Standards,* West Conshohocken, PA, 2001. Copyright, American Society for Testing and Materials, 100 Barr Harbor Drive, West Conshohocken, PA 19428-2959. Reprinted with permission.

[2] Joseph E. Bowles, *Engineering Properties of Soils and Their Measurement,* 2d ed., McGraw-Hill Book Company, New York, 1978.

[3] The Asphalt Institute, *The Asphalt Handbook,* College Park, MD, 1962.

Soils Testing Laboratory
California Bearing Ratio Test

Sample No. _____   Project No. _____

Location _____   Boring No. _____

Depth of Sample _____   Date of Test _____

Description of Soil _____

Tested by _____

Method Used for Preparation and Compaction of Specimen _____

[A] Moisture Content Determination

	Before Compaction	After Compaction	Top 1-in. Layer after Soaking	Average Moisture Content after Soaking
Can no.				
Mass of wet soil + can (g)				
Mass of dry soil + can (g)				
Mass of can (g)				
Mass of water (g)				
Mass of dry soil (g)				
Moisture content (%)				

Average moisture content before soaking _____ %
Average moisture content after soaking _____ %

[B] Unit Weight Determination

		Before Soaking	After Soaking
(1)	Mass of mold + compacted soil specimen (g)	_____	_____
(2)	Mass of mold (g)	_____	_____
(3)	Mass of compacted soil specimen (g) [(3) = (1) − (2)]	_____	_____
(4)	Diameter of mold (in.)	_____	_____
(5)	Area of soil specimen (in.^{2})	_____	_____
(6)	Height of soil specimen (in.)	_____	_____
(7)	Volume of soil specimen (in.^{3}) [(7) = (5) × (6)]	_____	_____
(8)	Wet unit weight (lb/ft^3) $\left[(8) = \dfrac{(3)}{453.6} \times \dfrac{1728}{(7)} \right]$	_____	_____
(9)	Moisture content (%) (from part [A])	_____	_____
(10)	Dry unit weight (lb/ft^3) $\left[(10) = \dfrac{(8)}{(9) + 100} \times 100 \right]$	_____	_____

[C] Swell Data

	Initial Swell Measurement	Final Swell Measurement
(1) Surcharge weight (lb)	_____	_____
(2) Time and date	_____	_____
	_____	_____
(3) Elapsed time (h)	_____	_____
(4) Dial reading (in.)	_____	_____
(5) Initial height of soil specimen (in.) [from part [B], (6)]	_____	_____
(6) Swell (% of initial height) $\left[(6) = \dfrac{(4)}{(5)} \times 100 \right]$	_____	_____

[D] Bearing Ratio Data

(1) ☐ Soaked ☐ Unsoaked (check one)

(2) Weight of surcharge _____ lb

(3) Proving ring calibration _____ lb/in.

Penetration (in.)	Proving Ring Dial Reading (in.)	Piston Load (lb)	Area of Piston (in.²)	Penetration Stress (psi)
(1)	(2)	(3) = (2) × proving ring calibration	(4)	(5) = $\dfrac{(3)}{(4)}$

$$\text{CBR at 0.10-in. penetration} = \frac{\left(\begin{array}{l}\text{corrected penetration stress}\\\text{for 0.10-in. penetration (from}\\\text{curve of penetration stress}\\\text{versus penetration)}\end{array}\right)}{1{,}000} \times 100 = \underline{\hspace{2cm}} \%$$

$$\text{CBR at 0.20-in. penetration} = \frac{\left(\begin{array}{l}\text{corrected penetration stress}\\\text{for 0.20-in. penetration (from}\\\text{curve of penetration stress}\\\text{versus penetration)}\end{array}\right)}{1{,}500} \times 100 = \underline{\hspace{2cm}} \%$$

Design CBR = \underline{\hspace{2cm}} %

Soils Testing Laboratory
California Bearing Ratio Test

Sample No. _____   Project No. _____

Location _____   Boring No. _____

Depth of Sample _____   Date of Test _____

Description of Soil _____

Tested by _____

Method Used for Preparation and Compaction of Specimen _____

[A] Moisture Content Determination

	Before Compaction	After Compaction	Top 1-in. Layer after Soaking	Average Moisture Content after Soaking
Can no.				
Mass of wet soil + can (g)				
Mass of dry soil + can (g)				
Mass of can (g)				
Mass of water (g)				
Mass of dry soil (g)				
Moisture content (%)				

Average moisture content before soaking _____ %
Average moisture content after soaking _____ %

[B] Unit Weight Determination

	Before Soaking	After Soaking
(1) Mass of mold + compacted soil specimen (g)	_____	_____
(2) Mass of mold (g)	_____	_____
(3) Mass of compacted soil specimen (g) $[(3) = (1) - (2)]$	_____	_____
(4) Diameter of mold (in.)	_____	_____
(5) Area of soil specimen (in.^{2})	_____	_____
(6) Height of soil specimen (in.)	_____	_____
(7) Volume of soil specimen (in.^{3}) $[(7) = (5) \times (6)]$	_____	_____
(8) Wet unit weight (lb/ft^3) $\left[(8) = \dfrac{(3)}{453.6} \times \dfrac{1728}{(7)}\right]$	_____	_____
(9) Moisture content (%) (from part [A])	_____	_____
(10) Dry unit weight (lb/ft^3) $\left[(10) = \dfrac{(8)}{(9) + 100} \times 100\right]$	_____	_____

	Initial Swell Measurement	Final Swell Measurement

[C] Swell Data

(1) Surcharge weight (lb) _____ _____

(2) Time and date _____ _____

                              _____ _____

(3) Elapsed time (h) _____ _____

(4) Dial reading (in.) _____ _____

(5) Initial height of soil specimen (in.) [from part [B], (6)] _____ _____

(6) Swell (% of initial height)

$$\left[(6) = \frac{(4)}{(5)} \times 100 \right]$$ _____ _____

[D] Bearing Ratio Data

(1) ☐ Soaked ☐ Unsoaked (check one)

(2) Weight of surcharge _____ lb

(3) Proving ring calibration _____ lb/in.

Penetration (in.)	Proving Ring Dial Reading (in.)	Piston Load (lb)	Area of Piston (in.^{2})	Penetration Stress (psi)
(1)	(2)	(3) = (2) × proving ring calibration	(4)	(5) = $\frac{(3)}{(4)}$

$$\text{CBR at 0.10-in. penetration} = \frac{\left(\begin{array}{l}\text{corrected penetration stress} \\ \text{for 0.10-in. penetration (from} \\ \text{curve of penetration stress} \\ \text{versus penetration)}\end{array}\right)}{1,000} \times 100 = \underline{\hspace{1cm}} \%$$

$$\text{CBR at 0.20-in. penetration} = \frac{\left(\begin{array}{l}\text{corrected penetration stress} \\ \text{for 0.20-in. penetration (from} \\ \text{curve of penetration stress} \\ \text{versus penetration)}\end{array}\right)}{1,500} \times 100 = \underline{\hspace{1cm}} \%$$

Design CBR = _____ %

Graph Papers

The following pages contain blank copies of graph papers that have been prepared for evaluating data from some of the test procedures herein. These graph papers are: semilogarithmic, 1 cycle; grain-size distribution curve; rectangular coordinate; semilogarithmic, 3 cycles; and semilogarithmic, 5 cycles.

Inasmuch as most of these will be needed more than once in carrying out the various tests presented in the text, it is suggested that they be photocopied as needed.

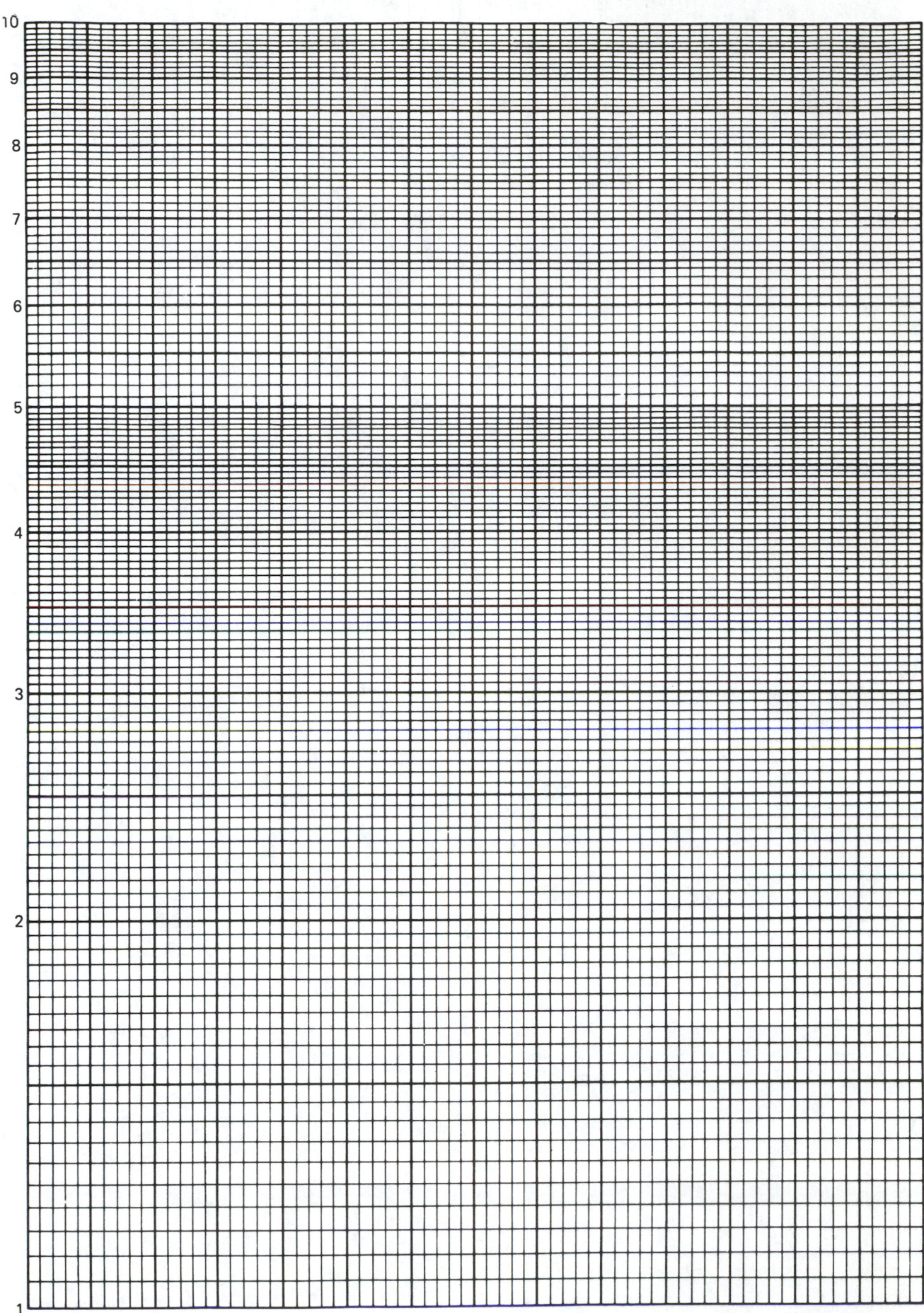

Semi-logarithmic, 1 cycle

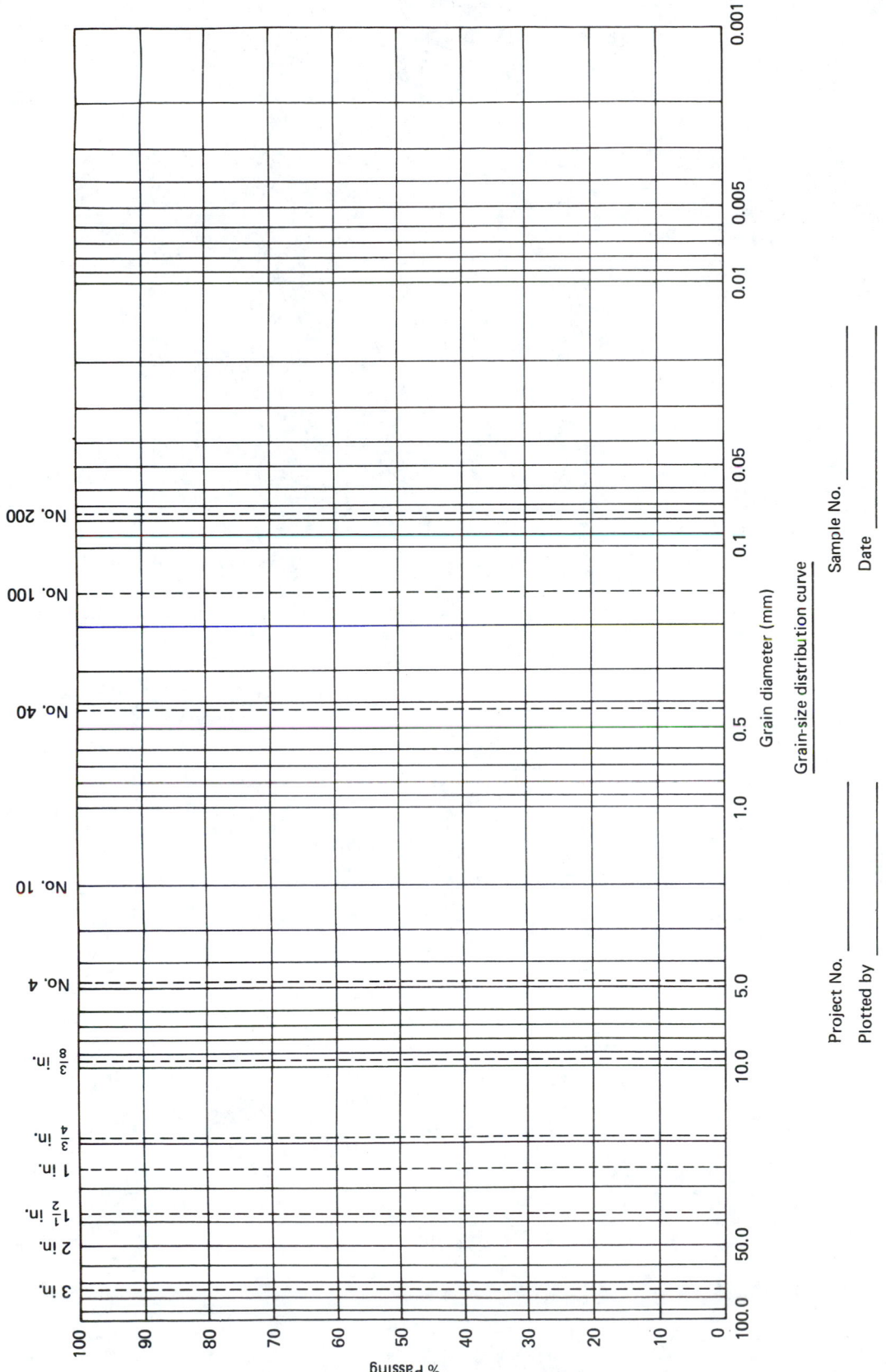

Grain-size distribution curve

Grain diameter (mm)

% Passing

Project No. _____

Plotted by _____

Sample No. _____

Date _____

Rectangular Coordinate

417

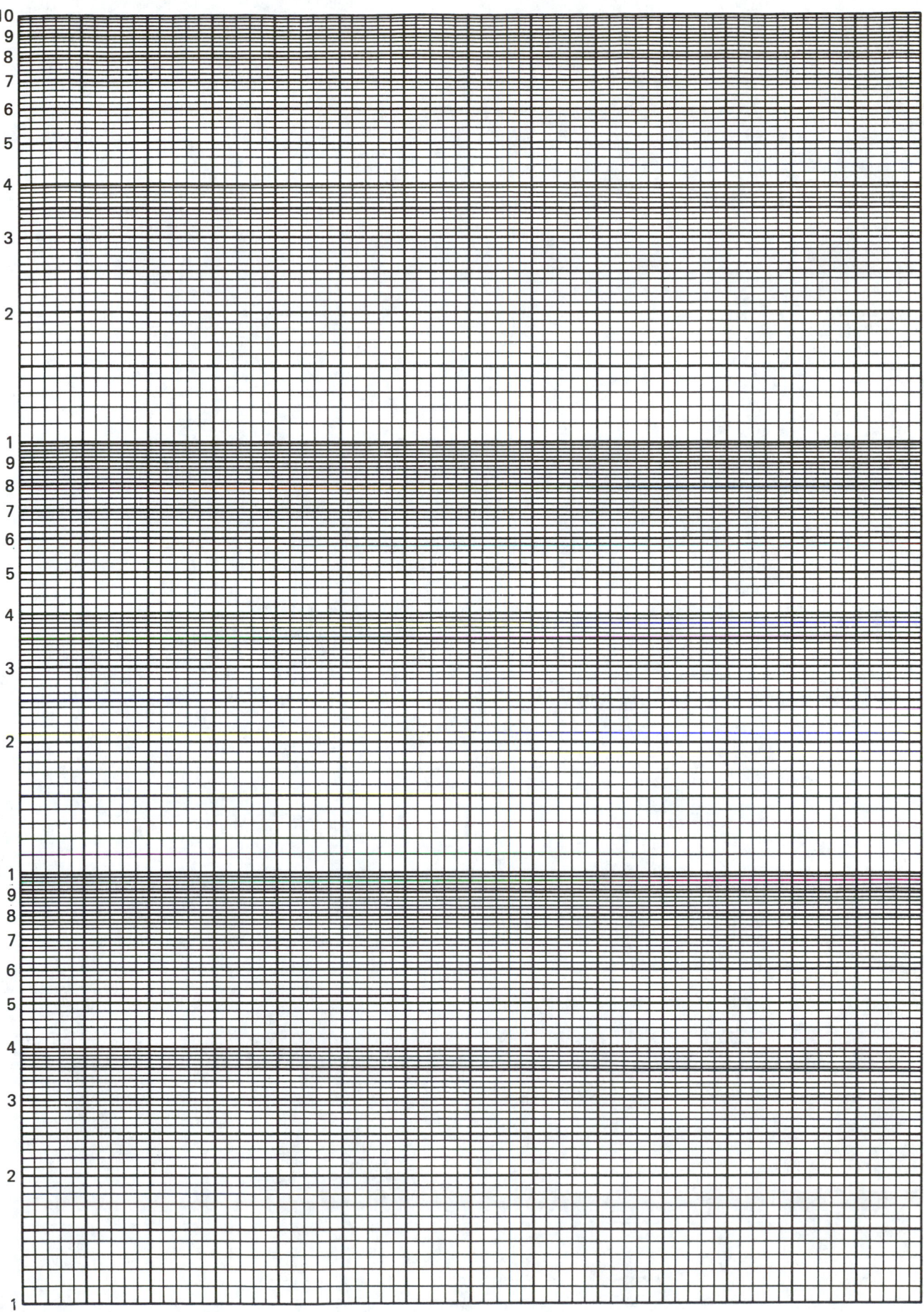

Semi-logarithmic, 3 cycles

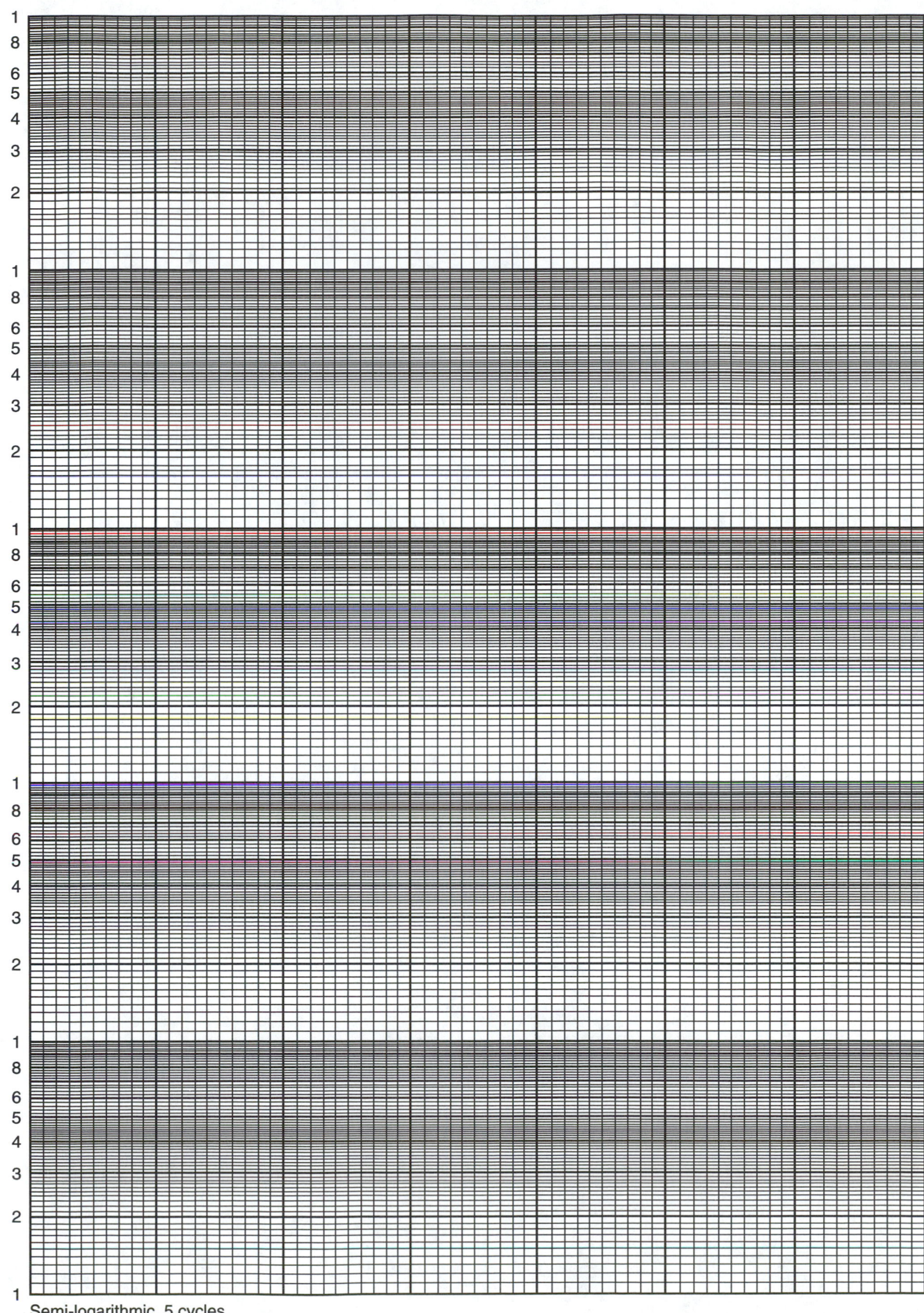

Semi-logarithmic, 5 cycles

CD Installation Tips
Soil Properties (5th Edition)
by Cheng Liu & Jack Evett

1. Log onto your hard drive (i.e. enter **C:** or **D:**)
2. Change to the root directory by entering CD\
3. Create a subdirectory by entering **MD ASEASY**
4. Log onto that subdirectory by entering **CD ASEASY**
5. Determine the drive letter of your CD drive.
6. Insert the CD into the computer (i.e. D: or E: typically)
7. Copy the following files from the CD to the hard drive: **COPY D:*.doc   COPY D:*.bat COPY D:*.diz**   (or COPY E: etc. if appropriate)
8. Uncompress the SOIL.WKS spreadsheet by entering **D:SOIL.EXE**

    Note:—You can 'reclaim' the original SOIL.WKS spreadsheet at any time by Logging onto the ASEASY subdirectory (i.e. enter C: then   CD \ASEASY   ) Insert the original CD and enter D:SOIL.EXE

9. Log onto the CD (i.e. enter D: or E:)
10. Run the Installation program by entering **INSTALL.** Press the **F10 key** to install, answer **Y** when it indicates the ASEASY subdirectory exists, press the **ESC** key when requested, and enter your **name** and the **specified number** when requested. Press Y when "Continue" is displayed. When it completes and indicates "Enter ASEASY" to run the program—**DO NOT** enter this, proceed to step #11 below.
11. Upon completion, log onto the hard drive, by entering **C:** then **CD \ASEASY**
12. Prepare a printer, with adequate paper.
13. Obtain hard copies of the information files by entering: **COPY *.DOC PRN**   and **COPY *.DIZ PRN**
14. To obtain a copy of the **120-page** AsEasyAs Manual, enter: **COPY ASEASY.MAN PRN**
15. To run the AsEasyAs Program, utilizing the prepared spreadsheet for the Soil Properties text, enter **SOIL.** When the ASEASYAS screen appears, press any key to continue, and the SOIL worksheet/menu will load. If display problems are experienced, refer to the documentation and modify the SOIL.BAT file accordingly.

Notes

1. To set up the proper video display and printer initially:
    Press the **ESC** key to get to the 'spreadsheet mode', enter **/UIV** then select the desired video display, **/UIP** then select the desired printer, then enter **/UIS** to save this new configuration and **/SS** will show you the system configuration.
2. The startup program could be copied to the root directory (COPY SOIL.BAT C:\) to eliminate the need for logging onto the ASEASY subdirectory each time the program is run.
3. Be sure to start the program by entering **SOIL** instead of **ASEASY,** because this batch file lets the hard drive provide extra memory space for the large spreadsheet.
4. Some computers may have problems displaying some of the graphic and Greek characters. This can often be fixed by adding the following to the CONFIG.SYS file: DEVICE=ANSI.SYS (after copying ANSI.SYS from your DOS disks to the hard drive).
5. Some printers may also have the same problems with the graphic and Greek characters. Consult your DOS and printer manuals for proper setup. (Laser printers may need additional font cartridges or a program to upload 'soft fonts', or one to put the printer in the proper mode.)
6. The AsEasyAs software provided with this manual is a shareware version, and its cost is NOT included in the price of the Soil Properties manual. In other words, its creator was gracious enough to allow its inclusion in the manual at no cost. Each individual user is expected to purchase their own registered copy upon their satisfactory usage of the software.
7. The spreadsheet data can be *cleared out* completely at any time by pressing **ALT Z** while in the spreadsheet command mode.

8. Ignore the 0.0's and ERR's in the blank spreadsheets; they will be re-placed by calculated results when data is entered.

9. If the Mohr circle plot does not plot with equal horizontal and vertical sizes, alter its plot parameters in cell AW19.

10. This shareware (free) version of ASEASYAS is a DOS program. It **does work** using a command prompt in Windows2000. However, if it does not run properly from a 'DOS prompt' under Windows or NT, simply create a bootable DOS disk and run the program normally. A 30-day Windows version of ASEASYAS is available from the ASEASYAS vendor. It retails for $60.

11. This spreadsheet is 100% compatible with the original Lotus program, but is not compatible with the Windows Excel spreadsheet program.

12. This disk was prepared by Ambrose Barry, who would appreciate comments and correction via e-mail at: abarry@uncc.edu or via the Engineering Technology HomePage on the World Wide Web at: **http://www.et.uncc.edu.**